Stellar Structure and Evolution

Stellar Structure and Evolution, the second volume in the Ohio State Astrophysics Series, takes advantage of our new era of stellar astrophysics, in which modern techniques allow us to map the interiors of stars in unprecedented detail. This textbook for upper-level undergraduate and graduate students aims to develop a broad physical understanding of the fundamental principles that dictate stellar properties. The study of stellar evolution focuses on the "life cycle" of stars: how they are born, how they live, and how they die. As elements ejected by one generation of stars are incorporated into the next generation, stellar evolution is intertwined with the chemical evolution of our galaxy. Focusing on key physical processes without going into encyclopedic depth, the authors present stellar evolution in a contemporary context, including phenomena such as pulsations, mass loss, binary interactions, and rotation, which contribute to our understanding of stars.

MARC PINSONNEAULT received his Ph.D. in astronomy from Yale University in 1988. He is a full professor of astronomy at The Ohio State University, where he has been teaching since 1994. He has an extensive research record in theoretical models of stellar structure and evolution, with an emphasis on stellar rotation and magnetism, rotationally induced mixing, helio- and asteroseismology, solar models, and solar neutrinos. He was elected a Fellow of the AAAS in 2010 and was recognized as a Distinguished University Scholar at Ohio State in 2017.

BARBARA RYDEN received her Ph.D. in astrophysical sciences from Princeton University. After postdocs at the Harvard-Smithsonian Center for Astrophysics and the Canadian Institute for Theoretical Astrophysics, she joined the astronomy faculty at The Ohio State University, where she is a full professor. She has 30 years of experience in teaching, at levels ranging from introductory undergraduate courses to advanced graduate seminars. She won the Chambliss Astronomical Writing Award for her textbook *Introduction to Cosmology*, and she is co-author, with Richard Pogge, of *Interstellar and Intergalactic Medium*.

"Pinsonneault and Ryden's book is a very welcome addition to the field of stellar evolution at a level appropriate to advanced undergraduate- or graduate-level study, since it manages to provide a clear, comprehensive overview of topics, without being intimidating in size or style. The textbook includes up-to-date results from contemporary missions such as Gaia and Kepler, with the final chapters discussing stellar rotation, pulsations, and binary evolution in depth. Most chapters include a few well-designed exercises, with a research-level reading list provided after the appendix. I would highly recommend it for Master's-level courses on stellar structure and evolution."

Professor Paul Crowther, University of Sheffield

"This text is a welcome addition to the pantheon of monographs and textbooks explaining the physical basics of stellar structure and evolution. Aimed primarily at an audience learning the material for the first time, this text explains the phases of the life of a star through a clear application of physical principles. Weaving together classical fluids, quantum mechanics, thermodynamics, and nuclear physics, it enables students and their instructors to gain the physical intuition needed for the study of stars in this time of their observational renaissance."

Professor Lars Bildsten, University of California, Santa Barbara

"This is a welcome addition to the literature, providing a comprehensive overview of stellar structure and evolution, and including insights from the latest data, techniques, and results."

Professor William Chaplin, University of Birmingham

Stellar Structure and Evolution

Marc Pinsonneault

The Ohio State University

Barbara Ryden

The Ohio State University

CAMBRIDGE
UNIVERSITY PRESS

Shaftesbury Road, Cambridge CB2 8EA, United Kingdom

One Liberty Plaza, 20th Floor, New York, NY 10006, USA

477 Williamstown Road, Port Melbourne, VIC 3207, Australia

314–321, 3rd Floor, Plot 3, Splendor Forum, Jasola District Centre, New Delhi – 110025, India

103 Penang Road, #05–06/07, Visioncrest Commercial, Singapore 238467

Cambridge University Press is part of Cambridge University Press & Assessment, a department of the University of Cambridge.

We share the University's mission to contribute to society through the pursuit of education, learning and research at the highest international levels of excellence.

www.cambridge.org
Information on this title: www.cambridge.org/highereducation/isbn/9781108835817
DOI: 10.1017/9781108869249

First published 2023

A catalogue record for this publication is available from the British Library

A Cataloging-in-Publication data record for this book is available from the Library of Congress

ISBN 978-1-108-83581-7 Hardback
ISBN 978-1-108-79882-2 Paperback

Additional resources for this publication at www.cambridge.org/osas_sse

For Julie, my Sun and Moon.   MP

*For Pat Westphal, who started me down
the inclined plane of physics.*   BR

Contents

Preface

This textbook is part of a series based on the curriculum for astronomy graduate students at The Ohio State University (OSU). In this curriculum, first-year graduate students take a five credit-hour course "Observed Properties of Astronomical Systems." This is followed by six courses, each of two or three credit-hours: "Atomic and Radiative Processes in Astrophysics," "Stellar Structure and Evolution," "Dynamics," "Cosmology," "Numerical and Statistical Methods in Astrophysics," and "The Interstellar Medium and the Intergalactic Medium." The philosophy of the OSU graduate program, however, is best encapsulated in the two credit-hour course "Order of Magnitude Astrophysics," which is offered every year to first- and second-year students. In this course, students work together to solve a wide range of astrophysical problems, using basic physical principles to find back-of-envelope solutions.

The Ohio State Astrophysics Series (OSAS), of which this is the second volume, is a projected series of books based on lecture notes for the six core courses and the first-year "Observed Properties" course. These textbooks will not be exhaustive monographs, but will instead adopt the back-of-envelope philosophy of the "Order of Magnitude" course to emphasize the most important physical principles in each subfield of astrophysics. The goal is to make our series a point of entry into the deeper and more detailed classic textbooks in our field. Although each volume in OSAS will stand on its own, care will be taken to unify notation and vocabulary as much as possible across volumes.

Stellar Structure and Evolution is based on the semester-long class of the same name. Stellar structure focuses on the underlying physics of stars. It naturally includes subjects that undergraduate astronomy and physics majors usually see in isolation: statistical mechanics, thermodynamics, electricity and magnetism, quantum mechanics, waves, fluid dynamics, and nuclear physics, among others. As such, the potential list of topics is vast, and we cannot address all of them. Our primary goal is to develop physical intuition for the fundamental principles that dictate the main properties of stars. A secondary goal is for students to see how disparate tools can be harnessed to understand a rich physical system like a star. Stellar evolution is the study of the "life cycle" of stars: how they are born, how they live, and how they die. Stellar evolution is intertwined with chemical evolution, the origin of the elements on the periodic table. Classical texts in the field have focused mainly on the evolution of spherical isolated stars. This does not reflect the current state of the art in a dynamic field. We therefore present stellar evolution in a modern context, including phenomena such as mass loss, binary interactions, and rotation where relevant.

This textbook uses the cgs (centimeter, gram, second) system of units commonly used in graduate education in astronomy. It also uses the most common astronomical distance units: the solar radius ($R_\odot$), the astronomical unit (au), and the parsec (pc). In addition, masses are given in units of the solar mass ($M_\odot$) and luminosities in units of the solar luminosity ($L_\odot$). On small scales, when we examine individual photons and other particles, the electron-volt (eV) will be a useful small unit of energy, with $1\,\mathrm{eV} = 1.602 \times 10^{-12}\,\mathrm{erg}$. Other helpful conversion factors, and the values of physical and astronomical constants, are included in the Appendices. Online resources for this book hosted by Cambridge University Press include ancillary materials such as a solutions manual and links to Jupyter notebooks for recreating and modifying figures in the textbook.

The text of this book was greatly improved by the careful reading and insightful recommendations of Jennifer Johnson (OSU). Many of the figures and images in this book are derived from works in the published astronomical literature. We are grateful to the authors and journals who promptly granted permission to use their figures. We are especially grateful to those of our colleagues who dug out their original data for us to replot for this volume. Particular thanks are due to Emily Griffith (OSU) for Figure 1.9, Zeki Eker (Akdeniz University) for Figures 1.12 and 1.13, Franck Delahaye (Observatoire de Paris) for Figure 4.2, Kohji Takahashi (GSI) for Figure 5.2, Patrick Vallely (OSU) for Figure 9.3, Jamie Tayar (Institute for Astronomy) for Figure 10.4, Gibor Basri (University of California, Berkeley) for Figure 10.7, Radek Poleski (University of Warsaw) for Figure 11.1, and Mathieu Vriard (OSU) for Figure 11.9. All original figures were created by Richard Pogge (OSU), in his role as technical editor of the Ohio State Astrophysics Series.

Properties of Stars

When quacks with pills political would dope us,
When politics absorbs the livelong day,
I like to think about that star Canopus,
So far, so far away.

Greatest of visioned suns, they say who list 'em;
To weigh it science almost must despair.
Its shell would hold our whole dinged solar system,
Nor even know 'twas there.

Bert Leston Taylor (1866–1921)
"Canopus" [1913]

A star can be defined as a self-gravitating ball of gas, usually spherical or spheroidal, that is powered by nuclear fusion in its interior. In this text, we will go slightly beyond the boundaries of this definition to discuss protostars and pre-main sequence stars (not yet powered by fusion), stellar remnants (no longer powered by fusion), and brown dwarfs (too small to be powered by fusion).

Less than 10% of the baryonic matter in the universe is contained in stars; less than 5% of the mass-energy of the universe is in the form of baryonic matter. Thus, stars make up less than 0.5% of the universe. Why do we devote an entire astrophysics textbook to such a small fraction of the universe? In part, we are simply following the well-trodden path of our astronomical ancestors. It wasn't until the twentieth century that astronomers were able to make extensive observations outside the visible range of the spectrum ($\lambda = 4000\,\text{Å}$–$7500\,\text{Å}$). Since human eyes evolved to take advantage of light emitted by a star, it isn't surprising that our eyes are quite good at detecting stars. Before the development of detectors that worked outside the visible range, astronomers had to study stars, and objects that reflect starlight, because that was what they could see.

However, even when we open our eyes (metaphorically) to the full spectrum of electromagnetic radiation, stars are well worth studying. The average mass density of baryonic matter today is $\rho_{\text{bary},0} = 4.2 \times 10^{-31}\,\text{g}\,\text{cm}^{-3}$. The average density of the Sun is $\rho_\odot = 1.410\,\text{g cm}^{-3}$, over 30 orders of magnitude greater than the baryonic density of the universe as a whole. When gas is compressed to such high densities, interesting physical processes, such as nuclear fusion, can occur. How does a self-gravitating fusion reactor regulate itself? How does the released energy escape from the self-gravitating fusion reactor? What happens when the self-gravitating fusion reactor runs out of fuel? How is the self-gravitating fusion reactor assembled from the low-density gas of interstellar space? How does rotation affect the self-gravitating fusion reactor's structure?

All these questions will be dealt with in this text; however, let's replace the phrase "self-gravitating fusion reactor" with the word "star," for the sake of brevity.

1.1 Observing the Sun

For those of us living on or near the Earth, the most easily observed star is the Sun. Many generations of astronomers have attempted to determine the length of the astronomical unit (au), originally defined as the average distance from the Earth's center to the Sun's center. Geometric methods, such as diurnal parallax, gave way to radar and to radio telemetry of interplanetary spacecraft. By the twenty-first century, the debate over how to correct for relativistic effects and for the gradually increasing size of the Earth's orbit (resulting from the Sun's mass-energy loss) became frustratingly tangled. Cutting the Gordian knot, the International Astronomical Union (IAU) resolved that the astronomical unit be defined as a conventional unit of length, with 1 au $\equiv 149\,597\,870.7\,\text{km}$. For our purposes, we can state that the length of the semimajor axis of the Earth's orbit is $a = 1\,\text{au}$, and that the perihelion and aphelion distances are $r_{\text{pe}} = 0.9833\,\text{au}$ and $r_{\text{ap}} = 1.0167\,\text{au}$.

The angular diameter of the Sun as seen from Earth ranges from $\theta_{\text{pe}} = 1951\,\text{arcsec}$ at perihelion to $\theta_{\text{ap}} = 1887\,\text{arcsec}$ at aphelion. The radius of the Sun in physical units is thus

$$R_\odot = r_{\text{ap}} \tan\left(\frac{\theta_{\text{ap}}}{2}\right) = r_{\text{pe}} \tan\left(\frac{\theta_{\text{pe}}}{2}\right) \approx 0.004\,65\,\text{au} \approx 696\,000\,\text{km}. \qquad (1.1)$$

This calculation assumes that the Sun has a well-defined radius, despite being a ball of gas rather than a ball of solid rock. In fact, a broadband optical image of the Sun, as seen in Figure 1.1, does have rather well-defined edges. This is because most of the visible light from the Sun comes from the thin **photosphere**, a layer that is only ~400 km thick. Quoting a single radius $R_\odot$ for the Sun also implicitly assumes that the Sun is spherical. Fortunately, this approximation is an excellent one. Although the Sun does have a measurable oblateness, its polar radius is smaller than its equatorial radius by only 5 km, representing a difference of less than one part in 10^5. Given that the Sun is continuously quivering (as a

Figure 1.1 An image of the Sun at visible wavelengths, taken on 2022 May 20, when the Sun had a sunspot group in one hemisphere. [Courtesy of NASA/SDO and HMI science team]

result of seismic waves), and that it is slowly expanding as it evolves, the IAU has recommended the use of a "nominal solar radius," defined as

$$1\,\mathrm{R}_\odot^{\mathrm{N}} \equiv 695\,700\,\mathrm{km}. \tag{1.2}$$

This is the value for the solar radius that we will use in this text.[1]

The solar image in Figure 1.1 shows a number of dark **sunspots** in the Sun's photosphere. Sunspots are regions where the Sun's magnetic field is much stronger than average. A typical magnetic field strength in the Sun's photosphere is $B_\odot \sim 3\,\mathrm{G}$; in a sunspot, however, the field strength can be as high as $B_{\mathrm{spot}} \sim 3000\,\mathrm{G}$. Sunspots appear dark because they are cooler than the surrounding photosphere. While the average photospheric temperature is $T \sim 5800\,\mathrm{K}$, the temperature at the center of a sunspot can be as low as $T \sim 3900\,\mathrm{K}$. (The resulting lower gas pressure within the sunspot compensates for its higher magnetic pressure, and the sunspot remains in pressure equilibrium with the surrounding photosphere.)

The existence of sunspots, which drift only gradually in solar latitude and longitude, permits us to measure the Sun's rotation period. Galileo, for instance, used his sunspot observations to estimate that the Sun's rotation period was roughly

---

[1] The use of a nominal solar radius enables us to make statements such as "When the Sun becomes a red giant, the Sun's radius will be 2.4 solar radii" without causing rampant confusion.

equal to one month ("mese lunare"), or 29 days. Subsequent observations revealed that the Sun is in differential rotation, with a period that ranges from $\mathcal{P}_{rot} = 24.5$ d at its equator to $\mathcal{P}_{rot} = 27.5$ d at latitude $\ell = \pm 45°$. (It's hard to use sunspots to determine the rotation period at higher latitudes, since spots stay fairly close to the equator.) The equatorial rotation period $\mathcal{P}_\odot = 24.5$ d corresponds to an angular speed of $\Omega_\odot = 2\pi/\mathcal{P}_\odot = 2.97 \times 10^{-6}$ s^{-1} and a rotation speed of $v_\odot = \Omega_\odot R_\odot = 2.07$ km s^{-1}.

From the size of the Earth's orbit, $a = 1$ au $= 1.495\,98 \times 10^{13}$ cm, and the length of the sidereal year, $\mathcal{P} = 365.256$ d $= 3.155\,81 \times 10^7$ s, we can use Kepler's third law, as modified by Newton, to find

$$GM_\odot = \frac{4\pi^2 a^3}{\mathcal{P}^2} = 1.3271 \times 10^{26} \text{ cm}^3 \text{ s}^{-2}, \tag{1.3}$$

where $M_\odot$ is the mass of the Sun and G is Newton's gravitational constant. (The mass of the Earth, which is three parts per million of the Sun's mass, can be ignored at this level of accuracy.) The product $GM_\odot$, known as the solar mass parameter, is better known than G and $M_\odot$ are individually. In fact, the IAU has recommended a "nominal solar mass parameter," defined as

$$1\,(GM_\odot)^N \equiv 1.327\,124\,4 \times 10^{26} \text{ cm}^3 \text{ s}^{-2}. \tag{1.4}$$

Using the best available value for the gravitational constant,[2] $G = 6.6743 \times 10^{-8}$ cm^3 g^{-1} s^{-2}, we find that the nominal solar mass parameter implies

$$1\,M_\odot = 1.9884 \times 10^{33} \text{ g}. \tag{1.5}$$

This is the value for the solar mass that we will use in this text.

The mass of the Sun is currently decreasing because of the **solar wind**, an outflow of charged particles from the Sun's extended hot corona. Satellites sent beyond the Earth's magnetosphere have studied the density, speed, and composition of the solar wind. The particles of the solar wind are mainly electrons and protons, with a smaller number of ^{4}He nuclei and other heavier ions. The solar wind is "gusty," with fluctuations in its density and speed. However, averaged over time, the Sun's mass loss rate from the solar wind is $\dot{M}_{wind} \approx 1.2 \times 10^{12}$ g s$^{-1} \approx 2.0 \times 10^{-8}$ M$_\odot$ Myr^{-1}.

At any given instant, the **solar irradiance** is the energy flux of sunlight, integrated over all frequencies, incident on a plane perpendicular to the Sun's rays at a distance of 1 au from the Sun. Since the Earth's atmosphere is very good at absorbing some frequencies of light, the solar irradiance can be measured accurately only by satellites. As shown in Figure 1.2, the solar irradiance varies with time over the solar activity cycle of ~ 11 yr. At solar minimum, when the Sun has few sunspots, flares, or plages (bright regions near sunspots), the solar irradiance is 1.3606×10^6 erg s^{-1} cm^{-2}, with little variation from day to day; at

---

[2] This is the 2018 CODATA recommended value for G, with a relative standard uncertainty of 22 parts per million.

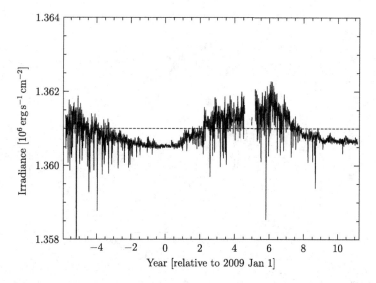

Figure 1.2 Solar irradiance from 2003 Feb to 2020 Feb, when the *Solar Radiation and Climate Experiment (SORCE)* satellite was in operation. During this time, solar minima occurred around 2008 Dec and 2019 Dec, while a solar maximum occurred around 2014 Apr. [Data from *SORCE*]

solar maximum, it averages 1.3614×10^6 erg s^{-1} cm^{-2}, with a larger variation. It may seem counterintuitive that the Sun produces more light when it is covered with cool sunspots at solar maximum; however, the increased light from flares and plages at solar maximum more than makes up for the decreased light from sunspots.

The solar irradiance averaged over a complete solar cycle is called the **solar constant**. A value of $S_\odot = 1.361 \times 10^6$ erg s^{-1} cm^{-2} is typically adopted for the solar constant. This value yields a computed **solar luminosity** of

$$1\,L_\odot = 4\pi a^2 S_\odot = 3.828 \times 10^{33} \text{ erg s}^{-1}. \tag{1.6}$$

This is equal to the IAU's recommended "nominal solar luminosity," and is the value for the solar luminosity that we will use in this text. In addition to emitting photons from its superficial photosphere, the Sun also emits neutrinos from its core. The Sun's *neutrino* luminosity is $L_{\nu,\odot} \approx 0.023\,L_\odot$. The equivalent mass loss rate of all the photons and neutrinos that the Sun tosses away into space is

$$\dot{M}_{\text{rad}} = \frac{L_\odot + L_{\nu,\odot}}{c^2} = 4.36 \times 10^{12} \text{ g s}^{-1} = 6.92 \times 10^{-8} \text{ M}_\odot \text{ Myr}^{-1}. \tag{1.7}$$

Thus, the loss of radiation provides more than three-quarters of the Sun's mass-energy loss, with the solar wind making only a minority contribution. The total mass loss rate of the Sun is

$$\dot{M}_\odot = \dot{M}_{\text{wind}} + \dot{M}_{\text{rad}} \approx 8.9 \times 10^{-8} \text{ M}_\odot \text{ Myr}^{-1}, \tag{1.8}$$

leading to a characteristic mass loss time $\text{M}_\odot / \dot{M}_\odot \approx 11\,000$ Gyr.

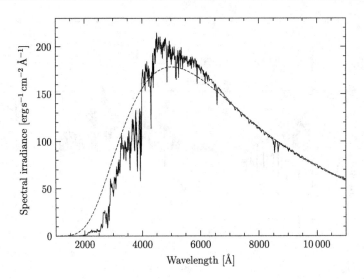

Figure I.3 The solar spectrum as seen 1 au from the Sun, without filtering by the Earth's atmosphere. The dashed line shows a blackbody spectrum with a temperature $T = 5772$ K. [ASTM E-490-00 solar spectrum]

An effective temperature $T_{eff,\odot}$ for the Sun's photosphere can be computed from the relation

$$L_\odot = 4\pi R_\odot^2 \sigma_{SB} T_{eff,\odot}^4, \tag{1.9}$$

where $\sigma_{SB} = 5.6704 \times 10^{-5}$ erg cm^{-2} s^{-1} K^{-4} is the Stefan–Boltzmann constant. (In other words, the Sun's effective temperature is the temperature of a perfect blackbody with the same luminosity and surface area as the Sun.) Using the IAU nominal values for the solar radius and solar luminosity, the effective temperature of the Sun is

$$T_{eff,\odot} = \left(\frac{L_\odot}{4\pi R_\odot^2 \sigma_{SB}} \right)^{1/4} = 5772 \text{ K}. \tag{1.10}$$

Although for some purposes the Sun may be safely approximated as a blackbody with $T = T_{eff,\odot} = 5772$ K, the detailed spectrum of the Sun, shown in Figure 1.3, is not extremely close to that of a blackbody. In particular, the Sun's spectrum is ultraviolet-deficient compared to a blackbody with $T = 5772$ K. The blackbody approximation is better in the near infrared, where there are fewer absorption lines in the Sun's spectrum.

The absorption lines in the solar spectrum give us information on the elements present in the photosphere. For instance, the existence of hydrogen is revealed by the presence of the hydrogen Balmer lines: Hα at $\lambda = 6563$ Å, Hβ at $\lambda = 4861$ Å, and so forth. However, determining the relative abundance of elements in the Sun is not a straightforward process. Some elements do not have detectable photospheric absorption lines; for instance, helium was first discovered from its emission lines in the Sun's chromosphere, the hotter layer just above the

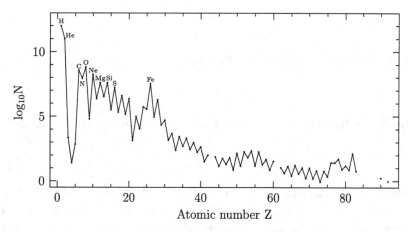

Figure 1.4 Abundances (by number) of elements present in the Sun at its formation; abundances are normalized to $N(H) = 10^{12}$ hydrogen atoms. The 10 most abundant elements are labeled. [Data from Lodders 2021]

photosphere. In addition, the abundances in the Sun's photosphere today are not the same as the abundances the Sun started with. Heavy elements settle slowly inward through diffusion, while unstable elements such as uranium undergo decay. Taking into account these effects, Figure 1.4 shows the reconstructed abundances of the protosolar nebula from which the Sun formed 4.57 Gyr ago.

Figure 1.4 doesn't display all the information that we have about protosolar abundances. In particular, it bins together all the different isotopes of each element. It is admittedly true that the most common elements are overwhelmingly made of a single isotope. For instance, in the protosolar nebula, hydrogen was 99.998% ordinary hydrogen (^1H) and only 0.002% deuterium (^2H) by number, while helium was 99.983% ordinary helium (^4He) and only 0.017% light helium (^3He). However, as we discuss in Section 5.3, the amount of ^2H and ^3He present in a star provides a useful probe of nuclear fusion conditions. Thus, we do sometimes care about scarce isotopes. (In addition, some elements do not have an overwhelmingly dominant isotope; for instance, bromine is 50.7% ^{79}Br and 49.3% ^{81}Br by number.)

By contrast, other astrophysical problems are largely indifferent to the exact details of elemental and isotopic abundances. For these problems, we need only three broad categories: (1) hydrogen, (2) helium, and (3) all other elements combined. Suppose that the mass density of gas is ρ, the number density of hydrogen nuclei is n_H, and the number density of helium nuclei is n_{He}. The mass of a hydrogen atom is $m_H = 1.674 \times 10^{-24}$ g; at solar abundance, the presence of ^2H raises the mean atomic mass by only 1 part in 25 000. The mass of a helium atom is $m_{He} = 6.646 \times 10^{-24}$ g; at solar abundance, the presence of ^3He lowers the mean atomic mass by only 1 part in 8000. Knowing the atomic masses, we can convert from the number density n_H to the mass fraction of hydrogen,

$$X \equiv \frac{n_H m_H}{\rho},$$ (1.11)

and from the number density n_{He} to the mass fraction of helium,

$$Y \equiv \frac{n_{He} m_{He}}{\rho}.$$ (1.12)

This means that the mass fraction of "metals," or all heavier elements combined,[3] is

$$Z \equiv 1 - X - Y.$$ (1.13)

The protosolar abundances are estimated to have been $X_{\odot,0} = 0.706$, $Y_{\odot,0} = 0.277$, and $Z_{\odot,0} = 0.017$. In the Sun's photosphere today, after diffusive settling of elements heavier than hydrogen, the abundances are $X_\odot = 0.739$, $Y_\odot = 0.246$, and $Z_\odot = 0.015$. The most abundant metal in the Sun is oxygen, which contributed about 43% of the protosolar metal mass. Carbon provided 18%, neon provided 14%, and iron provided 8% of the metallicity by mass. (For comparison with the solar values, the primordial abundances that came out of Big Bang Nucleosynthesis were $X_p \approx 0.753$ and $Y_p \approx 0.247$; the primordial mass fraction of metals was $Z_p \approx 3 \times 10^{-9}$, mostly in the form of ^7Li.)

Because we know the Sun better than we know any other star, our standard unit of length will be the solar radius, with $1\,R_\odot = 6.957 \times 10^{10}$ cm, our standard unit of mass will be the solar mass, with $1\,M_\odot = 1.9884 \times 10^{33}$ g, and our standard unit of power will be the solar luminosity, with $1\,L_\odot = 3.828 \times 10^{33}$ erg s^{-1}. (Since the Sun's volume is $V_\odot = 4\pi R_\odot^3/3 = 1.410 \times 10^{33}$ cm^3, we have the useful mnemonic that the Sun's most important properties are all $\sim 10^{33}$ in cgs units.)

1.2 Observing Other Stars

Determining the properties of stars other than the Sun is frequently made difficult by their large distance. The most reliable method of finding the distance to the nearest stars is trigonometric parallax. As seen from Earth over the course of one year, the apparent motion of a star on the sky can be fitted as a combination of linear proper motion (from the star's motion relative to the Sun) and a parallactic ellipse (from the Earth's orbital motion around the Sun). The semimajor axis of the ellipse, in angular units, is the parallax p of the star. The parallax is related to the star's distance in parsecs (pc) by the equation

$$\frac{d}{1\,pc} = \frac{1\,arcsec}{p}.$$ (1.14)

---

[3] Astronomers commonly use the term "metals" to mean "elements other than hydrogen or helium." A metallurgist might well be annoyed by this misuse of the word "metal," but we will accept it as a colorful metaphor and move onward.

Given this relation, $1 \, \mathrm{pc} = 206\,265 \, \mathrm{au}$, just as $1 \, \mathrm{radian} = 206\,265 \, \mathrm{arcsec}$. As an example, the nearby star Proxima Centauri has a parallax $p_{\mathrm{prox}} = 0.768\,07 \, \mathrm{arcsec}$. This parallax implies a distance $d_{\mathrm{prox}} = 1/p_{\mathrm{prox}} = 1.3020 \, \mathrm{pc}$; this can also be expressed as $d_{\mathrm{prox}} = 268\,550 \, \mathrm{au}$.[4]

Proxima Centauri is famously the Sun's nearest neighbor among the stars. Measuring accurate parallaxes for significantly more distant stars typically requires dedicated space-based missions. For instance, the *Gaia* spacecraft was launched in 2013 with the goal of measuring the parallax of $\sim 10^9$ stars, with an accuracy ranging from $\sigma_p \sim 10^{-5} \, \mathrm{arcsec} \sim 0.01 \, \mathrm{mas}$ for the apparently brightest stars in its sample to $\sigma_p \sim 0.3 \, \mathrm{mas}$ for the faintest.

Measuring the angular diameter of stars, given their large distances, typically requires interferometric techniques. The star (other than the Sun) that has the largest angular size as seen from Earth is R Doradus, a red variable star. Its angular diameter has been measured in the near infrared using aperture masking interferometry, with the result $\theta_{\mathrm{RD}} = 57 \pm 5 \, \mathrm{mas}$. Its parallax is $p_{\mathrm{RD}} = 18.31 \pm 0.99 \, \mathrm{mas}$, yielding a distance $d_{\mathrm{RD}} = 54.6 \pm 3.0 \, \mathrm{pc}$. Together, these values imply a physical radius $R_{\mathrm{RD}} = 1.56 \pm 0.16 \, \mathrm{au}$. Thus, R Doradus is physically much larger than the Sun, with $R_{\mathrm{RD}} \approx 330 \, R_\odot$.

R Doradus is far from being the largest star in our galaxy. Consider Betelgeuse (α Orionis), which is also a red variable star. The diameter of Betelgeuse varies with time; in the year 2019, measurements in the near infrared gave an angular diameter of $\theta_{\mathrm{bet}} = 42.61 \pm 0.05 \, \mathrm{mas}$.[5] The parallax of Betelgeuse is poorly known, in part because p_{bet} is small compared to the angular size θ_{bet}, and in part because Betelgeuse is variable in shape as well as size. If we assume $p_{\mathrm{bet}} = 5.5 \pm 1.0 \, \mathrm{mas}$, consistent with recent measurements, this implies a physical radius of $R_{\mathrm{bet}} = 3.9 \pm 0.7 \, \mathrm{au}$. Thus, the radius of Betelgeuse is $R_{\mathrm{bet}} \sim 2.5 R_{\mathrm{RD}} \sim 800 \, R_\odot$. Resolved images of Betelgeuse have been taken at visible and ultraviolet wavelengths, as seen in Figure 1.5; however, these images include the extended outer atmosphere of Betelgeuse (which is perceptibly non-spherical). The ultraviolet angular diameter of Betelgeuse, as shown in the right panel of Figure 1.5, is $\theta_{\mathrm{uv}} \approx 120 \, \mathrm{mas}$, nearly three times the size of the near-infrared photosphere of Betelgeuse; this corresponds to a physical radius of $R_{\mathrm{uv}} \sim 11 \, \mathrm{au}$ for the UV-emitting outer atmosphere.

Although the Sun is small compared to R Doradus and Betelgeuse, it is by no means a midget among stars. Consider, for instance, the stars of the α Centauri system. Within this triple star system, Proxima Centauri (at $d = 1.302 \, \mathrm{pc}$ from the Sun) is loosely bound to the tight binary α Centauri AB (at $d = 1.332 \, \mathrm{pc}$). Using interferometric techniques, the angular diameter of α Centauri A, the brighter star in the binary, is measured to be $\theta_{\mathrm{A}} = 8.512 \pm 0.022 \, \mathrm{mas}$, while that

---

[4] Observed parallaxes (and other properties) of some example stars are given in Table B.1 of Appendix B.

[5] The first interferometric measurement of the angular diameter of Betelgeuse was in 1920, when Michelson and Pease found $\theta_{\mathrm{bet}} = 47 \pm 5 \, \mathrm{mas}$ at visible wavelengths.

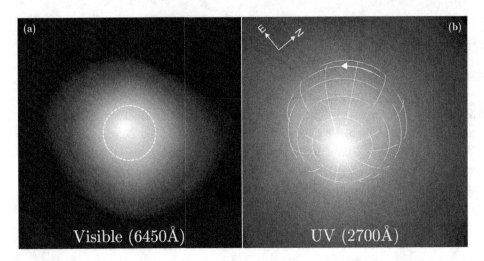

Figure 1.5 (a) Betelgeuse at $\lambda \sim 6450\,\text{Å}$ (*VLT*: 2019 Jan). Field of view is $180 \times 180\,\text{mas}$. Dashed circle = size of photosphere. North is up, east is left. [ESO/M. Montargès *et al.* 2021] (b) Betelgeuse at $\lambda \sim 2700\,\text{Å}$ (*HST*: 1995 Mar). Field of view is $180 \times 180\,\text{mas}$. Overlay indicates the orientation and rotation of Betelgeuse. [Uitenbroek *et al.* 1998]

of α Centauri B is $\theta_B = 6.002 \pm 0.048\,\text{mas}$ and that of Proxima Centauri is $\theta_{\text{prox}} = 1.02 \pm 0.08\,\text{mas}$. Given the distances to these stars, the physical radii of the stars in the binary system are $R_A = 1.22\,R_\odot$ and $R_B = 0.86\,R_\odot$, while little Proxima Centauri has $R_{\text{prox}} = 0.14\,R_\odot$. This means that the radius of Proxima Centauri is only $\sim 40\%$ bigger than that of the planet Jupiter, and is significantly smaller than the radius of puffy "hot Jupiter" exoplanets, swollen by absorbing radiative energy from their nearby parent star.

In addition to having a wide range of sizes, stars have a wide range of rotation speeds. One way to measure the rotation speed v_{rot} of a star is through the rotational broadening of the star's absorption lines. One problem with using rotational broadening to determine v_{rot} is that the line width tells you only $v_{\text{rot}} \sin i$, where i is the inclination of the star's rotation axis relative to the line of sight. Since the inclination is not known *a priori*, the line width generally gives only a lower limit on the value of v_{rot} for any particular star. Another problem with using rotational broadening is that the star's absorption lines also show thermal broadening. For atomic hydrogen, the root mean square thermal speed is

$$v_{\text{th}} = \left(\frac{3kT}{m_{\text{H}}}\right)^{1/2} \approx 12\,\text{km}\,\text{s}^{-1}\left(\frac{T}{5772\,\text{K}}\right)^{1/2}. \tag{1.15}$$

For slowly rotating stars like the Sun, thermal broadening is larger than the rotational broadening. However, for bright stars with high-resolution spectra, thermal broadening can be disentangled from rotational broadening since the thermal line profile is Gaussian, while the rotational line profile is non-Gaussian. (The exact shape of the rotational line profile depends on the limb darkening of the star; generally speaking, however, rotational line profiles lack the broad exponential wings of thermal line profiles.) For example, α Centauri A has $v_{\rm rot} \sin i = 2.7 \pm 0.7\,{\rm km\,s}^{-1}$; unless we happen to be looking at the star almost pole-on, this indicates that the rotation speed of α Centauri A is only slightly larger than that of the Sun. Since the radius of α Centauri A is also slightly larger than that of the Sun, this indicates that they have a similar rotation period $\mathcal{P}_{\rm rot}$. Some stars, however, have much greater rotation speeds. For instance, the star Altair (α Aquilae) has $v_{\rm rot} \sin i = 240\,{\rm km\,s}^{-1}$. Rotation at such high speeds, as we discuss in Chapter 10, profoundly affects the structure and evolution of a star.

The mass of a star is most readily determined if it is part of a binary system, with a companion that could be another star, a stellar remnant, a brown dwarf, or a planet. If the two bodies in the system have masses $M_{\rm A}$ and $M_{\rm B}$, then Newton's modification of Kepler's Third Law tells us the total mass of the system:

$$G(M_{\rm A} + M_{\rm B}) = \frac{4\pi^2 a^3}{\mathcal{P}^2}, \tag{1.16}$$

where $\mathcal{P}$ is the orbital period, and a is the semimajor axis of the relative orbit of the two bodies. In a visual binary like α Centauri AB, the motion of the two bodies relative to each other can be traced on the sky, as shown in Figure 1.6. We are actually seeing their orbit nearly edge-on, which explains why α Centauri A is so far from a focus of the projected ellipse in Figure 1.6. Using modern precision astrometry, the angular size of the orbit after deprojection is found to be $a_\alpha = 17.493 \pm 0.009$ arcsec. Given a parallax $p_\alpha = 750.8 \pm 0.4$ mas for the α Centauri system, this translates to a semimajor axis length of $a_\alpha = 23.30 \pm 0.02$ au. The orbital period of α Centauri A and B is $\mathcal{P}_\alpha = 79.76 \pm 0.02$ yr. The total mass of the α Centauri binary system is then

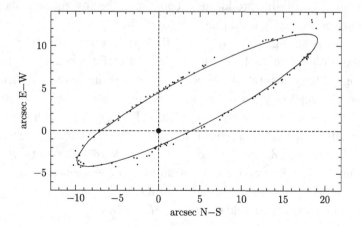

Figure 1.6 Measured position of α Centauri B (small dots) relative to α Centauri A (large dot) during the interval AD 1824–1893. As seen from Earth, α Cen B orbits counterclockwise. [Data from See 1893]

$$G(M_A + M_B) = \frac{4\pi^2 a_\alpha^3}{P_\alpha^2} = 2.639 \times 10^{26} \, \text{cm}^3 \, \text{s}^{-2} = 1.988 \, GM_\odot. \tag{1.17}$$

The allocation of mass between the two stars in a visual binary can be found by determining the barycenter (or center of mass) of the system, which lies along the line connecting the centers of the two stars, and which is at a distance

$$d_{\text{bary}} = a \left(\frac{M_B}{M_A + M_B} \right) \tag{1.18}$$

from star A. Over the course of a century or two, the proper motion of the barycenter will be well fitted by motion at constant speed along a straight line.[6] After taking out the annual wiggles from parallax and stellar aberration, it is found that the best fit for constant proper motion of the α Centauri system is $\mu_\alpha = 3.71 \, \text{arcsec} \, \text{yr}^{-1}$ with $d_{\text{bary}} = 0.457 a_\alpha$. This implies that α Centauri A is the more massive of the two stars, with $M_A = 1.079 \, M_\odot$ and $M_B = 0.909 \, M_\odot$.

The mass of a star can decrease with time if it has a **stellar wind**. The Sun's mass loss rate, $\dot{M}_\odot \approx 2 \times 10^{-8} \, M_\odot \, \text{Myr}^{-1}$, is tiny compared to that of some other classes of star. If a stellar wind leads to a high value of $\dot{M}$, it can be detected through the presence of a P Cygni line profile in the star's spectrum. P Cygni profiles, named after the star in whose spectrum they were first seen, combine blueshifted absorption with redshifted emission. Figure 1.7(a) shows the geometry of a stellar wind that gives rise to a P Cygni profile. The gas in the stellar wind is accelerated away from the star to an asymptotic velocity v_∞. The region labeled B_a contains gas seen in absorption against the star's photosphere; it thus produces a blueshifted absorption line, shown schematically as the dotted line in Figure 1.7(b). The region labeled B_e contains blueshifted gas seen in emission, while the region labeled R_e contains redshifted gas seen in emission. Together, these regions produce the emission line shown as the dashed line in Figure 1.7(b). (The region labeled O is occulted by the opaque star, and thus is unseen by the observer.) The P Cygni line produced by combining the absorption and emission lines is shown as the heavy solid line in Figure 1.7(b).

If a star with a strong stellar wind has P Cygni profiles in its spectrum, the width of the profile (Figure 1.7) tells us the asymptotic speed v_∞ of the wind. Determining the mass loss rate $\dot{M}$ associated with the wind is more difficult. Usually, it involves modeling spectral line formation in a spherically symmetric wind with constant $\dot{M}$, then comparing the computed P Cygni lines with the observed lines. For example, the spectral lines of the star P Cygni itself are consistent with a wind that has $v_\infty \approx 185 \, \text{km} \, \text{s}^{-1}$ and $\dot{M} \approx 30 \, M_\odot \, \text{Myr}^{-1}$, flowing away from a star of radius $R_\star \approx 75 \, R_\odot$ and mass $M_\star \sim 30 \, M_\odot$. The extravagant mass loss rate of P Cygni, more than a billion times that of the Sun, is enough to whittle away the star to nothing on a timescale of $t = M_\star / \dot{M} \sim 1 \, \text{Myr}$.

---

[6] The orbital period of α Centauri relative to Proxima Centauri is over 0.5 Myr; the orbital period of α Centauri relative to the galactic center is $\sim$240 Myr.

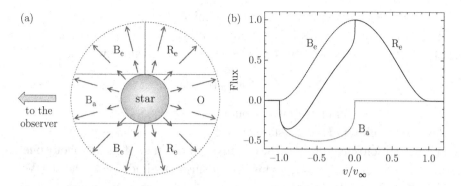

Figure 1.7 (a) Star with an accelerated stellar wind. (b) Resulting P Cygni profile in the star's spectrum, shown as a function of radial velocity relative to the star. Dotted line = blueshifted absorption. Dashed line = redshifted and blueshifted emission. Solid line = combined line profile.

Measuring the bolometric flux of a star (that is, its flux of photon energy integrated over all frequencies) is notoriously difficult. It's much simpler to measure the flux in one of the passbands for which the Earth's atmosphere is largely transparent. For instance, the Johnson–Cousins $UBVRI$ filters have long been in standard use; they stretch in effective wavelength from $\lambda_U \approx 3660\,\text{Å}$ for the U band to $\lambda_I \approx 7980\,\text{Å}$ for the I band. The Johnson–Cousins V band was originally devised to approximate the earlier "visual magnitude" system, based on the response of the human eye. The V band is centered at an effective wavelength of $\lambda_V \approx 5450\,\text{Å}$ (yellowish-green), and has a width of $\Delta\lambda \approx 840\,\text{Å}$. In the V band, the **apparent magnitude** m_V is related to the flux F_V by the relation

$$m_V = -2.5\log_{10}(F_V/F_{V,0}), \tag{1.19}$$

where the reference flux $F_{V,0}$ is chosen, for historical reasons, so that the star Vega has $m_V = 0.03$. At somewhat shorter wavelengths, the Johnson–Cousins B band is centered at $\lambda_B \approx 4360\,\text{Å}$ (blueish-indigo), and has a width of $\Delta\lambda \approx 890\,\text{Å}$. The apparent B magnitude can be defined as

$$m_B = -2.6\log_{10}(F_B/F_{B,0}), \tag{1.20}$$

where the reference flux is chosen so that Vega has $m_B = 0.03$.

Just as the apparent magnitude m is a logarithmic measure of a star's flux, the **absolute magnitude** M of a star is a logarithmic measure of its luminosity. The absolute magnitude is defined as the apparent magnitude that a star would have if it were at a distance $d = 10\,\text{pc}$. In the V band, for instance,

$$M_V \equiv m_V - 5\log_{10}\left(\frac{d}{10\,\text{pc}}\right) = m_V + 5\log_{10}\left(\frac{p}{100\,\text{mas}}\right). \tag{1.21}$$

We can define L_V as the portion of a star's luminosity that is detectable in the V band. For instance, the Sun's V band luminosity is $L_{V,\odot} = 4.6 \times 10^{32}$ erg s$^{-1} \approx$ $0.12\,L_\odot$. In terms of luminosity, the absolute magnitude can be written as

$$M_V = -2.5\log_{10}(L_V/L_{V,0}),\qquad(1.22)$$

where the reference luminosity works out to be $L_{V,0} = 85.4 L_{V,\odot}$.

The $B - V$ **color index** of a star, defined as $B - V \equiv m_B - m_V$, is useful to know since it is closely related to the effective temperature of a star's photosphere. Given the normalization of the Johnson–Cousins magnitude scale, the star Vega has $B - V = 0$; its effective temperature is $T_{\text{eff}} \approx 9600$ K. An empirical relation for solar-metallicity stars with colors in the range $-0.1 < B - V < 1.4$ is

$$T_{\text{eff}} \approx \frac{9630\,\text{K}}{1 + 1.05(B - V)}.\qquad(1.23)$$

For hotter stars ($B - V < -0.1$, $T_{\text{eff}} > 11\,000$ K), the $B - V$ color is a poor temperature diagnostic; at these high temperatures, both the B and V band are in the Rayleigh–Jeans tail of a blackbody, and the $B - V$ color index becomes nearly independent of temperature. For cooler stars ($B - V > 1.4$, $T_{\text{eff}} < 4000$ K), the $B - V$ color is also a poor temperature diagnostic; at these low temperatures, very little light emerges in the B band, and it becomes difficult to measure m_B. (For hot stars, the $U - B$ color index provides a useful temperature estimate, while for cool stars, $V - R$ or $V - I$ are often used.)

The effective temperature of a star is also closely related to its **spectral type**. In 1890, Edward Pickering and Williamina Fleming published a scheme in which each star's spectrum was assigned a letter from A through Q, based on the relative strength of various absorption lines, most notably the hydrogen Balmer lines. Further work eliminated erroneous spectral types: type C, for example, corresponding to stars with doubled Balmer lines, turned out to be due to "faulty focusing." In addition, some spectral types were merged: type E was folded into the G category, for instance, and type H was folded into K. By the year 1901, Annie Jump Cannon realized that if the remaining letters were arranged in the order OBAFGKM, they formed a logical sequence in which various absorption lines continuously increase or decrease in strength going from one letter to the next. For instance, the hydrogen Balmer lines increase steadily in strength going from O to A, then decrease from A to M. The spectra used by Cannon had sufficiently high resolution that she could further subdivide each letter in the OBAFGKM sequence into numbered subtypes. For example, G stars are arranged from G0 (closest in spectral appearance to F stars) to G9 (closest to K stars). Sometimes, fractional subtypes are used to produce finer subdivisions; for example, the star Proxima Centauri has spectral type M5.5. The OBAFGKM spectral types and their subdivisions became the standard system for classifying stellar spectra; they are sometimes referred to as the **Harvard spectral types**.

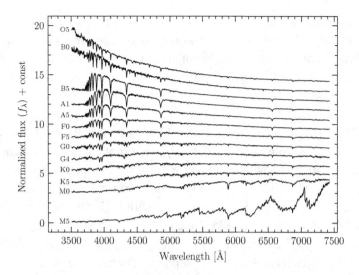

Figure 1.8 Spectra of stars with different spectral type, ranging from hottest (top) to coolest (bottom). Spectra are normalized to have $f_\lambda = 1$ at $\lambda = 5500\,\text{Å}$, then shifted vertically to avoid overlap. [Data from Jacoby *et al.* 1984]

Since O stars are seen to be blue ($B - V < -0.3$) and M stars are red ($B - V > 1.4$), it was correctly assumed that the OBAFGKM sequence is a temperature sequence from hot to cool photospheres. By the year 1921, the physicist Meghnad Saha was able to use his ionization equation to translate spectral types into quantitative temperatures on the Kelvin scale. An O star at $T > 22\,000\,\text{K}$, Saha calculated, has weak Balmer lines because the hydrogen in its photosphere is mostly ionized; conversely, an M star at $T < 4500\,\text{K}$ has weak Balmer lines because its hydrogen is mostly in the ground state. Saha found that $T \sim 12\,000\,\text{K}$ is the temperature at which the largest fraction of hydrogen atoms have electrons in the $n = 2$ energy level, ready to absorb Balmer photons and make the strong Balmer absorption lines characteristic of A stars. (Saha's 1921 calculations were slightly revised in the following decades; the temperature of maximum Balmer absorption, for instance, is now set at 9600 K.) Figure 1.8 shows spectra of stars with different effective temperature, and thus different spectral type; notice the strong Balmer absorption lines for type A1. It is also notable that the M5 star, which has $T_{\text{eff}} \sim 3000\,\text{K}$, is cool enough to display broad absorption bands from the molecules that exist in its atmosphere.

In the 1930s, astronomers realized that a two-dimensional spectral classification scheme would be a more informative way of describing the spectra of stars. The one-dimensional OBAFGKM scheme is useful because it correlates well with the effective temperature of a star. However, consider a pair of stars with the same effective temperature T_{eff} but with different values for their surface gravity $g \equiv GM_\star/R_\star^2$. The star with a higher value of g will have a higher pressure in its photosphere, in order to support its heavier atmosphere. This means that the higher-g star will show greater **pressure broadening** in its absorption lines. (Pressure broadening, also called collisional broadening, results from the frequent collisions between particles in a high-pressure gas; this is distinct from ordinary thermal broadening, which is the same for all stars with the same effective temperature.)

In a 1937 paper, W. W. Morgan pointed out that stars have a wide range of surface gravity g and thus have a detectably wide range of pressure broadening. For his example of a star with low surface gravity, Morgan adopted Betelgeuse. Given $M_{bet} \approx 18\,M_\odot$ and $R_{bet} \sim 800\,R_\odot$, the surface gravity of Betelgeuse is $g_{bet} \sim 3 \times 10^{-5} g_\odot$. For his example of a star with high surface gravity, Morgan used the nearby star Lalande 21185. Given $M_{lal} = 0.39\,M_\odot$ and $R_{lal} = 0.39\,R_\odot$, the surface gravity of Lalande 21185 is $g_{lal} = 2.5 g_\odot$, roughly 10^5 times the surface gravity of Betelgeuse. Thus, "supergiant" stars like Betelgeuse can be distinguished from "dwarf" stars like Lalande 21185 by the smaller pressure broadening of their absorption lines. Notice also that the surface area of Betelgeuse is $\sim 4 \times 10^6$ times that of Lalande 21185; since their effective temperatures are similar, this means that the luminosity of Betelgeuse must be $\sim 4 \times 10^6$ times that of Lalande 21185. Thus, Morgan's plan for classifying stars by the pressure broadening of their absorption lines was a way of sorting them by luminosity as well as by surface gravity.

Morgan and his collaborator Philip Keenan developed what is now called the MK system for spectral classification. The MK system combines the Harvard spectral types (OBAFGKM) with a system of **luminosity classes**, which might also be termed "surface gravity classes." Each luminosity class is designated by a Roman numeral, from I to V. Luminosity class I corresponds to stars with the narrowest lines at a given effective temperature; these are the highly luminous supergiant stars. Luminosity class III corresponds to less luminous giant stars, and luminosity class V, with the broadest lines, corresponds to dwarf stars of still lower luminosity.[7] The intermediate classes II and IV are called "bright giants" and "subgiants," respectively. Further refinements exist to the system of luminosity classes. For instance, it is useful to subdivide luminosity class I into two subclasses, with Ia representing brighter supergiants, and Ib the less extreme supergiants.

The complete MK **spectral class** of a star contains information about both a star's effective temperature and its luminosity. In Table B.1, for instance, the Sun and its neighbor α Cen A are both listed as spectral class G2V; that is, they are both dwarf (or main sequence) stars with effective temperature $T_{eff} \approx 5770\,\mathrm{K}$. Although Betelgeuse, R Doradus, and Proxima Centauri are all cool red stars of spectral type M, the width of their absorption lines reveals that Betelgeuse (M2Ib)

---

[7] Stars of luminosity class V are also called "main sequence" stars, for reasons explained in the next section.

is a red supergiant, R Doradus (M8III) is a red giant, and Proxima Centauri (M5.5V) is a red dwarf.

Determining the detailed elemental abundances of distant, faint stars is difficult. Thus, in place of the metallicity Z (the mass fraction of all metals added together), astronomers frequently use the abundance of a single metal, or a limited subset of metals, as a proxy for Z. One number that is often used to describe the metallicity of a star is the **iron abundance** [Fe/H]. If the number density of iron atoms in a star's photosphere is n_{Fe}, and the number density of hydrogen atoms is n_H, then

$$[Fe/H] \equiv \log_{10}(n_{Fe}/n_H) - \log_{10}(n_{Fe}/n_H)_\odot. \tag{1.24}$$

In the Sun's photosphere today,

$$\log_{10}(n_{Fe}/n_H)_\odot = -4.55, \tag{1.25}$$

or 1 iron atom for every 35 000 hydrogen atoms. The iron abundance is used as a proxy for Z, despite the fact that oxygen and carbon are the most abundant metals, because iron has a set of strong absorption lines visible over a wide range of stellar temperatures. For stars in our galaxy, the iron abundance usually lies in the range $-3 < [Fe/H] < +0.5$. However, metal-poor stars with $[Fe/H] < -3$ are known to exist; there have even been a few hyper-metal-poor stars discovered, with $[Fe/H] < -5$.

Knowing the value of [Fe/H] for a star doesn't tell you everything about its chemical composition. Just as the ratio of iron to hydrogen varies from one star to another, so does the ratio of other metals to iron. Consider magnesium, for instance, a relatively common metal whose most abundant isotope is ^{24}Mg. In the Sun, the magnesium-to-iron ratio is

$$\log_{10}(n_{Mg}/n_{Fe})_\odot = +0.08, \tag{1.26}$$

or six magnesium atoms for every five iron atoms. Figure 1.9(a) shows the value of

$$[Mg/Fe] \equiv \log_{10}(n_{Mg}/n_{Fe}) - \log_{10}(n_{Mg}/n_{Fe})_\odot \tag{1.27}$$

as a function of [Fe/H] for a sample of main sequence stars with $4200\,\mathrm{K} < T_{\mathrm{eff}} < 6700\,\mathrm{K}$.

The range of [Mg/Fe] for this sample is smaller than the range in [Fe/H]; it is generally true that stars that are poor in iron are also poor in magnesium. However, notice that stars with $[Fe/H] < -0.7$ have a tendency to be enhanced in magnesium relative to iron, with $[Mg/Fe] \sim +0.25$, on average. (To put this another way, a star whose iron abundance is 20% that of the Sun would most likely have a magnesium abundance ~35% that of the Sun.) Figure 1.9(a) would look very similar if, instead of magnesium, we used another one of the **alpha elements**. The alpha elements, ^{20}Ne, ^{24}Mg, ^{28}Si, ^{32}S, ^{36}Ar, and ^{40}Ca, can be thought of as

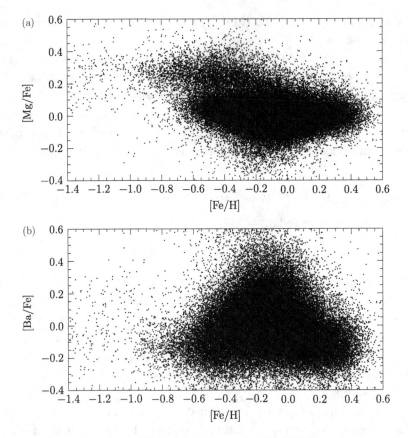

Figure 1.9 (a) [Mg/Fe] versus [Fe/H] for a sample of ∼85 800 main sequence stars. (b) [Ba/Fe] versus [Fe/H] for the same stars. [Data from GALAH Data Release 3]

being assembled from an integral number of ^{4}He nuclei.[8] The similarity of abundance patterns among the alpha elements suggests that they are made by the same process within stars.

Elements more massive than iron are relatively scarce within the Sun, as shown in Figure 1.4. As an example, consider barium. The most abundant isotope of barium is ^{138}Ba; this is far more massive than ^{56}Fe, the most abundant isotope of iron. In the Sun, the barium-to-iron ratio is

$$\log_{10}(n_{Ba}/n_{Fe})_\odot = -5.23, \tag{1.28}$$

or only 1 barium atom for every 170 000 iron atoms. Although barium and other heavy elements are scarce, they nonetheless exist, and we would like to have some understanding of how they formed. A plot of [Ba/Fe] versus [Fe/H], as shown in Figure 1.9(b), shows a distinctly different pattern than the equivalent

---

[8] To particle physicists, a ^{4}He nucleus is also known as an "alpha particle"; this explains how the "alpha elements" got their name.

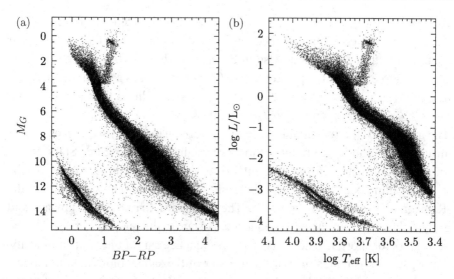

Figure 1.10 H–R diagram for $\sim$227 000 stars in *Gaia* Early Data Release 3 with measured parallax $p > 10$ mas, parallax uncertainty $\sigma_p/p < 0.01$, and apparent magnitude $m_G < 18$. (a) Absolute magnitude M_G versus $BP - RP$ color index. (b) Bolometric luminosity versus effective temperature. [Data from *Gaia Catalogue of Nearby Stars*]

plot for magnesium. This suggests that barium is formed by a process quite different from the mechanism for making alpha elements. In particular, note that stars with [Fe/H] < -0.7 have a slight tendency to be depleted in barium relative to iron; [Ba/Fe] ~ -0.05, on average, but with a large scatter. (To put this another way, a star whose iron abundance is 20% that of the Sun would most likely have a barium abundance $\sim$18% that of the Sun.)

1.3 Correlations among Properties

The Hertzsprung–Russell (H–R) diagram, named after Ejnar Hertzsprung and Henry Norris Russell, is a plot of stars' absolute magnitudes (or luminosities) as a function of their color indices (or effective temperatures, or Harvard spectral types). Figure 1.10, for instance, shows an H–R diagram for a sample of nearby stars viewed by the *Gaia* satellite. In Figure 1.10(a), the absolute magnitude M_G is given in the *Gaia* broad G passband, which has an effective wavelength of $\lambda_G \approx 6700$ Å, but whose transmissivity stretches at half-maximum from $\lambda \approx 4000$ Å to 8600 Å, embracing the entire visible spectrum plus a bit of the near infrared. The $BP - RP$ color in Figure 1.10 comes from comparing the *Gaia* blue photometer magnitude G_{BP} to the red photometer magnitude G_{RP}. The BP passband, roughly corresponding to the short-wavelength half of the broader G band, has an effective wavelength of $\lambda_{BP} \approx 5300$ Å. The RP passband, corresponding to the long-wavelength half of the G band, has $\lambda_{RP} \approx 8000$ Å. (To orient you in Figure 1.10, the Sun has $M_G \approx 4.6$ and $BP - RP \approx 0.8$ in the *Gaia* passbands.)

In Figure 1.10(b), the absolute magnitude M_G is converted to a luminosity L, using a bolometric correction to account for the fact that not all the light of a star is contained within the G band. Also in Figure 1.10(b), the $BP - RP$ color index is converted to effective temperature. The $BP - RP$ color is a useful temperature diagnostic over a wide range of T_{eff}; however, its usefulness fails for stars with $BP - RP > 4.5$, corresponding to $T_{eff} < 2500$ K.

The majority of the stars in this *Gaia* sample are on the **main sequence** running from upper left (hot and luminous) to lower right (cool and dim). Stars on the main sequence, such as Proxima Centauri, α Centauri A and B, Vega, and the Sun, are of luminosity class V. The intrinsic width of the main sequence on the H–R diagram is small, but non-zero. The **red giant branch** runs upward and slightly rightward from the main sequence. A star on the red giant branch, called a "red giant" for short, is redder than a main sequence star of the same luminosity, and more luminous than a main sequence star of the same color. Most red giants are of spectral type K or M and luminosity class III; the nearest red giant to the Sun, at $d = 10.4$ pc, is Pollux, of spectral class K0III. Pollux has $L \approx 33\,L_\odot$ and $T_{eff} \approx 4600$ K.

On the red giant branch, a concentration of stars can be seen at $M_G \approx 0.5$ and $BP - RP \approx 1.2$, corresponding to $L \sim 60\,L_\odot$ and $T_{eff} \sim 5000$ K; this concentration is known as the **red clump**. (The name "red clump" is something of a misnomer, since these stars are hot enough to be a yellow-orange color.) Red clump stars have spectral type from G8 through K2 and luminosity class III; the nearest red clump star to the Sun, at $d = 13.2$ pc, is Capella Aa, the brightest star in the Capella system. Capella Aa has $L \approx 80\,L_\odot$ and $T_{eff} \approx 5000$ K.

Below and to the left of the main sequence lie the **white dwarfs**, which are bluer than a main sequence star of the same luminosity, and less luminous than a main sequence star of the same color. The nearest white dwarf to the Sun, at $d = 2.67$ pc, is Sirius B, companion to the bright main sequence star Sirius A. The white dwarf Sirius B has $L \approx 0.03\,L_\odot$ and $T_{eff} \approx 26\,000$ K, implying a radius of $R \approx 0.009\,R_\odot$, about the same size as the Earth.

The H–R diagram of Figure 1.10 includes only stars that are our near neighbors, within $d = 100$ pc of the Sun. Since few stars within 100 pc are more luminous than $300\,L_\odot$, we have truncated Figure 1.10 at that maximum luminosity.[9] To get a reasonably large sample of highly luminous stars, we must go deeper into space. Figure 1.11 is an example of an H–R diagram that focuses primarily on bright stars. In this figure, the main sequence is seen to extend as far as the hottest, most luminous O stars at $T_{eff} \sim 45\,000$ K and $L \sim 600\,000\,L_\odot$. Across the top of the H–R diagram, supergiant stars are seen at both high and low temperatures. The nearest supergiant to the Sun, at $d \approx 96$ pc, is Canopus, with spectral class F0Ib. (Canopus has also been assigned the spectral class A9II; the distinction between a

---

[9] We apologize for this snub to R Doradus, which has $L \approx 5600\,L_\odot$ at $d \approx 55$ pc.

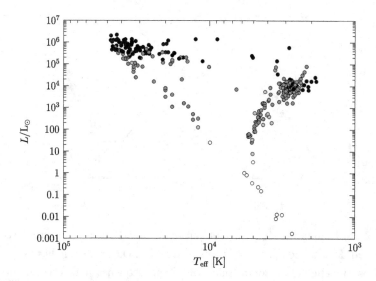

Figure 1.11 H–R diagram showing stars with well determined mass loss rates. White: $\log \dot{M} < -6$ (in units of $M_\odot \, \mathrm{Myr}^{-1}$). Light gray: $-6 < \log \dot{M} < -3$. Dark gray: $-3 < \log \dot{M} < 0$. Black: $\log \dot{M} > 0$. [Data from Cranmer and Saar 2011]

bright giant and a supergiant is not always clear-cut.) Canopus has $T_{\mathrm{eff}} \approx 7400 \, \mathrm{K}$ and $L \approx 11\,000 \, L_\odot$, implying a radius of $R \approx 65 \, R_\odot \sim 0.3 \, \mathrm{au}$. Although this is obviously not enough to hold the "whole dinged solar system," as the chapter epigraph says with poetic license, it would reach nearly to the orbit of Mercury.

The H–R diagram in Figure 1.11 also shows that highly luminous stars have high mass loss rates. Blue supergiants, with $T_{\mathrm{eff}} > 15\,000 \, \mathrm{K}$, can have mass loss rates as high as $\dot{M} \approx 30 \, M_\odot \, \mathrm{Myr}^{-1}$; the star P Cygni falls into this category. Red supergiants, with $T_{\mathrm{eff}} < 4500 \, \mathrm{K}$, can have even higher mass loss rates. For example, VY Canis Majoris is a highly variable red supergiant with $T_{\mathrm{eff}} \sim 3500 \, \mathrm{K}$ and $L \sim 300\,000 \, L_\odot$. It is surrounded by an irregular nebula of ejected matter; the mass and expansion speed of the nebula suggest that during the past $1200 \, \mathrm{yr}$, the average mass loss rate of VY CMa has been a staggering $\dot{M} \sim 600 \, M_\odot \, \mathrm{Myr}^{-1}$. Such a spendthrift loss of mass cannot last for long.

The luminosity of a main sequence star is not uniquely determined by its mass. The Sun, for instance, will spend a time $t_{\mathrm{MS},\odot} \sim 10 \, \mathrm{Gyr}$ as a main sequence star; during that time, its luminosity will roughly double. In addition, the luminosity of a main sequence star depends to some extent on its chemical composition. However, it is seen empirically that main sequence stars have a fairly close correlation between mass and luminosity, as shown in Figure 1.12. Overall, the mass–luminosity relation is moderately well fitted by the power law

$$L_{\mathrm{MS}} \sim 1 \, L_\odot \left(\frac{M_\star}{1 \, M_\odot} \right)^4, \tag{1.29}$$

which is plotted as the dashed line in Figure 1.12(a).

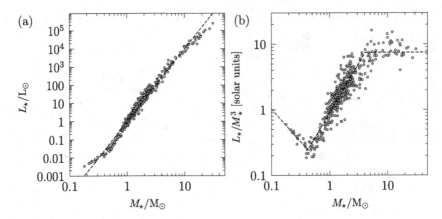

Figure 1.12 (a) Luminosity $L_\star$ as a function of mass $M_\star$ for a set of main sequence stars with well-determined mass, luminosity, and radius. Dashed line indicates $L_\star \propto M_\star^4$. (b) $L_\star/M_\star^3$ as a function of $M_\star$ for the same set of main sequence stars. Dashed line shows the broken power law of Equation 1.30. [Data from Eker *et al.* 2018]

To illustrate the deviations from a simple power law, Figure 1.12(b) shows a plot of $L_\star/M_\star^3$ as a function of $M_\star$. By taking out most of the steep dependence of luminosity on mass, this plot emphasizes the subtler features of the mass–luminosity relation, including the fairly abrupt change in slope at $M_\star \sim 0.4\,M_\odot$ and the more gradual flattening in slope at the high-mass end. For greater accuracy in fitting the mass–luminosity relation, it is often useful to employ a piecewise power law. For instance, a convenient fit to the data in Figure 1.12 is

$$L_{\mathrm{MS}}/L_\odot \approx \begin{cases} 0.091\,(M_\star/M_\odot)^2 & [M_\star < 0.4\,M_\odot], \\ 0.90\,(M_\star/M_\odot)^{4.5} & [0.4\,M_\odot < M_\star < 4\,M_\odot], \\ 7.2\,(M_\star/M_\odot)^3 & [4\,M_\odot < M_\star < 30\,M_\odot]. \end{cases} \qquad (1.30)$$

This function (divided by $M_\star^3$) is shown as the dashed line in Figure 1.12(b).

The steep dependence of luminosity on mass has serious implications for a star's main sequence lifetime,[10] t_{MS}. Stars are powered by nuclear fusion. However, fusion requires dense, hot conditions not found in the outer layers of stars; thus, only a fraction $f_{\mathrm{nuc}} < 1$ of a star's total mass is able to undergo fusion. The portion of the star that does undergo fusion converts mass to energy with an efficiency $\eta_{\mathrm{nuc}} \ll 1$. The maximum efficiency is attained when hydrogen is fused all the way to iron and nickel, the most tightly bound atomic nuclei. In this case, a fraction $\eta_{\mathrm{nuc}} \approx 0.009$ of the hydrogen's initial mass is converted to energy. The total energy released by nuclear fusion will then be

$$E_{\mathrm{nuc}} = \eta_{\mathrm{nuc}}(f_{\mathrm{nuc}}M_\star)c^2. \qquad (1.31)$$

---

[10] Astronomers commonly use the term "lifetime" to mean "the length of time a star can maintain its luminosity." A biologist might well be annoyed by this misuse of the term "life," but we will accept it as a colorful metaphor and move onward.

Scaling to a maximally efficient star, which fuses its entire mass from hydrogen to iron, the lifetime at a constant luminosity $L_\star$ will be

$$t_{\text{nuc}} = E_{\text{nuc}}/L_\star = \eta_{\text{nuc}} f_{\text{nuc}} c^2 (M_\star/L_\star) \tag{1.32}$$

$$= 133 \, \text{Gyr} \left(\frac{\eta_{\text{nuc}}}{0.009} \right) \left(\frac{f_{\text{nuc}}}{1.0} \right) \left(\frac{M_\star}{1 \, M_\odot} \right) \left(\frac{L_\star}{1 \, L_\odot} \right)^{-1}.$$

Using the empirical mass–luminosity relation for main sequence stars (Equation 1.30), we find that stars with $0.4 \, M_\odot < M_\star < 4 \, M_\odot$ have a main sequence lifetime of

$$t_{\text{MS}} \approx 150 \, \text{Gyr} \left(\frac{\eta_{\text{nuc}}}{0.009} \right) \left(\frac{f_{\text{nuc}}}{1.0} \right) \left(\frac{M_\star}{1 \, M_\odot} \right)^{-3.5}. \tag{1.33}$$

If stars in this mass range all have roughly similar values of f_{nuc}, this represents a steep decline in lifetime with mass. In the more massive range $4 \, M_\odot < M_\star < 30 \, M_\odot$, the dependence of main sequence lifetime on mass is

$$t_{\text{MS}} \approx 180 \, \text{Myr} \left(\frac{\eta_{\text{nuc}}}{0.009} \right) \left(\frac{f_{\text{nuc}}}{1.0} \right) \left(\frac{M_\star}{10 \, M_\odot} \right)^{-2}. \tag{1.34}$$

Since stars with $M_\star < 0.4 \, M_\odot$ have main sequence lifetimes much longer than the age of the universe, computing their lifetime is not yet of burning importance (as it were). At the high end of the mass range, Equation 1.34 implies that a $30 \, M_\odot$ star must have $t_{\text{MS}} < 20 \, \text{Myr}$, even if it has $f_{\text{nuc}} = 1$ and is able to use all its gas for nuclear fuel. An added complication, illustrated in Figure 1.11, is that very massive, luminous main sequence stars have high mass loss rates. Given $\dot{M} \sim 1 \, M_\odot \, \text{Myr}^{-1}$, typical for $30 \, M_\odot$ main sequence stars, the mass loss timescale $t = M_\star/\dot{M} \sim 30 \, \text{Myr}$ is not extremely long compared to the expected main sequence lifetime. Thus, we expect mass loss to have a significant effect on the evolution of the most massive main sequence stars; by discarding gas into interstellar space, a star with high mass loss rate is predicted by Equation 1.34 to increase its main sequence lifetime.

Main sequence stars also have a correlation between their mass and their photospheric radius, as shown in Figure 1.13. A useful fit to the mass–radius relation is the broken power law

$$R_{\text{MS}}/R_\odot \approx \begin{cases} 1.0 \, (M_\star/M_\odot) & [M_\star < 2 \, M_\odot], \\ 1.4 \, (M_\star/M_\odot)^{0.5} & [M_\star > 2 \, M_\odot]. \end{cases} \tag{1.35}$$

This is a steeper dependence of radius on mass than the $R_\star \propto M_\star^{1/3}$ dependence you would expect if all stars had the same density. More massive stars are less dense, with $\rho_\star \propto M_\star^{-2}$ for low-mass stars and $\rho_\star \propto M_\star^{-1/2}$ for high-mass stars. Figure 1.13 also shows that stars with $M_\star < 1 \, M_\odot$ have a small scatter in radius at a given mass, while higher-mass stars show a much larger scatter. This is because the radius of a star tends to increase during its main sequence

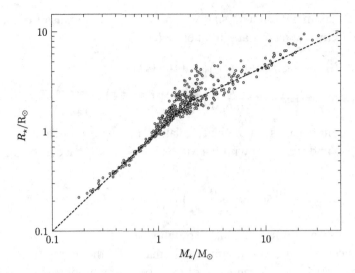

Figure 1.13 Radius $R_\star$ as a function of mass $M_\star$ for the same set of main sequence stars used in Figure 1.12. The dashed line shows the broken power law of Equation 1.35.

lifetime, as we discuss in Section 6.4. Since high-mass stars have a short main sequence lifetime, the sample of high-mass stars included in Figure 1.13 includes both very young stars ($t \ll t_{MS}$), which have not had time to swell in radius, and relatively old stars ($t \sim t_{MS}$), which have increased in radius by a factor $\sim$2 or more.

Equation 1.35 tells us that $M_\star \propto R_\star$ for main sequence stars with $M_\star < 2\,M_\odot$. Since a star's escape speed is $v_{esc} = (2GM_\star/R_\star)^{1/2}$, these low-mass main sequence stars all have roughly the same escape speed, $v_{esc} \sim v_{esc,\odot} \sim 600\,\mathrm{km\,s}^{-1}$. Equation 1.35 also tells us that $M_\star \propto R_\star^2$ for main sequence stars with $M > 2\,M_\odot$. Since a star's surface gravity is $g = GM_\star/R_\star^2$, these high-mass main sequence stars all have roughly the same surface gravity, $g \sim 0.5g_\odot \sim 14\,000\,\mathrm{cm\,s}^{-2}$. The surface gravity of stars is customarily written as $\log g$, where g is in units of $\mathrm{cm\,s}^{-2}$. The Sun, with $g_\odot = 27\,420\,\mathrm{cm\,s}^{-2}$, thus has $\log g_\odot = 4.438$, and high-mass main sequence stars have $\log g \approx \log g_\odot - 0.3 \approx 4.1$. (The custom of writing surface gravity as $\log g$ rather than g is useful because of the wide range of surface gravity found among stars and stellar remnants. For example, both the red giant R Doradus and the white dwarf Sirius B have a mass similar to the Sun. However, R Doradus, with $R_\star \approx 330\,R_\odot$, has $\log g \approx -0.6$; Sirius B, with $R_\star \approx 0.009\,R_\odot$, has $\log g \approx 8.6$.)

Because the luminosity of a main sequence star is closely correlated with its mass, and because its color is closely correlated with its luminosity, it is often said that "the main sequence is a *mass* sequence." If you know the mass of a main sequence star, you know, to first order, its luminosity, radius, and effective temperature. There is some dependence on chemical composition as well; since the composition of a star changes with time as a result of fusion, the observable properties of a star do gradually change while it is on the main sequence. However, the star's mass is of primary importance.

Keeping in mind the caveats about how metallicity and age effects smear out the main sequence, it is nevertheless useful to have a rough translation of the different observed spectral types into stellar luminosity and mass. For instance, Table 1.1 gives the range of T_{eff}, $L_\star$, and $M_\star$ for main sequence stars with solar abundance that are still near the beginning of their main sequence lifetime. The upper mass limit for O stars is poorly defined, so we have omitted it from Table 1.1. (There are very few stars known with $M_\star > 100\,M_\odot$, and they are all highly unstable.) The lower mass limit for M stars, discussed in Section 7.5, is related to the definition of "star"; if a gas ball has $M < 0.074\,M_\odot$, it isn't hot enough for fusion of hydrogen to occur.

1.4 Observing Clusters of Stars

The Hertzsprung–Russell diagram displayed in Figure 1.10 contains information from a variety of stars; the main attribute that they have in common is that they happen to find themselves within 100 pc of the Sun at the present moment. The stars of the solar neighborhood have a wide range of mass, effective temperature, and (especially) luminosity. In addition, they have a wide spread in age and a significant variation of chemical composition. From many points of view, it is useful to have such a diversity of stars in our immediate neighborhood, where they can be more easily studied. However, there are circumstances when we want to control some of the variables that can affect the structure and evolution of stars. For instance, the internal structure of a star depends on its metallicity; the elements that are present affect the star's opacity, which in turn regulates the transport of energy through the star. In addition, the fact that we talk at all about "stellar evolution" is a tacit admission that the properties of a star change with time.

Table 1.1 Properties of main sequence stars

Spectral type	T_{eff} [K]	$L_\star/L_\odot$	$M_\star/M_\odot$
O	32 000–	54 000–	19–
B	10 000–32 000	44–54 000	2.4–19
A	7300–10 000	8.5–44	1.65–2.4
F	6000–7300	1.58–8.5	1.13–1.65
G	5300–6000	0.51–1.58	0.88–1.13
K	3900–5300	0.065–0.51	0.56–0.88
M	2400–3900	0.0003–0.065	0.074–0.56

Data from Pecaut and Mamajek 2013

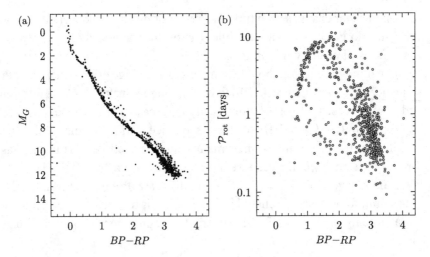

Figure 1.14 (a) H–R diagram for 1044 stars in the Pleiades which have *Gaia* EDR3 parallaxes with uncertainty $\sigma_p/p < 0.02$. [Data from Heyl *et al.* 2022] (b) Rotation period versus $BP - RP$ color index for 610 stars in the Pleiades. [Data from Godoy-Rivera *et al.* 2021]

Wouldn't it be useful if we could identify a population of stars that all had very nearly the same metallicity and age? Fortunately for astronomers, our galaxy contains numerous star clusters. The two basic types of cluster are **open clusters**, relatively young clusters of stars found in the disk of our galaxy, and **globular clusters**, older spherical clusters found in our galaxy's extended halo. Each star cluster is a gravitationally bound system, consisting of stars that formed at about the same time from the same cloud of interstellar gas. (Chapter 7 contains more details of how stars are assembled from lower-density gas.) The Pleiades, for instance, is a nearby example of an open cluster of stars, at a distance $d = 136\,\mathrm{pc}$. The famous "seven sisters" of the Pleiades are merely the most luminous members of a cluster containing well over 1000 stars. Figure 1.14(a) shows the H–R diagram for stars in the Pleiades whose parallax has been measured by *Gaia*.[11] Notice, in particular, how narrow the main sequence is when all the stars have the same age and metallicity. The stars lying just above and to the right of the Pleiades main sequence are actually unresolved binaries. The secondary star in the binary is dimmer and cooler than the primary star; by adding the light from the two stars together, we have a blended "star" that is brighter than (but also slightly redder than) the primary. A similar "binary main sequence" can be seen in Figure 1.10, somewhat smeared out by the spread in age and metallicity for stars in the immediate solar neighborhood.

[11] The naked-eye stars of the Pleiades are too annoyingly bright to have accurate *Gaia* parallaxes; the coolest, dimmest M dwarfs in the Pleiades are too frustratingly faint.

The metallicity of the stars in a cluster can be determined from their spectra; for the Pleiades, the metallicity turns out to be nearly solar, with $Z \approx Z_\odot$. Determining a cluster's age involves looking at the spectra of multiple stars in the cluster. Start by considering the bright (naked-eye) stars of the Pleiades; they have spectral type B6 through B8, with effective temperature $T_{eff} \sim 12\,000\,\text{K}$. However, the very brightest star, Alcyone (η Tauri), has a spectrum lacking the pressure broadening expected in a main sequence star; it is thus classified as a blue giant, of spectral class B7III. By contrast, the fainter star Asterope has pressure-broadened lines in its spectrum, and has the spectral class B8V. The primary difference between Alcyone and Asterope, two stars with the same age and same initial chemical composition, is their mass. For Alcyone, with $M_\star \approx 6\,M_\odot$, the calculated main sequence lifetime is (Equation 1.34)

$$t_{MS}[\text{Alcyone}] \approx 40\,\text{Myr}\left(\frac{f_{nuc}}{0.1}\right). \qquad (1.36)$$

Asterope is less massive, with $M_\star = 2.9\,M_\odot$, and thus has a longer main sequence lifetime (Equation 1.33):

$$t_{MS}[\text{Asterope}] \approx 290\,\text{Myr}\left(\frac{f_{nuc}}{0.1}\right). \qquad (1.37)$$

Thus, the observations are consistent with a cluster age of $t \sim 100\,\text{Myr}$, very broadly speaking. Asterope is still on the main sequence, but the more massive star Alcyone has exhausted the nuclear fuel in its core, and has moved on to the next stage of its existence. (Why that next stage involves swelling in radius and decreasing the surface gravity g is an interesting question that we discuss in Chapter 8.)

An accurate determination of the age of the Pleiades, or any other cluster of stars, requires detailed numerical modeling of the stars' evolution on the Hertzsprung–Russell diagram. This modeling is complicated by the fact that some of the stars in the Pleiades are rotating quite rapidly, which alters their evolution. In particular, rapid rotation mixes gas into a star's core from its surface layers; this increases the fraction f_{nuc} of a star's gas available for fusion, and extends its lifetime as a nuclear fusion reactor. Taking into account the effects of rotation, the age of the Pleiades is found to be $t \approx 125\,\text{Myr}$.

How do we know that stars in the Pleiades are rotating rapidly? In addition to measuring $v_{rot} \sin i$, there is another way to study the rotation properties of stars. The relatively young stars of the Pleiades have many starspots, analogous to the sunspots of the Sun. If starspots were extremely long-lived and were fixed in latitude and longitude, then the periodic variations in a star's brightness as it rotated would tell us $\mathcal{P}_{rot}$ in a straightforward manner. Since starspots appear and disappear, and change in size as they drift in latitude and longitude, determining the value of $\mathcal{P}_{rot}$ requires more sophisticated statistical analysis of a star's brightness as a function of time. Such an analysis (discussed in Section 10.5) yields the results in Figure 1.14(b).

All stars in the Pleiades have rotation periods shorter than the Sun's value of $\mathcal{P}_{rot,\odot} = 24.5$ d. In particular, stars in the Pleiades with a Sun-like color ($BP - RP \approx 0.8$) have rotation periods $\mathcal{P}_{rot} \approx 4$ d. This difference indicates that stars must spin down over the course of their main sequence existence. The hotter, higher-mass stars in the Pleiades are seen to have particularly large values for their specific rotational angular momentum, $j_{rot} \propto R_\star^2/\mathcal{P}_{rot}$. Consider, for instance, the star in Figure 1.14 that has $\mathcal{P}_{rot} = 0.18$ d and $BP - RP = -0.04$, corresponding to $T_{eff} \approx 12\,000$ K. This is the star HD 23923; like its sister Asterope, it has spectral class B8V. Because of its short rotation period and relatively large size ($R_\star \approx 2.4\,R_\odot$), HD 23923 has $j_{rot} \sim 800 j_{rot,\odot}$. (Although some cool stars in the Pleiades have a comparably short $\mathcal{P}_{rot}$, the smaller radii of these M dwarfs means their values of j_{rot} are not as startlingly large.) A successful theory of stellar rotation must explain a large number of observed features. Why are stars created with high angular momentum, and how are they spun down? Why do young high-mass stars have higher specific angular momentum than lower-mass stars? Why is there a range of rotation periods at a given mass? These are questions that we grapple with in Chapter 10.

Despite the fact that stars other than the Sun are vastly distant, we can coax out a great deal of observational information about stars. In the ideal case, we can find a star's luminosity, mass, photospheric radius, effective temperature, chemical composition, rotation period, and age. However, using the observed and deduced properties of stars to constrain the structure and evolution of stars requires combining many seemingly diverse fields of physics: Newtonian gravity, nuclear fusion, radiative transfer, thermodynamics, statistical mechanics, and quantum mechanics, to name a few. Having to combine so many fields of physics makes the study of stellar interiors challenging. However, it can also yield the deep satisfaction of understanding how the different branches of physics combine to form something as lovely as a star.

Exercises

1.1 The nominal solar mass parameter as given by the IAU is $GM_\odot = 1.327\,124\,4 \times 10^{26}$ cm^3 s^{-2}. Assuming the current mass-energy loss rate for the Sun remains constant, how long will it be until the IAU has to change the last significant digit of the solar mass parameter from a "4" to a "3"?

1.2 During its main sequence existence, the Sun is powered by the fusion of hydrogen into helium. This process releases 6.40×10^{18} erg per gram of hydrogen.

(a) What is the efficiency η_{nuc} of this fusion process?

(b) Detailed numerical models, discussed in Chapter 6, reveal that the Sun has a main sequence lifetime $t_{MS} = 10\,\text{Gyr}$. What is the fraction f_{nuc} of the Sun's mass that undergoes fusion on the main sequence? Assume a constant luminosity $L_\odot$.

(c) Given an initial hydrogen mass fraction $X_{\odot,0} = 0.706$, what fraction of the Sun's initial hydrogen undergoes fusion on the main sequence?

1.3 Assume that the mass–radius relation of Equation 1.35 holds for all main sequence stars.

(a) What is the mean density $\rho_\star$ for main sequence stars with $M_\star = 0.1, 1, 10$, and $100\,\text{M}_\odot$?

(b) At what rotation period $\mathcal{P}_{rot}$ would the centrifugal acceleration at the photosphere equal the surface gravity g for main sequence stars with $M_\star = 0.1, 1, 10$, and $100\,\text{M}_\odot$?

2

Equations of Stellar Structure

"A star is basically a pretty simple structure."
"You'd look pretty simple, Fred, at a distance of ten parsecs."
Fred Hoyle (1915–2001) and Roderick Redman (1905–1975)
at Cambridge Observatory [1954]

Since all stars other than the Sun are at a distance that is large compared to their diameter, discerning their detailed structure is challenging. In this chapter, we start with the Hoyle-ish assumption that a star is a pretty simple structure: a static, isolated sphere. In later chapters we will acknowledge in a Redman-esque way that stars evolve with time, that they are not necessarily spherical, and that they are frequently found in binary systems (or systems of even greater multiplicity).

2.1 Stars Conserve Mass

Consider a static spherical star of mass $M_\star$, photospheric radius $R_\star$, and photon luminosity $L_\star$. When we talk of a static spherical star, we mean that the star's internal properties are a function only of r, the distance from the star's center. There are, for instance, the mass density $\rho(r)$, the enclosed mass $M(r)$, the enclosed luminosity $L(r)$, the pressure $P(r)$, the temperature $T(r)$, and so forth. We assume that the star is not rotating (since that would break the spherical symmetry), that it is isolated (since tidal distortions would break the spherical symmetry), and that it has no mass loss (since that would introduce a time dependence in the star's mass). You may think that you've caught us out in a contradiction, since any non-zero luminosity $L_\star$ produces an equivalent mass loss rate $L_\star/c^2$. However, the timescale for mass-energy loss by radiation, $t = c^2(M_\star/L_\star)$, is necessarily much longer than the star's nuclear fusion lifetime $t_{\mathrm{nuc}} = \eta_{\mathrm{nuc}} f_{\mathrm{nuc}} c^2(M_\star/L_\star)$, given $\eta_{\mathrm{nuc}} \ll 1$ and $f_{\mathrm{nuc}} < 1$. Thus, we can treat mass-energy loss by radiation as being negligibly small. Our static spherical star is permitted to have internal convection, as long as the convection properties are constant with time.

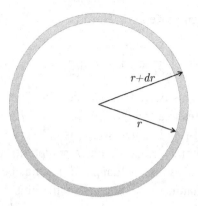

Figure 2.1 A thin shell centered on a star's central point.

Imagine a thin spherical shell of thickness dr, centered on the star's center, as illustrated in Figure 2.1. If $dr \ll r$, where r is the distance of the shell from the star's center, the mass of the shell is

$$dM = 4\pi\rho r^2 dr. \tag{2.1}$$

Thus, the *first* equation of stellar structure is a simple statement of mass conservation:

$$\frac{dM}{dr} = 4\pi\rho r^2. \tag{2.2}$$

In Equation 2.2, we are using the radius r as the independent variable. However, under some conditions, it's more convenient to use the enclosed mass M as the independent variable, yielding

$$\frac{dr}{dM} = \frac{1}{4\pi\rho r^2}. \tag{2.3}$$

For instance, if a star undergoes radial pulsations, the radius r of a thin shell of material changes with time, but the amount of mass M interior to the shell remains constant; this makes M the more convenient variable in this case.

Inside a spherically symmetric star, the gravitational potential is

$$\Phi(r) = -\frac{GM}{r}, \tag{2.4}$$

yielding a gravitational acceleration of

$$g(r) = \frac{GM}{r^2} \tag{2.5}$$

pointing toward the star's center. If gravity were the only force at work, with

$$\frac{d^2r}{dt^2} = -\frac{GM}{r^2}, \tag{2.6}$$

then, using Newton's form of Kepler's Third Law, the **freefall time** would be

$$t_{ff} = \left(\frac{\pi^2 R_\star^3}{8GM_\star}\right)^{1/2} = \left(\frac{3\pi}{32G\rho_\star}\right)^{1/2}, \tag{2.7}$$

where $\rho_\star$ is the mean density of the star. For the Sun's mean density $\rho_\odot = 1.41\,\text{g cm}^{-3}$, this translates to a freefall time $t_{ff,\odot} = 1770\,\text{s}$, or about half an hour.

We've assumed a non-rotating star, which is not very physical. How fast can a star rotate without blatantly violating our assumption of spherical symmetry? Speaking generally, a star can be well approximated as being spherical when its rotation speed at the equator, $v_{rot} = \Omega R_\star$, is small compared to its escape speed, $v_{esc} = (2GM_\star/R_\star)^{1/2}$. More quantitatively, the equatorial radius of a star will be larger than its polar radius by a fraction

$$f \sim \left(\frac{v_{rot}}{v_{esc}}\right)^2 \sim \frac{\Omega^2 R_\star^3}{2GM_\star} \sim \Omega^2 t_{ff}^2. \tag{2.8}$$

In Chapter 10, we discuss how this relation results from a star's tendency to stratify itself on equipotential surfaces in a rotating frame of reference. For the Sun, with a rotation speed of $v_{rot} = 2.07\,\text{km s}^{-1}$ and an escape speed of $v_{esc} = 618\,\text{km s}^{-1}$, we expect $f \sim 10^{-5}$; this is, in fact, comparable to the observed oblateness of the Sun.

2.2 Stars Are in Hydrostatic Equilibrium

The Sun has existed for longer than half an hour. Thus, there must be an outward force that almost exactly counterbalances the inward force of gravity. This outward force is provided by a pressure gradient. Consider, once again, our thin shell of thickness dr. The isotropic pressure at the inner surface of the shell is P; the pressure at the outer surface is slightly different, $P + dP$. We cut a cylinder of cross-sectional area dA out of the thin shell, as shown in Figure 2.2. The total force on the cylinder from pressure forces working in the radial direction is the sum of the pressure forces on the two end caps of the cylinder:

$$F_P = PdA - (P + dP)dA = -dPdA, \tag{2.9}$$

where the choice of sign tells us that the net force is inward if $dP > 0$ and outward if $dP < 0$. The total mass of the little cylinder is $\rho(r)dr\,dA$, so the gravitational force on the cylinder is

$$F_g = -\frac{GM}{r^2}\rho\,dr\,dA. \tag{2.10}$$

If the pressure force (Equation 2.9) is equal in magnitude but opposite in direction to the gravitational force (Equation 2.10), thus keeping the cylinder in

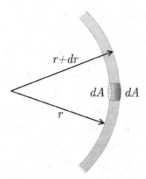

Figure 2.2 Pressure acting on the inner end cap and outer end cap of a small cylinder embedded in a thin shell.

equilibrium, we find that

$$dPdA = -\frac{GM}{r^2}\rho\,dr\,dA, \qquad (2.11)$$

or

$$\frac{dP}{dr} = -\frac{GM\rho}{r^2}. \qquad (2.12)$$

Equation 2.12 is the hydrostatic equilibrium equation for a spherical system, and is the *second* equation of stellar structure. Using the enclosed mass M as the independent variable, the equation of hydrostatic equilibrium becomes

$$\frac{dP}{dM} = \frac{dP}{dr}\frac{dr}{dM} = -\frac{GM}{4\pi r^4}. \qquad (2.13)$$

The Sun's age is $t_\odot = 4.57\,\mathrm{Gyr} \sim 10^{14} t_{\mathrm{ff},\odot}$. The fact that the Sun has existed for so long without collapsing or exploding tells us that over the long run, the gravitational force balances the pressure gradient force to better than one part in 10^{14}.

The equation of hydrostatic equilibrium merely assumes that there exists a pressure. Determining the dominant source of pressure, and calculating the resulting value of P, requires understanding the microphysics of what is happening on the scale of individual atoms, atomic nuclei, and electrons. In Chapter 3, we examine the possible **equations of state** that apply within stellar interiors, where an "equation of state" is the relation that gives the pressure P in terms of the density ρ, temperature T, and chemical composition of a gas. Although a simple ideal gas law, with $P \propto \rho T$, is often applicable within stars, other contributions, such as radiation pressure and degeneracy pressure, must sometimes be taken into consideration.

2.3 Stars Are in Thermal Equilibrium

So far, we've been talking in general terms about spheres of gas in hydrostatic equilibrium. We haven't yet addressed one of the most interesting attributes of stars: they glow in the dark, with a total photon luminosity L_*. All the photon energy that the star tosses away has to come from somewhere. At an immediate level, the answer to the question "Why do stars shine?" is "Stars shine because they are hot." That is, the photon energy is drawn from the star's reservoir of thermal energy. However, we then must ask the obvious follow-up question: "Why don't stars cool down?" Stars can maintain thermal equilibrium, thus keeping a constant temperature profile $T(r)$, if they have a non-zero specific energy generation rate ϵ.[1] The most obvious energy source in a star is nuclear fusion; however, there are circumstances where the gravitational potential energy of a star also contributes significantly to the specific energy generation rate.

If a thin shell has mass dM and luminosity dL, its temperature remains constant if $\epsilon \, dM = dL$, or

$$\frac{dL}{dM} = \epsilon. \tag{2.14}$$

This is the *third* equation of stellar structure; in terms of the radial variable r, it can be written as

$$\frac{dL}{dr} = 4\pi \rho r^2 \epsilon. \tag{2.15}$$

Equation 2.15 is also known as the equation of thermal equilibrium, since the balance between photon luminosity and energy generation keeps the temperature T constant.

For the Sun as a whole, the average specific energy generation rate is

$$\epsilon_\odot = \frac{L_\odot}{M_\odot} = 1.925 \, \text{erg s}^{-1} \, \text{g}^{-1}. \tag{2.16}$$

This is often expressed as a mass-to-light ratio Υ, with

$$\Upsilon_\odot \equiv \frac{M_\odot}{L_\odot} = 0.519 \, \text{g s erg}^{-1}, \tag{2.17}$$

or about 4000 metric tons per horsepower. For comparison, a draft horse, with $M \sim 0.8$ ton, can produce one horsepower; however, a working draft horse can easily consume ~ 8 tons of dry food per year. Thus, an immortal draft horse working for 4.57 Gyr would require a heap of hay with a mass of $\sim 4 \times 10^{10}$ tons. Obviously, the Sun is not powered by the digestion of hay, but by something 10^7 times more efficient (and producing far less manure).

---

[1] In the field of stellar interiors, ϵ stands for specific luminosity or specific power ($\text{erg s}^{-1} \, \text{g}^{-1}$), and *not* specific energy (erg g^{-1}). This is a conventional usage that can be traced back to Arthur Eddington.

In the nineteenth century, it was thought that the Sun's sole source of energy was its gravitational potential energy. For a spherically symmetric star, the gravitational potential energy is

$$W = -\int_0^{M_\star} \frac{GMdM}{r}. \tag{2.18}$$

For a star of uniform density, with $M \propto r^3$, the gravitational potential energy would be

$$W_\star = -\frac{3}{5} \frac{GM_\star^2}{R_\star}. \tag{2.19}$$

More generally, we can write

$$W_\star = -\alpha \frac{GM_\star^2}{R_\star}, \tag{2.20}$$

with $\alpha \approx 1$ for stars with plausible, centrally concentrated density profiles. If we assume that a star's luminosity has been fueled by gravitational potential energy, this gives a lifetime equal to the **Kelvin–Helmholtz** time,

$$t_{\text{KH}} \equiv \frac{-W_\star}{L_\star} = \frac{\alpha GM_\star}{R_\star} \Upsilon_\star, \tag{2.21}$$

where $\Upsilon_\star$ is the star's mass-to-light ratio. For the Sun, assuming $\alpha \approx 1$, the Kelvin–Helmholtz time is

$$t_{\text{KH},\odot} \approx 1.0 \times 10^{15} \, \text{s} \approx 30 \, \text{Myr}. \tag{2.22}$$

In 1899, when Lord Kelvin gave an address entitled "The Age of the Earth as an Abode fitted for Life," he concluded that the time required for the Earth to cool to its current state was "more than 20 and less than 40 million years." Since this range agreed with the Kelvin–Helmholtz time for the Sun, Kelvin thought he had a self-consistent model for the evolution of a solar system that was only ~30 Myr old. As it turns out, Kelvin had neglected the effect of convective heat transport when doing his calculations of the cooling time for the Earth. The Earth and Sun are actually 4.57 Gyr old, and the Sun must have an additional, non-gravitational, energy source.

In the 1920s, Arthur Eddington proposed that stars are powered by the fusion of four hydrogen nuclei into a single helium nucleus. In 1939, Hans Bethe showed in more detail how main sequence stars fuse hydrogen into helium. This fusion process, called "hydrogen burning,"[2] has an efficiency $\eta_{\text{nuc}} = 0.007\,12$, about 80% of the maximal efficiency yielded by fusing hydrogen all the way to iron.

---

[2] Astronomers commonly use the term "hydrogen burning" to mean "the fusion of hydrogen into helium." A chemist might well be annoyed by this misuse of the term "burning," but we will accept it as a colorful metaphor and move onward.

Thus, from Equation 1.32, a star powered by hydrogen burning has a lifetime equal to the nuclear time

$$t_{\text{nuc}} = 105\,\text{Gyr}\left(\frac{f_{\text{nuc}}}{1.0}\right)\left(\frac{M_\star}{1\,\text{M}_\odot}\right)\left(\frac{L_\star}{1\,\text{L}_\odot}\right)^{-1}, \qquad (2.23)$$

where f_{nuc} is the fraction of the star's mass that undergoes fusion from hydrogen to helium. For the Sun, with $f_{\text{nuc}} \approx 0.1$, the nuclear time is $t_{\text{nuc},\odot} \approx 10\,\text{Gyr}$. Thus, the Sun has $t_{\text{nuc}} \sim 300 t_{\text{KH}} \sim 2 \times 10^{14} t_{\text{ff}}$. Similarly, other stars on the main sequence have a hierarchy of timescales in which the nuclear time is long compared to the Kelvin–Helmholtz time, which in turn is very long compared to the freefall time.

Computing the specific energy generation rate ϵ for nuclear fusion processes requires a fairly sophisticated grasp of nuclear physics. We postpone nuclear sophistication until Chapter 5, and now simply point out that ϵ is a function of ρ, T, and the elemental composition of the gas undergoing fusion. (The composition is usually expressed in terms of the mass fraction X_i of each element.) Typically, the dependence of $\epsilon(\rho, T, X_i)$ on temperature is extremely steep, meaning that energy generation by fusion takes place in the central hot, dense regions of stars.

2.4 Stars Transport Energy

So far, we have three differential equations of stellar structure, plus an equation of state $P(\rho, T, X_i)$ and a specific energy generation rate $\epsilon(\rho, T, X_i)$. Even if we assert that we know the elemental composition $X_i(r)$ exactly, we are still left with four unknowns: $M(r)$, $\rho(r)$, $T(r)$, and $L(r)$. Our three equations of stellar structure (Equations 2.2, 2.12, and 2.15) will have to be joined by a fourth equation. Don't panic, though: an end to the long procession of equations of stellar structure is in sight.

Our necessary fourth equation can be extracted from one essential piece of physics that we have neglected so far. Energy is generated in the star's core; the fusion reactions produce gamma rays and positrons that heat the core. (Neutrinos are also produced, but they escape promptly; in the case of the Sun, neutrinos carry away 2.3% of the generated energy.) Although the energy is generated in the hot core of the star, it is radiated from the cooler surface of the star. Somehow, the energy has to get from a point near the star's center to a point in the star's photosphere. This is not an impossible task; after all, heat has a natural tendency to flow from hot regions, like a star's center, to cooler regions, like a star's photosphere. There are three basic mechanisms for carrying energy within a star. **Conductive** transport is the transfer of heat by the collision of particles (molecules, atoms, ions, and free electrons) due to their random thermal motions. **Radiative** transport is the transfer of heat by the motion of photons through a medium. **Convective** transport is the transfer of heat by the bulk flow of hot fluid. Because conductive and radiative transport can both be thought of as diffusion processes, we will treat them analogously in this section of the text. Since convective transport tends to occur through chaotic, turbulent flow, it's a bit trickier to describe; our grappling with the complexity of convection will be delayed until Section 4.2.

First consider the transport of energy by conduction. Since the heat flow, in this case, results from random thermal motion of particles, the energy flux $\vec{F}$ (energy per unit time per unit area) can be described by a diffusion equation:

$$\vec{F} = -D_{th}\vec{\nabla}U_{th}. \tag{2.24}$$

In Equation 2.24, U_{th} is the thermal energy density of the hot material, and D_{th} is the thermal diffusivity. The negative sign on the right-hand side of Equation 2.24 reminds us that heat flows from regions of high thermal energy density to regions of low thermal energy density. Engineers have measured the thermal diffusivity of many substances at room temperature, $T \sim 300\,\mathrm{K}$. Dense (honest-to-god) metals like gold, silver, and copper have $D_{th} \sim 1\,\mathrm{cm^2\,s^{-1}}$. Water has $D_{th} \approx 0.0014\,\mathrm{cm^2\,s^{-1}}$ at room temperature, and air has $D_{th} \approx 0.19\,\mathrm{cm^2\,s^{-1}}$. If the Sun were made of solid gold, the time for heat to be conducted from the core to the surface would be

$$t_{cond} \sim \frac{R_\odot^2}{D_{th}} \sim 5 \times 10^{21}\,\mathrm{s} \sim 10^5\,\mathrm{Gyr}. \tag{2.25}$$

Conduction, therefore, will not be significant in the Sun unless its thermal diffusivity is many orders of magnitude greater than that of gold.

For simplicity, let's treat the Sun's interior as a dense gas of pure ionized hydrogen. (The presence of helium and metals slightly complicates the situation, but doesn't fundamentally alter the underlying physics.) In gaseous ionized hydrogen, the thermal conduction is done mainly by the fast-moving free electrons rather than by the sluggish free protons. The thermal diffusivity then takes the simple value

$$D_{th} = \frac{1}{3}v_{th}\lambda_{elec}, \tag{2.26}$$

where v_{th} is the typical random thermal speed of the free electrons and λ_{elec} is their mean free path; that is, the typical distance they travel between collisions. At a temperature T, the root mean square thermal speed of a free electron can be written as

$$\frac{v_{th}}{c} = \left(\frac{3kT}{m_e c^2}\right)^{1/2} = 0.050\left(\frac{T}{5 \times 10^6\,\mathrm{K}}\right)^{1/2}. \tag{2.27}$$

We have scaled Equation 2.27 to a temperature of $T = 5 \times 10^6\,\mathrm{K}$ because the fusion reactions that power the Sun, and other main sequence stars, take place at a temperature $T_c \approx 10^7$ K. Thus, for a typical temperature in the solar interior, we adopt something halfway between the central temperature and the much lower photospheric temperature. The mean free path of the free electrons is

$$\lambda_{\text{elec}} = \frac{1}{n\sigma}, \tag{2.28}$$

where n is the number density of objects with which a free electron can collide, and σ is the cross section for collisions. A free electron can interact equally well with attractive protons (which have number density n_p) as with repulsive electrons ($n_e = n_p$ in ionized hydrogen). Thus, the number density of possible colliders is

$$n = n_e + n_p = \frac{2\rho}{m_p} = 1.69 \times 10^{24} \text{ cm}^{-3} \left(\frac{\rho}{\rho_\odot}\right), \tag{2.29}$$

where $\rho_\odot = 1.41 \text{ g cm}^{-3}$ is the mean density of the Sun. In general, we expect the cross section σ to depend on kinetic energy. A free electron in an ionized gas doesn't undergo physical impacts; instead, it undergoes electrostatic interactions with electrons and protons that deflect the electron's motion through some angle. Let's state that an electron undergoes a "collision" when it is deflected through an angle of 90 degrees or more. Such a large deflection can occur only when the magnitude of the electrostatic potential energy at closest approach is comparable to the initial kinetic energy of the electron; that is, when the closest approach distance r_e obeys

$$\frac{e^2}{r_e} \approx \frac{1}{2} m_e v_{\text{th}}^2. \tag{2.30}$$

This gives an effective cross section of

$$\sigma \approx \pi r_e^2 \approx 4\pi \left(\frac{e^2}{m_e v_{\text{th}}^2}\right)^2. \tag{2.31}$$

This cross section for electrostatic "collisions" can be rewritten as

$$\sigma \approx \frac{3}{2}\sigma_e \left(\frac{v_{\text{th}}}{c}\right)^{-4} \approx 1.6 \times 10^{-19} \text{ cm}^2 \left(\frac{v_{\text{th}}/c}{0.05}\right)^{-4}, \tag{2.32}$$

where

$$\sigma_e = \frac{8\pi}{3} \left(\frac{e^2}{m_e c^2}\right)^2 = 6.652 \times 10^{-25} \text{ cm}^2 = 0.6652 \text{ barn} \tag{2.33}$$

is the Thomson cross section of the electron.

Consider the plight of a free electron in the Sun's interior. It is moving at $\sim$5% the speed of light, in a crowded neighborhood with $\sim 10^{24}$ electrons and protons per cubic centimeter; these neighbors typically have collisional cross sections of more than 100 kilobarns. The mean free path of the electron is therefore quite short (Equation 2.28):

$$\lambda_{elec} = \frac{m_p}{2\rho} \frac{1}{\sigma} \approx 3.7 \times 10^{-6} \, \text{cm} \left(\frac{\rho}{\rho_{\odot}}\right)^{-1} \left(\frac{v_{th}/c}{0.05}\right)^4. \tag{2.34}$$

Combining this result with the thermal speed of an electron, from Equation 2.27, we find that an ionized gas has thermal diffusivity

$$D_{th} \approx 1900 \, \text{cm}^2 \, \text{s}^{-1} \left(\frac{\rho}{\rho_{\odot}}\right)^{-1} \left(\frac{v_{th}/c}{0.05}\right)^5. \tag{2.35}$$

Although this is greater than the thermal diffusivity of gold, it's not large enough to cool the Sun's interior effectively.[3]

Conductive transport can be thought of as moving energy with a poorly organized bucket brigade; each gas particle hands on energy to a neighbor in a random direction. The bucket brigade can be made more effective if the buckets are handed on at a faster speed (increasing v_{th}) and if the bucket handlers are placed further apart (increasing the mean free path λ). In stars, it is found that radiative transport is more effective than conductive transport at moving energy. The interior of a star is filled with photons. Since stellar interiors are not transparent, the mean free path of a photon will be

$$\lambda_{phot} = \frac{1}{n\sigma}, \tag{2.36}$$

where n is the number density of absorbers and scatterers, and σ is their typical cross section for interaction with photons. The mean free path can also be written as

$$\lambda_{phot} = \frac{1}{\rho\kappa}, \tag{2.37}$$

where κ is the **opacity** of the stellar material; it has dimensionality of area per unit mass, and can be thought of as the sum of the cross sections of all the absorbers or scatterers in a given volume divided by the total mass within that volume. In general, κ is a function of ρ, T, and X_i.

At a radius r, the photons have a blackbody energy density

$$U_{\gamma}(r) = aT(r)^4, \tag{2.38}$$

where the radiation density constant is $a = 7.566 \times 10^{-15} \, \text{erg cm}^{-3} \, \text{K}^{-4}$. Expressed in terms of the energy density U_{γ} of photons, the diffusion equation for radiative transport is

$$\vec{F} = -D_{\gamma} \vec{\nabla} U_{\gamma}, \tag{2.39}$$

---

[3] In the Sun's corona, however, with a very low density and fairly high temperature, conduction does play an important role in determining the thermal structure.

where $\vec{F}$ is the energy flux and D_γ is the radiative diffusion coefficient. By analogy with Equation 2.26, which gives the diffusion coefficient for massive gas particles, we can write the diffusion coefficient for photons as

$$D_\gamma = \frac{1}{3}c\lambda_{\text{phot}} = \frac{c}{3\kappa\rho}. \tag{2.40}$$

Radiative transport is more effective than conductive transport within stars partly because photons travel faster than electrons; at $T \sim 5 \times 10^6\,\text{K}$, an electron's thermal speed is $\sim 5\%$ the speed of a photon. However, the main reason why radiative transport is more effective is that the mean free path of a photon is much longer than that of a free electron. At a temperature $T \sim 5 \times 10^6\,\text{K}$ and density $\rho \sim \rho_\odot$, we have seen that the mean free path of an electron in ionized hydrogen is $\lambda_{\text{elec}} \sim 4 \times 10^{-6}\,\text{cm}$. At the same temperature and density, the mean free path of a photon between scatterings from a free electron is

$$\lambda_{\text{phot}} = \frac{1}{n_e\sigma_e} = \frac{m_p}{\rho\sigma_e} = 1.8\,\text{cm}\left(\frac{\rho}{\rho_\odot}\right)^{-1}, \tag{2.41}$$

where σ_e is the Thomson cross section of the electron (Equation 2.33). Although other sources of opacity exist, as discussed in Section 4.1, electron scattering is always present in ionized gas. Thus, electron scattering provides a lower bound on the opacity κ of ionized gas; this in turn implies that Equation 2.41 gives an upper bound on the photon mean free path λ_{phot} for hot gas with density ρ. Scaling to $\lambda_{\text{phot}} = 2\,\text{cm}$, the radiative diffusion coefficient is

$$D_\gamma \approx 2 \times 10^{10}\,\text{cm}^2\,\text{s}^{-1}\left(\frac{\lambda_{\text{phot}}}{2\,\text{cm}}\right). \tag{2.42}$$

The time for energy to be carried from the Sun's center to its photosphere by radiative transport is then

$$t_\gamma \sim \frac{R_\odot^2}{D_\gamma} \sim 8000\,\text{yr}\left(\frac{\lambda_{\text{phot}}}{2\,\text{cm}}\right)^{-1}. \tag{2.43}$$

Even when additional sources of opacity decrease the typical photon mean free path below $\lambda_{\text{phot}} \sim 2\,\text{cm}$, this still remains a comfortably short timescale compared to the age of the Sun.

In a spherically symmetric star, the general radiative diffusion equation (Equation 2.39) for blackbody radiation reduces to

$$\frac{L(r)}{4\pi r^2} = -D_\gamma \frac{dU_\gamma}{dr} = -D_\gamma \frac{d}{dr}\left(aT^4\right). \tag{2.44}$$

Using Equation 2.40 for the diffusion coefficient, this becomes

$$\frac{L(r)}{4\pi r^2} = -\frac{4ac}{3\kappa\rho}T^3\frac{dT}{dr}. \tag{2.45}$$

In a star where energy is transported through radiation rather than convection, Equation 2.45 is the *fourth* equation of stellar structure for static spherical stars.[4]
In summary,

$$\frac{dM}{dr} = 4\pi\rho r^2, \tag{2.46}$$

$$\frac{dP}{dr} = -\frac{GM\rho}{r^2}, \tag{2.47}$$

$$\frac{dL}{dr} = 4\pi\rho r^2 \epsilon, \tag{2.48}$$

$$\frac{dT}{dr} = -\frac{3\kappa\rho L}{16\pi a c T^3 r^2} \tag{2.49}$$

are the four coupled differential equations of stellar structure for a spherical static star with radiative heat transport. Combined with relations for $P(\rho, T, X_i)$, $\kappa(\rho, T, X_i)$, and $\epsilon(\rho, T, X_i)$, they are a set of four equations for the four unknowns $M(r)$, $\rho(r)$, $L(r)$, and $T(r)$. This assumes that you have a complete knowledge of the chemical composition of the star at each radius; in general, X_i changes with time as fusion and convective mixing occur.

Now that we have an elegant set of differential equations for stellar structure, we must investigate the (occasionally less elegant) details. In Chapter 3, we examine the microphysics that determines the pressure P in the star's interior. In Chapter 4, we consider the sources of opacity κ within a star, and determine under what circumstances radiative transport breaks down and convective transport takes over. In Chapter 5, we then examine the nuclear physics that determines the energy generation rate ϵ from fusion and the resulting rate of change for the star's composition X_i. Finally, after all these considerations, we will be well situated to attempt solutions of our elegant foursome of differential equations.

Exercises

2.1 Show that, for any density profile $\rho(r)$, the gravitational potential energy of a spherical object with mass $M_\star$ and radius $R_\star$ must satisfy the relation

$$|W| > 0.5\frac{GM_\star^2}{R_\star}. \tag{2.50}$$

2.2 For the Sun, the Kelvin–Helmholtz time is $t_{\mathrm{KH}} \approx 30\,\mathrm{Myr}$, while the nuclear timescale is $t_{\mathrm{nuc}} \approx 10\,\mathrm{Gyr}$. What are the corresponding timescales for a $0.1\,\mathrm{M_\odot}$ star and a $10\,\mathrm{M_\odot}$ star? In the standard scenario for star formation, gas collapses and heats up on the Kelvin–Helmholtz timescale until it is hot

---

[4] Full-disclosure footnote: In deriving Equation 2.45, we implicitly assumed that the opacity κ is the same for all photons. In general, however, κ can depend on frequency; in such a case, as we discuss in Section 4.1, an appropriate frequency-averaged value of κ must be used.

enough to ignite fusion. Given the timescales you have calculated, is it easier to detect high-mass stars or low-mass stars prior to fusion ignition?

2.3   What was the Kelvin–Helmholtz time of Lord Kelvin? You may assume a spherical physicist; state explicitly any other assumptions you make.

3

Equations of State

The Sun is in stable hydrostatic equilibrium, as are other main sequence stars. As we found in Section 2.2, the Sun has existed for $\sim 10^{14}$ times its freefall time. The condition of hydrostatic equilibrium is the single most crucial governing condition in stars, and is in large part responsible for why we can understand so much about them. Because stars are large and massive compared to a rocky planet like the Earth, we expect that a balance between pressure gradients and gravity inside a star will require very high internal pressure. However, there can be very different ways in which high pressure can be achieved, as two examples from the Earth make clear. Both the atmosphere and the oceans are in hydrostatic equilibrium; air pressure thus decreases with altitude above sea level, while pressure in the ocean increases with depth. If you climb Mount Everest, the temperature and density of the air both tend to decrease with altitude; this is how the atmosphere generates a pressure gradient to balance gravity. However, if you dive deep into the ocean, the temperature and density of the water are both nearly constant once you are below the solar-heated surface layers. Thus, the high pressure in the liquid ocean must be achieved by a different mechanism than in the gaseous atmosphere. The relation among pressure, temperature, density, and chemical composition, called the **equation of state**, is different for the air and for the ocean.

The equation of state is determined by processes on microscopic scales, but has profound macroscopic consequences. If the equation of state dictates that high pressure requires high temperature, then a large pressure gradient implies a large temperature gradient, and thus a large heat flow. However, a large heat flow can be maintained only if there exists a powerful energy source. By contrast, if a high pressure can be produced solely by increasing the density, then a large pressure gradient can be produced even in a relatively cold object. Writ large, this is the difference between the Sun and Jupiter. The Sun is a billion times more luminous than Jupiter, in part because the Sun requires a much higher central pressure, produced by a high central temperature. Jupiter, by contrast, can produce its required central pressure without a high central temperature, and hence without a large heat flow toward its surface.

To explore equations of state in detail, we will begin by setting a lower bound on the central pressure of the Sun, and see how this minimum pressure scales with mass $M_\star$ and radius $R_\star$. We will then turn to the broader question of how this necessary pressure is generated in stellar interiors.

3.1 Central Pressure of Stars

Directly measuring the central pressure of the Sun would be a remarkably difficult task; a probe capable of withstanding the high temperature and pressure in the Sun's interior is not something we can manufacture now, or in the foreseeable future. However, we can derive a surprisingly firm lower bound on the Sun's central pressure (or the central pressure of any star) using only the basic physical concepts of hydrostatic equilibrium and mass conservation.

The equation of hydrostatic equilibrium (Equation 2.12) can be written as

$$\frac{dP}{dr} = -\frac{GM(r)\rho(r)}{r^2}. \tag{3.1}$$

Integrating from a star's center ($r = 0$, $P = P_c$) to its surface ($r = R_\star$, $P \ll P_c$), hydrostatic equilibrium implies

$$-\int_{P_c}^{0} dP = P_c = G \int_{0}^{R_\star} \frac{M(r)\rho(r)}{r^2} dr, \tag{3.2}$$

where we have set the surface pressure equal to zero, since it is much lower than the central pressure. To proceed further, we need to know the density profile $\rho(r)$ within the star. Knowing $\rho(r)$, we can use the equation of mass conservation (Equation 2.2) to find $M(r)$, then use the equation of hydrostatic equilibrium, in the form of Equation 3.2, to determine the star's central pressure P_c.

It would be simple to compute the star's central pressure if the star were of uniform density, with $\rho(r) = \rho_\star = $ constant. Although we don't expect a self-gravitating gas ball to have a uniform density (this would be a better approximation for a rocky planet), it does provide a useful, analytically tractable starting point. In the case of a uniform-density star, the enclosed mass is

$$M(r) = \frac{4\pi}{3}\rho_\star r^3 \tag{3.3}$$

and the hydrostatic equilibrium equation yields a central pressure of

$$P_c = \frac{2\pi}{3}G\rho_\star^2 R_\star^2 = \frac{3}{8\pi}\frac{GM_\star^2}{R_\star^4}. \tag{3.4}$$

Using the solar mass and solar radius implies

$$P_{c,\odot} = \frac{3}{8\pi}\frac{GM_\odot^2}{R_\odot^4} = 1.345 \times 10^{15}\,\text{dyn\,cm}^{-2} \tag{3.5}$$

for a uniform-density Sun. Compared to the standard air pressure at sea level on Earth, 1 atm $= 1.013 \times 10^6$ dyn cm^{-2}, this is an astonishingly high pressure, with $P_{c,\odot} \sim 10^9$ atm.

Although the central pressure of a uniform-density Sun (Equation 3.5) is high by terrestrial standards, it provides only a lower bound on the central pressure of the real Sun. Stars are made of compressible gas, so their density is not uniform; instead, a star's density $\rho(r)$ is compressed to a higher value at the star's center than in its outer layers. As a simple analytic model for a star whose density decreases with radius, consider a star with mass $M_\star$, radius $R_\star$, and a density profile $\rho \propto r^{-\eta}$, where $\eta > 0$. Inside such a star, the pressure gradient must be, from Equation 3.1, $dP/dr \propto r^{1-2\eta}$. Thus, although the pressure gradient for a uniform-density star vanishes at the center, a star with $\rho \propto r^{-0.5}$ will have a constant pressure gradient throughout the star, while a steeper density profile (with $\eta > 0.5$) must be accompanied by a pressure gradient that steepens as you approach the star's center. Integrating the equation of hydrostatic equilibrium reveals that the central pressure of a star with $\rho \propto r^{-\eta}$ is

$$P_c = \frac{3}{8\pi}\frac{1-\eta/3}{1-\eta}\frac{GM_\star^2}{R_\star^4} \tag{3.6}$$

when $0 < \eta < 1$. Thus, as long as the density decreases with radius, the central pressure P_c must be greater than in a uniform-density star of the same mass and radius.

Detailed solar models, examined in Section 6.4, yield a central pressure of $P_{c,\odot} = 2.51 \times 10^{17}$ dyn cm^{-2}, nearly 200 times the central pressure of a uniform-density Sun. The central pressure of the Sun is associated with a central temperature of $T_{c,\odot} = 1.57 \times 10^7$ K and a central density of $\rho_{c,\odot} = 153$ g cm^{-3}. The central temperature of the Sun is therefore $\sim 50\,000$ times Earth's average air temperature at sea level, while its central density is $\sim 130\,000$ times the density of Earth's air at sea level. The mean mass of the gas particles at the Sun's center (mostly free electrons, protons, and He nuclei) is also quite a bit lower than the mean mass of molecules in the Earth's atmosphere; this also plays a role in determining pressure. If the mean particle mass in the Sun were equal to that in the

Earth's air, the temperature at the center of the Sun would have to be even higher
to maintain hydrostatic equilibrium.

Before we proceed to investigating the relation among pressure, density,
temperature, and mean particle mass, we can draw an additional lesson from
Equation 3.6, which gives the central pressure of a star with a density profile
$\rho \propto r^{-\eta}$. Note that the central pressure always scales as $P_c \propto M_\star^2/R_\star^4$, with the
exponent η only entering the numerical normalization. Thus, in a family of stars
with a similar density profile, the more massive stars must have either higher central pressure or a larger radius, or both. It is found empirically that main sequence
stars with $M_\star > 2\,\mathrm{M}_\odot$ have $R_\star \propto M_\star^{1/2}$, as shown in Figure 1.13. Thus, if these
high-mass main sequence stars all have similar density profiles, they should have
similar central pressures. Using the normalization of Equation 1.35, this pressure
is $P_c \sim P_{c,\odot}/4 \sim 6 \times 10^{16}\,\mathrm{dyn\,cm^{-2}}$.

We can also consider another consequence of the relation $P_c \propto M_\star^2/R_\star^4$; if the
Sun were to expand (without greatly changing the shape of its density profile), its
central pressure would drop. Conversely, if the Sun were to contract, its central
pressure would increase. Thus, stars can adjust their central pressure (and density
and temperature) by expanding or contracting their radius; this is an important
driver of the stately process of stellar evolution.

3.2 Quantum Statistics

The physical conditions within stars cover a wide range of temperatures, densities, and chemical compositions. Thus, the full equations of state used by stellar
evolution software include a number of physical effects. In practice, stellar evolution software uses lookup tables of laboriously pre-computed results. However, in
many circumstances, a much simpler equation of state can be assumed. Inside the
Sun, for instance, the ideal gas law provides a remarkably good approximation to
the equation of state.

To examine equations of state within stellar interiors, we begin with some simplifying assumptions. First, we assume that we can treat the material in stellar
interiors as a **compressible gas**. This assumption does not hold true to arbitrarily
high density. Consider, for instance, an isolated hydrogen atom in its ground state;
it is surrounded by an electron cloud with radius comparable to the Bohr radius,

$$a_0 = \frac{1}{2\pi\alpha}\frac{h}{m_e c} = 21.8\frac{h}{m_e c} = 5.29 \times 10^{-9}\,\mathrm{cm}, \tag{3.7}$$

where $\alpha \approx 1/137$ is the fine structure constant. In a gas of atomic hydrogen, as
long as the number density is much less than the Bohr density,

$$n_0 = \frac{3}{4\pi a_0^3} = 1.61 \times 10^{24}\,\mathrm{cm^{-3}} \tag{3.8}$$

atoms can be treated as isolated particles and the gas is compressible. The Bohr density corresponds to a mass density of $\rho_0 = m_H n_0 \approx 2.7\,\mathrm{g\,cm^{-3}}$; this density is actually exceeded in the central regions of most stars. However, stellar interiors also have very high temperatures, reaching $T \sim 10^7\,\mathrm{K}$ or more at the star's center. At such high temperatures, the gas within a star is highly ionized (in other words, it is a plasma). In this highly ionized state, the density of the material can approach that of an atomic nucleus ($\rho \sim 10^{14}\,\mathrm{g\,cm^{-3}}$) while still remaining a compressible gas.

Our next assumption is that the particles in the star's interior are **non-interacting**. Initially, this assumption may sound ridiculous: in Section 2.4, after all, we pointed out that a free electron in the Sun's interior can travel only $\sim$40 nm before being strongly deflected by another charged particle. How can we soberly claim that an electron undergoing half a million deflections per nanosecond is a "non-interacting" particle? In the context of equations of state, the assumption that a particle is "non-interacting" is equivalent to the approximation that it behaves like a billiard ball; although it does undergo brief collisions, between those collisions it moves without significant influence from long-range interactions. The assumption of non-interacting particles, in this sense, implies that the total pressure P is the sum of the pressures from all components: photons, free electrons, naked atomic nuclei, ions, atoms, and molecules. Corrections due to long-range electrical repulsion and attraction are included in full equation of state calculations, but in most circumstances, these corrections are small. At the Sun's center, for example, the deviation from an ideal gas law is at the 1% level.

The final assumption that we use while examining equations of state is the assumption of **local thermodynamic equilibrium**. As we discussed in Section 2.4, the mean free paths of electrons ($\lambda_{elec} \sim 40\,\mathrm{nm}$) and photons ($\lambda_{phot} \sim 2\,\mathrm{cm}$) in a star's interior are very small compared to a star's radius. The mean free paths of nuclei, ions, atoms, and molecules are also very small. Thus, at a given location within a star, the energy distribution of all these particles can be described with the same temperature T. (Neutrinos, with a mean free path much longer than the radius of a star, do not share in this local thermodynamic equilibrium.) The temperature T, moreover, also describes the ionization state of the gas; the population of electron energy levels in atoms, ions, and molecules; and the population of rotational and vibrational levels in molecules.

We can now take up the tools of statistical mechanics to specify the distribution of particle momentum (p) and energy (E). Consider a volume element within the star that is large compared to the mean free path of all the particles it contains,[1] but small compared to the radius of the star. There are N particles within the volume V in question. We assume that the volume element is small enough to be approximated as having a uniform temperature T and number density $n = N/V$.

---

[1] We ignore the neutrinos passing through the volume.

In terms of particle momentum, we may write the number density as

$$n = \int_0^\infty n_{\text{state}}(p) f_{\text{state}}(p) dp, \tag{3.9}$$

where $n_{\text{state}}(p)dp$ is the number density of states with momenta in the range $p \to p + dp$, and f_{state} is the fraction of states in that momentum range that are actually occupied by particles. (In this context, a "state" is any distinct, permissible combination of position and momentum for a particle in the volume element.) The pressure P can be inferred by imagining a wall enclosing the volume element, and computing the momentum imparted by elastic collisions of particles with the wall. For a given particle momentum $\vec{p}$, the imparted momentum will range from 0 to $2p$, depending on the relative orientation of the wall and the momentum vector. The rate of collisions will depend on the average velocity of the particles and the number density of particles. Assuming isotropy, and doing appropriate angle averages, we find the pressure integral

$$P = \int_0^\infty \frac{pv}{3} n_{\text{state}}(p) f_{\text{state}}(p) dp. \tag{3.10}$$

In terms of particle energy, rather than momentum, we may write the number density of particles as

$$n = \int_0^\infty n_{\text{state}}(E) f_{\text{state}}(E) dE, \tag{3.11}$$

where $n_{\text{state}}(E)dE$ is the number density of states with energy in the range $E \to E + dE$, and f_{state} is the fraction of states in that energy range that are occupied. Using Equation 3.11, we may write the energy density U of the particles as

$$U = \int_0^\infty E\, n_{\text{state}}(E) f_{\text{state}}(E) dE. \tag{3.12}$$

Computing the number density of states requires a bit of thought. Formally, the number of continuum states is infinite in a classical system, where there are an infinite number of possible momentum vectors (p_x, p_y, p_z). However, we can avoid an awkward encounter with infinity by applying the Heisenberg uncertainty principle. In a given physical volume V, the uncertainty principle sets an upper limit on the number of distinct states, through the relation

$$\Delta p_x \Delta p_y \Delta p_z \Delta x \Delta y \Delta z > h^3, \tag{3.13}$$

where the deltas represent the uncertainty in momentum and position. Since we constrain the particles to lie in the volume V, this leads to a minimum uncertainty in the momentum:

$$\Delta p_x \Delta p_y \Delta p_z > \frac{h^3}{V}. \tag{3.14}$$

In (p_x, p_y, p_z) momentum space, the volume corresponding to the range $p \to p + dp$ is $4\pi p^2 dp$. Equation 3.14 then implies that if particles are confined within a physical volume V, the maximum number of distinct states that can be contained within a momentum volume $4\pi p^2 dp$ is

$$N_{\text{state}} dp = \frac{4\pi p^2 dp}{\Delta p_x \Delta p_y \Delta p_z} = \frac{4\pi p^2 V}{h^3} dp. \tag{3.15}$$

This leads to a number density of states

$$n_{\text{state}}(p) = \frac{N_{\text{state}}}{V} = \frac{4\pi p^2}{h^3}. \tag{3.16}$$

However, the laws of quantum mechanics permit us to increase this number density for particles with two spin states (this includes photons, electrons, and protons). In this case, two particles with opposite spin can occupy the same momentum state and still be distinguishable; this doubles the number density of states to

$$n_{\text{state}}(p) = \frac{8\pi p^2}{h^3} \qquad \text{[two spin states]} \tag{3.17}$$

and yields a total number density of particles (Equation 3.9)

$$n = \frac{8\pi}{h^3} \int_0^\infty p^2 f_{\text{state}}(p) dp \tag{3.18}$$

and pressure (Equation 3.10)

$$P = \frac{1}{3} \frac{8\pi}{h^3} \int_0^\infty p^3 v f_{\text{state}}(p) dp. \tag{3.19}$$

The fraction of states that are occupied, f_{state}, is determined by the type of quantum statistics that the particle obeys. There are two main particle families: fermions (particles with half-integral spin, such as electrons and protons) and bosons (particles with integral spin, such as photons). Fermions follow a **Fermi–Dirac** distribution of occupied states,

$$f_{\text{state}}(E) = \frac{1}{e^\alpha e^{E/kT} + 1} \qquad \text{[F–D]}. \tag{3.20}$$

By contrast, bosons follow a **Bose–Einstein** distribution of occupied states,

$$f_{\text{state}}(E) = \frac{1}{e^\alpha e^{E/kT} - 1} \qquad \text{[B–E]}. \tag{3.21}$$

In Equations 3.20 and 3.21, the $e^{E/kT}$ term is the same for all particles regardless of mass, because of our assumption of local thermodynamic equilibrium; at a given temperature T, this term indicates that lower energy states are preferred. The e^α term is a normalization factor required to produce the correct total number density n; if the density n is low, this requires f_{state} to be small, and hence implies

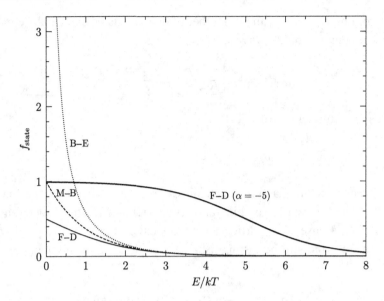

Figure 3.1 Fermi–Dirac (solid line), Bose–Einstein (dotted), and Maxwell–Boltzmann (dashed) distributions with $\alpha = 0$, plus the Fermi–Dirac distribution for $\alpha = -5$ as the heavy solid line.

a large value of e^{α}. At sufficiently low density, $e^{\alpha}e^{E/kT} \gg 1$; in this limit, a good approximation for f_{state} is the **Maxwell–Boltzmann** distribution,

$$f_{state}(E) = e^{-\alpha}e^{-E/kT} \qquad [\text{M–B}], \qquad (3.22)$$

which is indifferent to whether particles are fermions or bosons.

The basic differences among the Fermi–Dirac, Bose–Einstein, and Maxwell–Boltzmann distributions are shown in Figure 3.1. The lighter lines show the three different distributions assuming a normalization of $\alpha = 0$. Note that the Bose–Einstein distribution can have $f_{state} > 1$ at low energies; that is, multiple bosons can occupy the same state. The heavier line in Figure 3.1 shows the Fermi–Dirac distribution for $\alpha = -5$, representing a high total density n; in the limit $E \ll -\alpha kT$, the Fermi–Dirac distribution approaches $f_{state} = 1$. The striking differences among the distributions will manifest as very different dependencies of the pressure P on the physical conditions within a star.

3.3 Ideal Gas Pressure

The Maxwell–Boltzmann distribution (Equation 3.22) is the simplest of the three possible distributions we have considered. When particle velocities are non-relativistic, we can write $E = mv^2/2$, $p = mv$, and therefore $E = p^2/2m$. Assuming a Maxwell–Boltzmann distribution for non-relativistic particles, Equation 3.18

becomes

$$n = e^{-\alpha} \frac{8\pi}{h^3} \int_0^\infty p^2 \exp\left(-\frac{p^2}{2mkT}\right) dp. \tag{3.23}$$

Switching to the dimensionless variable $\beta \equiv p^2/(2mkT)$, the number density can be written as

$$n = e^{-\alpha} \frac{4\pi(2mkT)^{3/2}}{h^3} \int_0^\infty \beta^{1/2} e^{-\beta} d\beta. \tag{3.24}$$

In this equation, the definite integral is $\Gamma(3/2) = \sqrt{\pi}/2$.

Similarly, the assumption of a Maxwell–Boltzmann distribution for non-relativistic particles yields a pressure (Equation 3.19) of

$$P = e^{-\alpha} \frac{8\pi}{3mh^3} \int_0^\infty p^4 \exp\left(-\frac{p^2}{2mkT}\right) dp. \tag{3.25}$$

Again using the dimensionless variable $\beta \equiv p^2/(2mkT)$, the pressure becomes

$$P = e^{-\alpha} \frac{4\pi(2mkT)^{5/2}}{3mh^3} \int_0^\infty \beta^{3/2} e^{-\beta} d\beta. \tag{3.26}$$

Here, the definite integral is $\Gamma(5/2) = 3\sqrt{\pi}/4$. Dividing Equation 3.26 by Equation 3.24, we find that the relation between n and P is

$$\frac{P}{n} = \frac{2mkT}{3m} \frac{\Gamma(5/2)}{\Gamma(3/2)}. \tag{3.27}$$

However, since $\Gamma(5/2) = (3/2)\Gamma(3/2)$, this simplifies to

$$P_{\text{ideal}} = nkT, \tag{3.28}$$

which is the familiar **ideal gas law**.

Although its derivation required a number of simplifying assumptions, the ideal gas law turns out to be an excellent description for most stars in most stages of stellar evolution. In an ideal gas, the pressure depends on both density and temperature, implying that pressure gradients will be produced by a combination of density and temperature gradients. Within a star, the ideal gas pressure depends on the number density n of gas particles, while the pressure gradient in hydrostatic equilibrium depends on the mass density ρ. Thus, to build a complete stellar model, we need to know the relation among the number density, the mass density, and the chemical composition of a gas. This requires a discussion of the **mean molecular mass**.

A star generally contains many elements, with each element represented by one or more isotopes. In turn, each isotope can have multiple ionization states. Suppose that the total number of species in the star is N_{spe}, where in this context a "species" means a particular ionization state of a particular isotope of a particular element.[2] For species i, the density is given by the number density of

---

[2] We are considering the hot interior of a star, where the number of molecules is negligible; in the atmospheres of cool stars, molecular species must also be included.

atomic nuclei, n_i. Considering only species i, the number of free particles (including free electrons) per atomic nucleus is $N_{z,i}$. A species consisting of neutral atoms thus has $N_{z,i} = 1$, a singly ionized species has $N_{z,i} = 2$, and so forth to higher ionization states. For species i, the total mass per atomic nucleus is μ_i, expressed in atomic mass units. The atomic mass unit, $m_{\text{AMU}} = 1.6605 \times 10^{-24}$ g, is defined as one-twelfth the mass of a ^{12}C atom. However, the proton mass ($m_p = 1.0073 m_{\text{AMU}}$) and neutron mass ($m_n = 1.0087 m_{\text{AMU}}$) are both nearly equal to an atomic mass unit, while the electron mass ($m_e \approx 0.0005 m_{\text{AMU}}$) and the binding energy per nucleon[3] in an atomic nucleus ($B/A < 0.0095 m_{\text{AMU}} c^2$) are both much smaller than an atomic mass unit. Thus, for any species, we can approximate μ_i as the mass number A_i; that is, the number of nucleons in the atomic nucleus. For example, ^{1}H has $\mu_i = 1.0078 \approx 1$, ^{56}Fe has $\mu_i = 55.9354 \approx 56$, and ^{208}Pb has $\mu_i = 207.9767 \approx 208$.

The total mass density, summing over all species in a gas, is then

$$\rho = m_{\text{AMU}} \sum_{i=1}^{N_{\text{spe}}} n_i \mu_i \approx m_{\text{AMU}} \sum_{i=1}^{N_{\text{spe}}} n_i A_i. \tag{3.29}$$

The total number density n of all free particles in the gas is

$$n = \sum_{i=1}^{N_{\text{spe}}} n_i N_{z,i}. \tag{3.30}$$

The mass fraction of species i is $X_i = n_i \mu_i m_{\text{AMU}} / \rho$. Thus, we can rewrite the number density n as

$$n = \frac{\rho}{m_{\text{AMU}}} \sum_{i=1}^{N_{\text{spe}}} X_i \frac{N_{z,i}}{\mu_i} \approx \frac{\rho}{m_{\text{AMU}}} \sum_{i=1}^{N_{\text{spe}}} X_i \frac{N_{z,i}}{A_i}. \tag{3.31}$$

We can now recast the ideal gas law of Equation 3.28 in terms of ρ, T, and the mean molecular mass, designated by the symbol $\bar{\mu}$:

$$P_{\text{ideal}} = \frac{k}{m_{\text{AMU}}} \frac{\rho T}{\bar{\mu}}, \tag{3.32}$$

where the mean molecular mass is given by the formula

$$\bar{\mu} = \left[\sum_{i=1}^{N_{\text{spe}}} X_i \frac{N_{z,i}}{\mu_i} \right]^{-1} \approx \left[\sum_{i=1}^{N_{\text{spe}}} X_i \frac{N_{z,i}}{A_i} \right]^{-1}. \tag{3.33}$$

Note, by the way, that astronomers customarily refer to the mean particle mass as the mean "molecular" mass even when there are no molecules present. This is an eccentric, but long-established, usage.

---

[3] The word "nucleon," introduced by the physicist Christian Møller in 1941, is simply a shorthand term for "proton or neutron."

For a gas of neutral atomic ^{1}H, the mean molecular mass is $\bar{\mu} = 1$, while for a gas of fully ionized ^{1}H, $\bar{\mu} = 1/2$. Similarly, for neutral atomic ^{4}He, $\bar{\mu} = 4$, while for fully ionized ^{4}He, $\bar{\mu} = 4/3$. Heavier elements that are fully ionized typically have[4] $\bar{\mu} \approx 2$. More broadly, the mean molecular mass can be computed for any mixture of elements, if the mass fractions and ionization states are known. For instance, given the abundances in the protosolar nebula, we can compute a mean molecular mass of $\bar{\mu} = 1.29$ if the gas consists of neutral atoms, and $\bar{\mu} = 0.614$ if the gas is completely ionized.

Since H and He are far more abundant than heavier elements in a typical star, they largely determine the mean molecular mass. If a star converts its ionized ^{1}H ($\bar{\mu} = 0.5$) into ionized ^{4}He ($\bar{\mu} \approx 1.33$) by a process of nuclear fusion, there will be an inevitable drift toward higher mean molecular mass. If the density and temperature remained constant, this would result in a pressure decrease (Equation 3.32). Thus, stars that undergo nuclear fusion must increase their central temperature or central density (or both) in order to remain in hydrostatic equilibrium. This result follows from the ideal gas law alone.

3.4 Degeneracy Pressure

We have stated that we can use the simple Maxwell–Boltzmann distribution,

$$f_{\text{state}}(E) \propto \exp\left(-\frac{E}{kT}\right), \tag{3.34}$$

when the density n is "sufficiently low." At a more quantitative level, we can compute the maximum density n_{MB} at which the Maxwell–Boltzmann distribution still applies, permitting us to ignore the complexities of Fermi–Dirac or Bose–Einstein statistics. To simplify matters, assume that the temperature T is low enough that particle motions are non-relativistic (for instance, $T \ll m_e c^2/k \sim 6 \times 10^9$ K for free electrons). In this case, $p^2 = 2mE$, and the number density of states becomes (using Equation 3.17)

$$n_{\text{state}}(E)dE = n_{\text{state}}(p)\frac{dp}{dE}dE = \frac{8\pi\sqrt{2}}{h^3}m^{3/2}E^{1/2}dE. \tag{3.35}$$

Thus, at a given energy E, the most massive particles have the most states available for occupation. Equation 3.35 can also be written as

$$n_{\text{state}}(E)dE = \frac{8\pi\sqrt{2}}{\lambda_C^3}\left(\frac{E}{mc^2}\right)^{1/2}\frac{dE}{mc^2}, \tag{3.36}$$

---

[4] Among the most common heavy elements (or "metals"), the mean molecular mass at full ionization ranges from $\bar{\mu} = 12/7 \approx 1.71$ for ^{12}C to $\bar{\mu} = 56/27 \approx 2.07$ for ^{56}Fe.

where

$$\lambda_C \equiv \frac{h}{mc} \qquad (3.37)$$

is the Compton wavelength of the gas particle in question.

For non-relativistic particles obeying the Maxwell–Boltzmann distribution, the number density of particles in the energy range $E \rightarrow E + dE$ is

$$n(E)dE = n_{state}(E)f_{state}(E)dE \propto E^{1/2}e^{-E/kT}dE, \qquad (3.38)$$

from Equations 3.34 and 3.35. Normalized to the total number density of particles, $n = \int n(E)dE$, Equation 3.38 becomes

$$n(E)dE = nF_M(E)dE, \qquad (3.39)$$

where

$$F_M(E)dE = \frac{2}{\sqrt{\pi}} \left(\frac{E}{kT} \right)^{1/2} \exp\left(-\frac{E}{kT} \right) \frac{dE}{kT} \qquad (3.40)$$

is the **Maxwellian distribution** of particle energies.[5] Although the Maxwellian energy distribution is useful in low-density gases, fermions (such as free electrons) can't have an actual number density $n(E)$ that is greater than $n_{state}(E)$, the number density of available states. Taking the Maxwellian number density $n(E)$ from Equations 3.39 and 3.40, and comparing it to n_{state} from Equation 3.36, we find that when $E \ll kT$, the requirement that $n(E) \leq n_{state}(E)$ is equivalent to requiring $n \leq n_{MB}$, where

$$n_{MB}(m, T) = \frac{4\pi\sqrt{2\pi}}{\lambda_C^3} \left(\frac{kT}{mc^2} \right)^{3/2}. \qquad (3.41)$$

In a gas containing particles of different mass m, this upper density limit goes as $n_{MB} \propto \lambda_C^{-3}m^{-3/2} \propto m^{3/2}$. In a gas of ionized hydrogen, the electrons will thus switch from Maxwell–Boltzmann to Fermi–Dirac statistics at a much lower density than the protons, since $(m_e/m_p)^{3/2} \sim 10^{-5}$. If pure hydrogen is at $T \sim 10^7$ K, comparable to the temperature at the Sun's center, the maximum density at which electrons obey Maxwell–Boltzmann statistics is

$$\rho_{MB}(T) = m_H n_{MB} = 256 \, g \, cm^{-3} \left(\frac{T}{10^7 \, K} \right)^{3/2}. \qquad (3.42)$$

In Figure 3.2, the light dashed lines show the Maxwellian energy distribution for electrons at $T = 10^7$ K, scaled to mass densities ranging from $\rho = 100 \, g \, cm^{-3}$ for the lowest curve (comparable to the density in the Sun's core) to $\rho = 1000 \, g \, cm^{-3}$ for the highest curve. The heavy solid line in Figure 3.2 is the number density $n_{state}(E)$, which represents the maximum density

---

[5] James Clerk Maxwell derived this distribution in 1860, when Ludwig Boltzmann was a teenager and Max Planck was just a toddler.

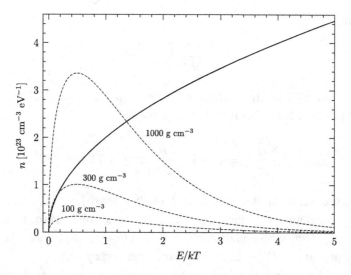

Figure 3.2 Light dashed lines: Number density of electrons in a Maxwellian distribution, assuming a gas of pure hydrogen at $T = 10^7$ K ($kT = 862$ eV). Heavy line: Maximum number density permitted for electrons by Fermi–Dirac statistics.

permitted by Fermi–Dirac statistics. Although the Maxwellian distribution for $\rho = 100 \, \text{g cm}^{-3}$ remains below the n_{state} line, and thus avoids the region forbidden by quantum mechanics, electrons in the higher-density gases cannot have a Maxwell–Boltzmann distribution for $f_{\text{state}}(E)$.

Since $n_{\text{MB}} \propto T^{3/2}$, in the limit that $T \to 0$, a gas of any density must start to deviate from a Maxwell–Boltzmann distribution. For a gas of bosons, the Bose–Einstein distribution permits $f_{\text{state}} > 1$; in the limit $T \to 0$, this implies that all the bosons pile into the lowest-energy state. For a gas of fermions, by contrast, the Fermi–Dirac distribution has an upper limit of $f_{\text{state}} = 1$. In the limit $T \to 0$, this implies that every low-energy state will be occupied by a single fermion up to a maximum energy E_{F}, known as the **Fermi energy**. The momentum corresponding to the Fermi energy is the **Fermi momentum**, p_{F}. The relation between the Fermi momentum and the density can be found from Equation 3.18, assuming that $f_{\text{state}} = 1$ for $p \le p_{\text{F}}$ and $f_{\text{state}} = 0$ for $p > p_{\text{F}}$. This yields

$$n = \frac{8\pi}{h^3} \int_0^{p_{\text{F}}} p^2 dp = \frac{8\pi p_{\text{F}}^3}{3h^3}, \tag{3.43}$$

implying

$$p_{\text{F}} = \left(\frac{3}{8\pi}\right)^{1/3} h n^{1/3}. \tag{3.44}$$

In the limit $T \to 0$, the fermions are referred to as being **degenerate**.[6] In the case of full degeneracy ($T = 0$), the pressure computed from Equation 3.19 takes the simple form

$$P_{\text{deg}} = \frac{8\pi}{3h^3} \int_0^{p_F} p^3 v\, dp. \tag{3.45}$$

In the non-relativistic limit, we can use the relation $v = p/m$ to obtain the fully degenerate pressure for a non-relativistic system:

$$P_{\text{deg,nr}} = \frac{8\pi p_F^5}{15mh^3} = \left(\frac{3}{\pi}\right)^{2/3} \frac{h^2}{20m} n^{5/3} \approx 0.0485 \frac{h^2}{m} n^{5/3}. \tag{3.46}$$

This pressure depends on the mass m and number density n of the fermions, but not on the temperature T, which is assumed to be absolute zero. In an electrically neutral system, with equal numbers of electrons and protons, the electron degeneracy pressure is 1836 times the proton degeneracy pressure.

If hydrostatic equilibrium is a balance between electron degeneracy pressure and gravity, then the outward pressure gradient force is supplied almost entirely by the electrons, while the inward gravitational force is supplied almost entirely by the nucleons. In this situation, a useful parameter is the mean molecular mass per electron, μ_e, defined as

$$\mu_e = \left[\sum_{i=1}^{N_{\text{spe}}} X_i \frac{N_{e,i}}{\mu_i}\right]^{-1}, \tag{3.47}$$

where $N_{e,i}$ is the number of electrons per atom in a species. Electrical neutrality requires that $N_{e,i} = Z_i$, where Z_i is the atomic number of species i (that is, the number of protons in the atomic nucleus). Since, as we have seen, μ_i is well approximated by the mass number A_i, we can write

$$\mu_e \approx \left[\sum_{i=1}^{N_{\text{spe}}} X_i \frac{Z_i}{A_i}\right]^{-1}, \tag{3.48}$$

and μ_e for any pure species can be thought of as the number of nucleons per electron. For ^{1}H, we can take $\mu_i = A_i = 1$ and $N_{e,i} = Z_i = 1$; thus, pure hydrogen has $\mu_e = 1$. By contrast, ^{4}He and the most abundant metals (^{16}O, ^{12}C, and ^{20}Ne) have $A_i = 2Z_i$; thus, these elements have $\mu_e = 2$. In terms of μ_e, the number density of electrons in an arbitrary mixture can be written as

$$n_e = \frac{\rho}{m_{\text{AMU}}\mu_e}. \tag{3.49}$$

---

[6] The word "degenerate" comes from a Latin word meaning "to depart from the ways of your *genus*, or ancestral stock." When electrons become degenerate, they depart from the ancestral ways of classical physics, and follow the newfangled ways of quantum mechanics.

Thus, the Fermi momentum (Equation 3.44) for electrons can be expressed in terms of ρ as

$$p_F = \left(\frac{3}{8\pi}\right)^{1/3} h \left(\frac{\rho}{m_{AMU}\mu_e}\right)^{1/3} \tag{3.50}$$

and the electron degeneracy pressure can be written as (compare to Equation 3.46)

$$P_{deg,nr} = \left(\frac{3}{\pi}\right)^{2/3} \frac{h^2}{20m_e} \left(\frac{\rho}{m_{AMU}\mu_e}\right)^{5/3} \tag{3.51}$$

$$= 1.004 \times 10^{13} \text{ dyn cm}^{-2} \, \mu_e^{-5/3} \left(\frac{\rho}{1\,\text{g cm}^{-3}}\right)^{5/3}. \tag{3.52}$$

The electron degeneracy pressure given in Equation 3.52 is computed at absolute zero temperature, and thus represents a strict minimum to the true pressure at a given density ρ. As the temperature starts to creep above $T = 0$, the distribution f_{state} grows an exponential tail of occupied states with $E > E_F$ and $p > p_F$; this boosts the value of $P_{deg,nr}$ at a given value of ρ.

The degeneracy pressure computed in Equation 3.52 assumes that the non-relativistic approximation holds true. The assumption of non-relativistic electrons fails badly when the Fermi momentum reaches $p_F = m_e c$. From Equation 3.44, this criterion holds true at an electron density n_{rel} given by the relation

$$\left(\frac{3}{8\pi}\right)^{1/3} h n_{rel}^{1/3} = m_e c, \tag{3.53}$$

or

$$n_{rel} = \frac{8\pi}{3} \lambda_{C,e}^{-3} = 5.865 \times 10^{29} \text{ cm}^{-3}, \tag{3.54}$$

where

$$\lambda_{C,e} \equiv \frac{h}{m_e c} = 2.426 \times 10^{-10} \text{ cm} \tag{3.55}$$

is the Compton wavelength of the electron. The number density n_{rel} at which degenerate electrons become relativistic corresponds to a mass density of

$$\rho_{rel} = \frac{8\pi}{3} \frac{m_{AMU}}{\lambda_{C,e}^3} \mu_e = 9.74 \times 10^5 \text{ g cm}^{-3} \, \mu_e. \tag{3.56}$$

In the highly relativistic limit, we may use the approximation $v = c$ in the relation for degeneracy pressure (Equation 3.45), yielding

$$P_{deg,rel} = \frac{8\pi c}{3h^3} \int_0^{p_F} p^3 dp = \frac{2\pi c p_F^4}{3h^2}. \tag{3.57}$$

Using the equation for the Fermi momentum of electrons (Equation 3.50), we find the relation between pressure and mass density for a gas supported by relativistic electron degeneracy pressure:

$$P_{\text{deg,rel}} = \frac{1}{8}\left(\frac{3}{\pi}\right)^{1/3} hc \left(\frac{\rho}{m_{\text{AMU}}\mu_e}\right)^{4/3} \tag{3.58}$$

$$= 1.243 \times 10^{23} \text{ dyn cm}^{-2} \mu_e^{-4/3} \left(\frac{\rho}{10^6 \text{ g cm}^{-3}}\right)^{4/3}. \tag{3.59}$$

Thus, degenerate electrons become relativistic at a density $\rho \sim 10^6$ g cm^{-3} (with the exact value depending on the number of nucleons per electron) and pressure $P_{\text{deg}} \sim 10^{23}$ dyn cm$^{-2} \sim 10^{17}$ atm.

When degeneracy pressure is the main source of support for a star, there are several direct and important consequences:

(a) Degeneracy pressure does not depend on temperature. This permits the existence of cold objects in hydrostatic equilibrium. Such objects include brown dwarfs (substellar gaseous spheres without nuclear fusion), as well as old, cold white dwarfs and neutron stars (former stars that have exhausted their nuclear fuel).

(b) When a star is supported by degeneracy pressure, it no longer has a mechanical response (expansion or contraction) when its temperature changes. This affects the stability of the star, especially when a new nuclear fuel source is ignited, driving up the temperature.

(c) The same condition that makes a system degenerate – the fact that particles are forced to occupy high-momentum states because lower states are occupied – also increases the mean free path of the degenerate particles. Thus, thermal conduction in degenerate systems can become very efficient, driving down temperature gradients.

(d) The first two equations of stellar structure (mass conservation and hydrostatic equilibrium), combined with the degenerate relation for $P(\rho)$, can be solved for $\rho(r)$ and $P(r)$, without knowing anything about the luminosity and temperature within a star. In Section 9.1 we therefore develop a remarkably simple theory for white dwarf structure. (Neutron stars, as we see in Section 9.2, are more difficult to explain because of their extraordinarily high density.)

(e) Finally, degeneracy pressure switches from $P_{\text{deg,nr}} \propto \rho^{5/3}$ at low densities to $P_{\text{deg,rel}} \propto \rho^{4/3}$ at higher densities. In other words, degenerate matter becomes *softer* at higher densities because of relativistic effects, with a given change in pressure being associated with a larger change in density. This softening has important consequences for the mass range of objects that can be supported by degeneracy pressure.

3.5 Radiation Pressure

So far, we have considered the ideal gas pressure contributed by massive particles at relatively high temperature (Equation 3.32) and the degeneracy pressure contributed by fermions independent of temperature (Equations 3.52 and 3.59). Under some conditions, however, the **radiation pressure** provided by photons can be significant. For an individual photon moving at $v = c$, we can write the energy $E = h\nu$ and momentum $p = h\nu/c$ in terms of the frequency ν.

The Bose–Einstein distribution for photons (Equation 3.21) can be written as

$$f_{\text{state}}(\nu) = \frac{1}{e^{h\nu/kT} - 1}. \tag{3.60}$$

Notice that we have set $e^{\alpha} = 1$ in Equation 3.60. This is because photon number is not conserved, and as photons are created and destroyed in their interactions with other particles, they reach an equilibrium state in which the normalization equals unity. Expressed in terms of the frequency $\nu = (c/h)p$, the number density of available states is (Equation 3.17)

$$n_{\text{state}}(\nu) = n_{\text{state}}(p)\frac{dp}{d\nu} = \frac{8\pi \nu^2}{c^3}. \tag{3.61}$$

The total number density of photons at a temperature T can then be written as

$$n = \int_0^\infty n_{\text{state}}(\nu)f_{\text{state}}(\nu)d\nu = \frac{8\pi}{c^3} \int_0^\infty \frac{\nu^2 d\nu}{\exp(h\nu/kT) - 1}. \tag{3.62}$$

Using the substitution $\beta \equiv h\nu/kT$, Equation 3.62 becomes

$$n = \frac{8\pi}{h^3 c^3}(kT)^3 \int_0^\infty \frac{\beta^2 d\beta}{e^\beta - 1}. \tag{3.63}$$

The definite integral in Equation 3.63 has the numerical value $2\zeta(3)$, where $\zeta(3) \approx 1.2021$ is the Riemann zeta function computed at $s = 3$. This leads to the relation

$$n = bT^3, \tag{3.64}$$

where

$$b = \frac{16\pi(1.2021)k^3}{h^3 c^3} \approx 20.29 \, \text{cm}^{-3} \, \text{K}^{-3}. \tag{3.65}$$

Similarly, the energy density of photons at a temperature T can be written as

$$U = \int_0^\infty (h\nu)n_{\text{state}}(\nu)f_{\text{state}}(\nu)d\nu = \frac{8\pi h}{c^3} \int_0^\infty \frac{\nu^3 d\nu}{\exp(h\nu/kT) - 1}. \tag{3.66}$$

Again using the substitution $\beta \equiv h\nu/kT$, Equation 3.66 becomes

$$U = \frac{8\pi(kT)^4}{h^3 c^3} \int_0^\infty \frac{\beta^3 d\beta}{e^\beta - 1}. \tag{3.67}$$

The definite integral in Equation 3.67 has the numerical value $6\zeta(4)$, where $\zeta(4) = \pi^4/90 \approx 1.0823$. This leads to the relation

$$U = aT^4, \tag{3.68}$$

where

$$a = \frac{8\pi^5 k^4}{15h^3 c^3} \approx 7.566 \times 10^{-15} \, \text{erg cm}^{-3} \, \text{K}^{-4} \tag{3.69}$$

is the same radiation density constant that we encountered when discussing energy transport in Section 2.4. Combining Equations 3.63 and 3.67, we find that the mean photon energy is

$$h\bar{\nu} = \frac{U}{n} = 2.701kT = 2.328 \, \text{keV} \left(\frac{T}{10^7 \, \text{K}}\right). \tag{3.70}$$

Starting from the pressure integral given in Equation 3.19, the radiation pressure provided by the photons is calculated to be

$$P_{\text{rad}} = \frac{1}{3}U = \frac{1}{3}aT^4. \tag{3.71}$$

Since energy per unit volume has the same dimensionality as force per unit area, the energy density U and radiation pressure P_{rad} are related by a pure numerical factor, which is simply the geometric factor of $1/3$ that enters into the basic pressure integral (Equation 3.10).

When radiation pressure is the dominant source of pressure, the presence of a pressure gradient requires a temperature gradient. The necessity of temperature gradients in stars supported by radiation pressure enables us to make a strong claim: given radiative energy transport and a radiation pressure equation of state, there exists a unique luminosity L for an enclosed mass M. To support this bold claim, suppose a star has

$$P = P_{\text{rad}} = \frac{a}{3}T^4. \tag{3.72}$$

Hydrostatic equilibrium then implies (from Equation 2.47)

$$\frac{dP}{dr} = \frac{4aT^3}{3}\frac{dT}{dr} = -\frac{GM\rho}{r^2}. \tag{3.73}$$

However, the equation of radiative transport (Equation 2.49) gives us another relation involving the temperature gradient:

$$\frac{dT}{dr} = \frac{3\kappa\rho L}{16\pi acT^3 r^2}. \tag{3.74}$$

Combining Equations 3.73 and 3.74, we can solve for a unique luminosity $L_{\text{Edd}}(M)$, called the **Eddington luminosity**:

$$L_{\text{Edd}} = \frac{4\pi cGM}{\kappa}. \tag{3.75}$$

If the mass is pure ionized hydrogen, and if the opacity is provided entirely by scattering from free electrons, this relation becomes

$$L_{\mathrm{Edd}} = 3.29 \times 10^4 \, L_\odot \left(\frac{M}{1 \, M_\odot} \right). \tag{3.76}$$

Although this luminosity is high, it is approached in the most massive stars.[7] Furthermore, since $L_{\mathrm{Edd}} \propto M$, every star in which pressure support and energy transport are supplied by radiation has the same mass-to-light ratio, $\Upsilon_{\mathrm{Edd}} = M_\star/L_{\mathrm{Edd}} \approx 3.0 \times 10^{-5} \, M_\odot/L_\odot$, assuming electron scattering opacity. This value of Υ implies a main sequence lifetime, from Equation 1.32, of

$$t_{\mathrm{Edd}} \approx 0.32 \, \mathrm{Myr} \left(\frac{f_{\mathrm{nuc}}}{0.1} \right) \tag{3.77}$$

for a star radiating at its Eddington luminosity. Although main sequence stars with $M_\star < 30 \, M_\odot$ have a mass-luminosity dependence much steeper than $L_\star \propto M_\star$, a linear mass-luminosity relation is in fact observed for the most massive, shortest-lived stars.

In summary, we now have three possible relations for the equation of state. The ideal gas pressure, $P_{\mathrm{ideal}} \propto \rho T$, depends on both temperature and density. The degeneracy pressure, $P_{\mathrm{deg,nr}} \propto \rho^{5/3}$ or $P_{\mathrm{deg,rel}} \propto \rho^{4/3}$, depends on density but not on temperature. Finally, the radiation pressure, $P_{\mathrm{rad}} \propto T^4$, depends on temperature but not on mass density. The different dependencies on ρ and T mean that each of the three sources of pressure dominates in a particular region of the ρ–T plane. Comparison of the ideal gas pressure (Equation 3.32) and the non-relativistic degeneracy pressure (Equation 3.52) tells us that these two pressures are equal at a temperature-dependent critical density

$$\rho_{\mathrm{i\text{-}d}} = \frac{40\pi\sqrt{5}}{3} \frac{m_{\mathrm{AMU}}}{\lambda_{C,e}^3} \bar{\mu}^{-3/2} \mu_e^{5/2} \left(\frac{kT}{m_e c^2} \right)^{3/2}. \tag{3.78}$$

For ionized hydrogen, the critical density $\rho_{\mathrm{i\text{-}d}}$ between ideal gas pressure and degeneracy pressure becomes

$$\rho_{\mathrm{i\text{-}d}} = 2130 \, \mathrm{g\,cm^{-3}} \left(\frac{T}{10^7 \, \mathrm{K}} \right)^{3/2}. \tag{3.79}$$

This is an order of magnitude larger than the density ρ_{MB} at which the Maxwell-Boltzmann distribution and the resulting ideal gas law first start to break down (Equation 3.42); in the intervening density range, the ideal gas pressure gradually gives way to electron degeneracy pressure. The critical density $\rho_{\mathrm{i\text{-}d}}$ is plotted as the solid line in Figure 3.3.

---

[7] Using the empirical mass-luminosity relation of Equation 1.30, we expect that a main sequence star with $M_\star = 30 \, M_\odot$ will have $L_\star \sim 0.2 L_{\mathrm{Edd}}$. If we daringly extrapolate to higher stellar mass, we find that $L_\star = L_{\mathrm{Edd}}$ at $M_\star \sim 65 \, M_\odot$.

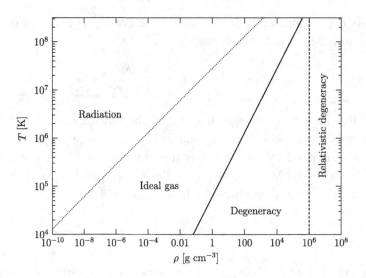

Figure 3.3 Regions in the density–temperature plane where the pressure is dominated by radiation pressure, ideal gas pressure, non-relativistic degeneracy pressure, and relativistic degeneracy pressure. A gas of pure hydrogen is assumed.

Comparison of the ideal gas pressure (Equation 3.32) and the radiation pressure (Equation 3.71) tells us that these two pressures are equal at a temperature-dependent critical density

$$\rho_{\text{r-i}} = \frac{8\pi^5}{15} \frac{m_{\text{AMU}}}{\lambda_{C,e}^3} \bar{\mu} \left(\frac{kT}{m_e c^2} \right)^3. \tag{3.80}$$

For ionized hydrogen, the critical density $\rho_{\text{r-i}}$ between radiation pressure and ideal gas pressure becomes

$$\rho_{\text{r-i}} = 0.0455 \,\text{g cm}^{-3} \left(\frac{T}{10^7 \,\text{K}} \right)^3. \tag{3.81}$$

This relation is plotted as the dotted line in Figure 3.3. Notice that radiation pressure dominates at high temperature and low density, while degeneracy pressure dominates at low temperature and high density. The ideal gas law dominates the pressure only in an intermediate strip. Because of the different rules for pressure in each region, we realize that some of the diversity in stellar behavior results from the different pressure sources upon which they draw.

Exercises

3.1 Consider a spherical star with mass $M_\star$, radius $R_\star$, and density profile $\rho \propto r^{-\eta}$, where $\eta > 0$.

(a) For what range of η is the gravitational potential energy W of the star a finite negative number?

(b) Suppose that the star has $M_\star = 1\,M_\odot$ and $R_\star = 1\,R_\odot$. For what range of η is the star's Kelvin–Helmholtz time greater than its hydrogen burning lifetime, as given in Equation 2.23? Express the lower limit on η as a function of f_{nuc}, the fraction of the star that undergoes hydrogen burning.

3.2 At the Sun's center, the temperature is $T_{c,\odot} = 1.57 \times 10^7\,K$ and the density is $\rho_{c,\odot} = 153\,g\,cm^{-3}$.

(a) What is the root mean square speed of a proton and of an electron at the Sun's center? Assume a non-relativistic ideal gas.

(b) What is the average distance between protons at the center of the Sun if we approximate it as a fully ionized gas of hydrogen?

(c) Repeat the calculations for parts (a) and (b), assuming $T = 2 \times 10^6\,K$ and $\rho = 0.1\,g\,cm^{-3}$, values appropriate for the base of the Sun's convection zone.

(d) Using the non-relativistic approximation, at what temperature is the average speed of an electron comparable to the speed of light? At what temperature is the average speed of a proton comparable to the speed of light?

3.3 Numerical models give the following values for the central temperature T_c and central density ρ_c of main sequence stars that have just begun to fuse hydrogen:

Mass	$\log T_c$	$\log \rho_c$
$[M_\odot]$	[K]	$[g\,cm^{-3}]$
50	7.6	0.3
10	7.5	0.8
1	7.2	2.0
0.1	6.6	2.8

For each mass, compute and compare the three components of the pressure (radiation, ideal gas, and full electron degeneracy), assuming a fully ionized gas of hydrogen.

4

Stellar Energy Transport

Deep Throat: Follow the money.
Woodward: What do you mean? Where?
Deep Throat: Oh, I can't tell you that.
Woodward: But you could tell me that.

Carl Bernstein, Bob Woodward, and William Goldman,
All the President's Men screenplay [1976]

When investigating a political scandal, the standard advice is "follow the money." When investigating stellar structure, a comparably useful piece of advice is "follow the energy." Since energy cannot be created or destroyed (if we regard mass as a sort of congealed energy), forensic investigation of a star's energy content will uncover whatever physical processes are hidden in a star's opaque interior. In this chapter, we will investigate the sources of a star's opacity, showing that when the opacity is sufficiently large, a star turns to convection rather than radiation to carry energy from core to photosphere.

4.1 Opacity

Determining the opacity κ starts with the simple question, "Why is the Sun opaque?" First, we realize that although the Sun is opaque to some particles (such as photons), it is *not* opaque to other particles (such as neutrinos). The typical energy of a neutrino produced by nuclear fusion in the Sun is $E \sim 0.3\,\text{MeV}$. At this neutrino energy, a baryon's cross section for interaction with neutrinos is $\sigma_{\text{neu}} \sim 10^{-42}\,\text{cm}^2$, or $\sigma_{\text{neu}} \sim 1$ attobarn, if we use the nuclear physicists' favorite unit of area, 1 barn $= 10^{-24}\,\text{cm}^2$. Traveling through material with density equal to the Sun's mean density of $\rho_\odot = 1.41\,\text{g\,cm}^{-3}$, the mean free path of a solar neutrino before interacting is

$$\lambda_{\text{mfp}} \approx \frac{m_{\text{AMU}}}{\rho \sigma_{\text{neu}}} \approx \frac{1.66 \times 10^{-24}\,\text{g}}{(1.41\,\text{g\,cm}^{-3})(10^{-42}\,\text{cm}^2)} \sim 10^{18}\,\text{cm} \sim 10^7\,\text{R}_\odot. \tag{4.1}$$

Thus, the vast majority of neutrinos stream straight through the Sun without interacting.

Photons, however, are much less antisocial than neutrinos; they frequently interact, particularly with electrons. In general, opacity can result from scattering of photons or from absorption of photons (with the eventual re-emission of one or more photons). In an ionized gas, most scattering is due to free electrons, whose cross section for photon scattering is the Thomson cross section, $\sigma_e = 6.652 \times 10^{-25}\,\mathrm{cm}^2$, or about two-thirds of a barn (see Equation 2.33). The number density of free electrons depends on the mass density ρ, the chemical composition X_i, and the degree of ionization of the gas. For simplicity, let's start by assuming a gas of pure hydrogen, with a number density $n_\mathrm{H} = \rho/m_\mathrm{H}$ of hydrogen nuclei, of which n_HI are in neutral hydrogen atoms and $n_p = n_\mathrm{H} - n_\mathrm{HI}$ are free protons. To conserve charge neutrality, the number density of free electrons is $n_e = n_p$. We can describe the degree of ionization of the hydrogen gas by the **fractional ionization**

$$x \equiv \frac{n_p}{n_p + n_\mathrm{HI}} = \frac{n_e}{n_\mathrm{H}}. \tag{4.2}$$

If the hydrogen satisfies the conditions of being a compressible gas of non-interacting particles in local thermodynamic equilibrium, then its ionization state is well described by the **Saha equation**,

$$\frac{n_p n_e}{n_\mathrm{HI}} = \left(\frac{2\pi m_e kT}{h^2}\right)^{3/2} \exp(-I_\mathrm{H}/kT), \tag{4.3}$$

where $I_\mathrm{H} = 13.60\,\mathrm{eV}$ is the ionization energy of hydrogen.

In terms of the fractional ionization x, the Saha equation for hydrogen can be rewritten as

$$\frac{x^2}{1-x} = \frac{1}{n_\mathrm{H}} \left(\frac{2\pi m_e kT}{h^2}\right)^{3/2} \exp(-I_\mathrm{H}/kT). \tag{4.4}$$

At a given temperature T, the density at which the hydrogen is half-ionized ($x = 1/2$) is given by the relation

$$\rho_{1/2} = 0.50\,\mathrm{g\,cm}^{-3} \left(\frac{kT}{13.6\,\mathrm{eV}}\right)^{3/2} \exp\left(-\frac{13.6\,\mathrm{eV}}{kT}\right). \tag{4.5}$$

Figure 4.1 shows the line of half-ionization in the ρ–T plane, as predicted by the Saha equation. In the solar photosphere (the square data point at lower left), the density and temperature are in the neutral regime, where hydrogen is less than half ionized. However, most of the Sun is well within the ionized regime.

We must be careful, though, to respect the limits of the Saha equation. Consider, for instance, the conditions at the Sun's center, where $kT = 1350\,\mathrm{eV}$ and $\rho = 153\,\mathrm{g\,cm}^{-3}$. The Saha equation (Equation 4.4) predicts that the ionization fraction of hydrogen at these conditions should be $x = 0.70$. It certainly

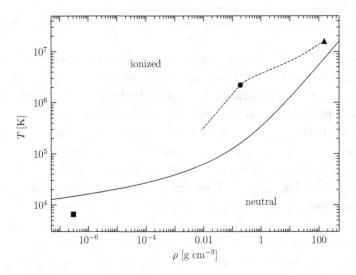

Figure 4.1 Solid line: division between mostly ionized hydrogen ($x > 1/2$) and mostly neutral hydrogen, according to the Saha equation. Triangle: conditions at the Sun's center. Circle: conditions at the base of the Sun's convective zone. Square: conditions at the Sun's photosphere. Dashed line: run of density and temperature in the Sun's interior.

seems odd that hydrogen at a temperature of $kT \approx 100 I_{\mathrm{H}}$ should be so mildly ionized. In fact, the true ionization fraction of hydrogen at such high temperature and density is extremely close to $x = 1$. The Saha equation is not truly applicable at the Sun's center because the density there is much higher than the value $\rho_0 \sim 3\,\mathrm{g\,cm^{-3}}$ at which the separation between nuclei becomes smaller than the Bohr radius. If 30% of the hydrogen were actually in the form of neutral atoms, then their electron clouds would overlap, and the assumption of a compressible gas of non-interacting particles would break down. In general, at densities greater than $3\,\mathrm{g\,cm^{-3}}$, hydrogen cannot have electrons bound to individual nuclei; at relatively low temperature, it becomes an incompressible metal (like the layer of metallic hydrogen in the cool interior of Jupiter), while at the high temperature found in the Sun's core, it becomes a fully ionized compressible gas.

In the low-density regime where the Saha equation strictly applies ($\rho \ll 3\,\mathrm{g\,cm^{-3}}$), the transition temperature from neutral to ionized is $T_{1/2} \sim 0.1 I/k$, where I is the relevant ionization energy. Thus, for hydrogen, with $I_{\mathrm{H}} = 13.6\,\mathrm{eV}$, the transition temperature is $T_{1/2} \sim 15\,000\,\mathrm{K}$. The first ionization energy of helium is $I_{\mathrm{He,1}} = 24.6\,\mathrm{eV}$, yielding $T_{1/2} \sim 30\,000\,\mathrm{K}$ in the low-density regime; the second ionization energy of helium is $I_{\mathrm{He,2}} = 54.4\,\mathrm{eV}$, yielding $T_{1/2} \sim 60\,000\,\mathrm{K}$. Thus, when you dive deep enough below the photosphere to reach a temperature of $T \gg 60\,000\,\mathrm{K}$, you reach the realm where hydrogen and helium, providing 98.5% of the Sun's mass, are almost completely ionized.

In a fully ionized gas, the number density of free electrons can be written as (Equation 3.47)

$$n_e = \frac{\rho}{m_{AMU}\mu_e},\tag{4.6}$$

where ρ is the mass density (contributed almost entirely by nucleons of mass $\sim m_{AMU}$) and μ_e is the number of nucleons per electron. For hydrogen $\mu_e = 1$, and for helium $\mu_e = 2$. Although the most abundant metals have equal numbers of neutrons, protons, and electrons, and thus have $\mu_e = 2$, some metals are neutron-rich, and thus have $\mu_e > 2$. In the Sun, the most abundant isotope with more neutrons than protons is ^{56}Fe, which has $\mu_e = 56/26 = 2.15$. However, ^{56}Fe and other neutron-rich isotopes like ^{22}Ne, ^{25}Mg, and ^{26}Mg provide only $\sim 12\%$ by mass of the Sun's metals. Thus, we can make a good estimate of μ_e for the Sun by using the approximation that all metals have $\mu_e = 2$. This approximation yields

$$\frac{1}{\mu_e} \approx \frac{X}{1} + \frac{Y}{2} + \frac{Z}{2} = X + \frac{1-X}{2} = \frac{1+X}{2},\tag{4.7}$$

and thus

$$n_e \approx \frac{1+X}{2}\frac{\rho}{m_{AMU}}\tag{4.8}$$

for a fully ionized gas. To compute the number density of free electrons in the Sun's highly ionized interior, the most important things to know are thus the mass density ρ and the hydrogen mass fraction X.

In a crude way, we can compute a mean free path for photons in the Sun, assuming electron scattering is the only source of opacity:

$$\lambda_{\mathrm{mfp}} \sim \frac{1}{\bar{n}_e\sigma_e} \sim \frac{2}{1+X}\frac{m_{AMU}}{\rho_\odot\sigma_e} \sim 2\,\mathrm{cm} \sim 3 \times 10^{-11}\,R_\odot.\tag{4.9}$$

Thus, although the mean free path for photons is much larger than the mean free path for electrons, it is much smaller than the solar radius. This has important implications for the thermal structure of the Sun. The mean temperature gradient in the Sun is

$$\frac{\Delta T}{\Delta r} = \frac{T_c - T_{\mathrm{eff}}}{0 - R_\odot} \approx \frac{1.57 \times 10^7\,\mathrm{K}}{-6.96 \times 10^{10}\,\mathrm{cm}} \approx -2.3 \times 10^{-4}\,\mathrm{K\,cm^{-1}}.\tag{4.10}$$

Even if a photon happens to be moving exactly in the radial direction, the change in ambient temperature between scatterings will be only

$$\Delta T \sim \frac{\Delta T}{\Delta r}\lambda_{\mathrm{mfp}} \sim 0.5\,\mathrm{mK}\left(\frac{\lambda_{\mathrm{mfp}}}{2\,\mathrm{cm}}\right).\tag{4.11}$$

Now consider a spherical shell of radius r centered on the Sun's central point (Figure 2.1), with a thickness dr equal to several times the local mean free path λ_{mfp} for photons. A photon that enters the shell will scatter many times from free electrons before exiting the shell. Since the mean free path for encounters among electrons, protons, and ions is shorter than λ_{mfp} for photons, this implies that all particles in the shell (except for neutrinos) are in local thermodynamic equilibrium at a temperature $T(r)$, to within a few millikelvin.

The opacity due to scattering from free electrons in a highly ionized gas is, making use of Equation 4.8,

$$\kappa_e = \frac{n_e \sigma_e}{\rho} = \frac{1+X}{2} \frac{\sigma_e}{m_{AMU}} = 0.200(1+X)\,\text{cm}^2\,\text{g}^{-1}. \tag{4.12}$$

For $X_\odot = 0.706$, this represents a scattering opacity of $\kappa_{e,\odot} = 0.342\,\text{cm}^2\,\text{g}^{-1}$. However, scattering does not represent the only source of opacity. Photons can be absorbed as well as scattered; in fully ionized hydrogen, the main absorption process is **free–free absorption**, also known as inverse bremsstrahlung. Free–free *emission* occurs when a free electron is accelerated in passing a proton or ion. However, free–free emission is a reversible process, so the inverse process of free–free *absorption* can also occur in an ionized gas.

An obvious question to ask is whether the opacity of highly ionized gas is dominated by scattering or by absorption. Unsurprisingly, the answer to the question is "it depends." Whether scattering or absorption dominates depends on the local density ρ and temperature T. It also depends on the frequency ν at which you are observing, since the free–free opacity $\kappa_{ff,\nu}$, unlike the electron scattering opacity κ_e, is a function of frequency; in the limit $h\nu \gg kT$, the free–free opacity drops steeply, with $\kappa_{ff,\nu} \propto \nu^{-3}$. Let's start, then, by asking what the free–free opacity κ_{ff} might be for the typical frequency $h\nu \sim 3kT$ that we expect for blackbody radiation.

Our thin spherical shell of radius r and thickness dr (Figure 2.1) has a temperature $T(r)$ that describes its photons, electrons, protons, and ions. The energy density of photons in the frequency range $\nu \to \nu + d\nu$ is

$$U_\nu d\nu = h\nu\, n_{state}(\nu) f_{state}(\nu) d\nu. \tag{4.13}$$

Using the values of n_{state} and f_{state} appropriate for photons, from Equations 3.60 and 3.61 we find the blackbody formula

$$U_\nu = \frac{8\pi h}{c^3} \frac{\nu^3}{\exp(h\nu/kT) - 1}. \tag{4.14}$$

The blackbody spectrum peaks at a frequency given by the relation $h\nu_{peak} = 2.82kT$; this is close to, but not identical with, the mean photon energy $h\bar{\nu} = 2.70kT$ for a blackbody spectrum. For photons at the very peak of the blackbody spectrum, with $\nu_{peak} = 2.82kT/h$, Equation 4.14 becomes

$$U_\nu \approx \frac{8\pi h}{e^{2.82} - 1} \left(\frac{2.82kT}{hc}\right)^3 \tag{4.15}$$

$$\approx 7.9 \times 10^{-8}\,\text{erg cm}^{-3}\,\text{Hz}^{-1} \left(\frac{T}{10^6\,\text{K}}\right)^3. \tag{4.16}$$

The photons providing this energy density have a brief and precarious life, since they are surrounded by electrons ready to absorb them in a free–free interaction. During a short time interval dt, the probability that a photon with frequency ν will be absorbed is

$$d\tau_\nu = \rho \kappa_{\mathrm{ff},\nu} c \, dt. \tag{4.17}$$

The energy density of photons with frequency ν thus decreases during the interval dt by an amount

$$dU_\nu = -U_\nu d\tau_\nu = -U_\nu \rho \kappa_{\mathrm{ff},\nu} c \, dt. \tag{4.18}$$

At the frequency $\nu_{\mathrm{peak}} = 2.82 kT/h$, the shell loses photon energy through free–free absorption at the rate

$$\frac{dU_\nu}{dt} = -U_\nu \rho \kappa_{\mathrm{ff},\nu} c \tag{4.19}$$

$$\approx -2400 \, \mathrm{erg \, s^{-1} \, cm^{-3} \, Hz^{-1}} \left(\frac{\rho}{1 \, \mathrm{g \, cm^{-3}}} \right) \left(\frac{T}{10^6 \, \mathrm{K}} \right)^3 \left(\frac{\kappa_{\mathrm{ff},\nu}}{1 \, \mathrm{cm^2 \, g^{-1}}} \right),$$

where we have used Equation 4.16 for the value of U_ν at the frequency ν_{peak} of peak emission. The characteristic time for absorbing photon energy is

$$t_{\mathrm{loss}} = -U_\nu \left(\frac{dU_\nu}{dt} \right)^{-1} = \frac{1}{\rho \kappa_{\mathrm{ff},\nu} c} \tag{4.20}$$

$$\approx 3.3 \times 10^{-11} \, \mathrm{s} \left(\frac{\rho}{1 \, \mathrm{g \, cm^{-3}}} \right)^{-1} \left(\frac{\kappa_{\mathrm{ff},\nu}}{1 \, \mathrm{cm^2 \, g^{-1}}} \right)^{-1}. \tag{4.21}$$

Thus, even if the opacity had the minuscule value $\kappa_{\mathrm{ff},\nu} \sim 10^{-10} \, \mathrm{cm^2 \, g^{-1}}$, photons in the Sun's dense interior would be absorbed in the blink of an eye (given $t_{\mathrm{blink}} \sim 0.3 \, \mathrm{s}$ and an assumed density $\rho \sim \rho_\odot$).

To maintain the blackbody energy density U_ν, photons must be added to the shell at the same rate as they are removed. In fully ionized gas, just as free–free absorption is an effective way to destroy photons, free–free emission is an effective way to create photons. For free–free emission in an ionized gas, the emissivity (power per unit volume per unit frequency) is a well-known function. In fully ionized hydrogen, the free–free emissivity is

$$\frac{dU_\nu}{dt} = \frac{32\pi}{3} \left(\frac{2\pi}{3} \right)^{1/2} \frac{e^6}{m_e^2 c^3} \left(\frac{m_e}{kT} \right)^{1/2} n_e n_p e^{-h\nu/kT} g_{\mathrm{ff}}, \tag{4.22}$$

where $g_{\mathrm{ff}}(\nu)$ is the Gaunt factor, which is of order unity. At the frequency ν_{peak}, the emissivity is

$$\frac{dU_\nu}{dt} \approx 1.4 \times 10^6 \, \mathrm{erg \, s^{-1} \, cm^{-3} \, Hz^{-1}} \left(\frac{T}{10^6 \, \mathrm{K}} \right)^{-1/2} \left(\frac{\rho}{1 \, \mathrm{g \, cm^{-3}}} \right)^2 g_{\mathrm{ff}}. \tag{4.23}$$

In order to balance the rate at which photon energy is lost (Equation 4.19) and the rate at which it is gained (Equation 4.23), the free–free opacity at ν_{peak} must be

$$\kappa_{\text{ff},\nu} \approx 600 \, \text{cm}^2 \, \text{g}^{-1} \left(\frac{\rho}{1 \, \text{g cm}^{-3}} \right) \left(\frac{T}{10^6 \, \text{K}} \right)^{-3.5}. \tag{4.24}$$

An opacity that has the dependence $\kappa \propto \rho T^{-3.5}$ is said to follow **Kramers' law**, after the physicist Hendrik Kramers.[1] By extension, an opacity that has the dependence $\kappa \propto \rho^\alpha T^{-\beta}$, for arbitrary values of α and β, follows a **generalized Kramers' law**.

Equation 4.24 gives the opacity at the peak of a blackbody spectrum. More usually, we want an opacity that is averaged over *all* frequencies of light. The most useful way of averaging over ν for a blackbody spectrum was developed by the astrophysicist Svein Rosseland in 1924. Let's follow in his footsteps, and reconstruct his definition of the **Rosseland mean opacity**. Consider a spherical surface of radius $r < R_\star$ centered upon a star's center. Let $L_\nu d\nu$ be the enclosed luminosity in the frequency range $\nu \to \nu + d\nu$. Then, at frequency ν, the radiative diffusion equation (compare to Equation 2.44) becomes

$$\frac{L_\nu}{4\pi r^2} = -\frac{c}{3\kappa_\nu \rho} \frac{d}{dr}(U_\nu). \tag{4.25}$$

Assuming a blackbody spectrum $U_\nu(T)$, this equation can be rewritten as

$$\frac{3\rho L_\nu}{4\pi c r^2} = -\frac{1}{\kappa_\nu} \frac{dU_\nu}{dT} \frac{dT}{dr}. \tag{4.26}$$

Now we integrate over frequency to find the relation between the bolometric luminosity L and the temperature gradient dT/dr:

$$\frac{3\rho L}{4\pi c r^2} = -\frac{dT}{dr} \int_0^\infty \frac{1}{\kappa_\nu} \frac{dU_\nu}{dT} d\nu. \tag{4.27}$$

Rosseland realized he could define a weighted harmonic mean opacity, now called the Rosseland mean opacity κ_R, using the relation

$$\frac{1}{\kappa_R} \equiv \int_0^\infty \frac{1}{\kappa_\nu} \frac{dU_\nu}{dT} d\nu \left/ \int_0^\infty \frac{dU_\nu}{dT} d\nu \right. . \tag{4.28}$$

However, the denominator on the right-hand side of Equation 4.28 can be written as

$$\int_0^\infty \frac{dU_\nu}{dT} d\nu = \frac{d}{dT} \int_0^\infty U_\nu d\nu = \frac{d}{dT}(aT^4) = 4aT^3. \tag{4.29}$$

Thus, the Rosseland mean opacity is given by the relation

$$\frac{1}{\kappa_R} = \frac{1}{4aT^3} \int_0^\infty \frac{1}{\kappa_\nu} \frac{dU_\nu}{dT} d\nu. \tag{4.30}$$

---

[1] In 1923, Kramers did pioneering quantum mechanical studies of absorption and emission; the next year, Arthur Eddington showed that Kramers' work implied an opacity $\kappa \propto \rho T^{-3.5}$.

Using the Rosseland mean opacity as given in Equation 4.30, the equation of radiative diffusion (Equation 4.27) can be expressed more simply as

$$\frac{dT}{dr} = -\frac{3\rho L}{4\pi cr^2}\frac{\kappa_R}{4aT^3} = -\frac{3\rho\kappa_R L}{16\pi acT^3 r^2}. \tag{4.31}$$

However, this is the standard equation for radiative energy transport within a star; see, for instance, Equation 2.49. Thus, although inverting κ and weighting it with the temperature derivative of U_ν might initially seem like an eccentric thing to do, the Rosseland mean opacity defined in this way is the correct opacity to use when tracking the radiative transport of energy within a star.

The Rosseland mean opacity for a gas of metallicity Z, taking only free–free absorption into account, can be approximately computed as

$$\kappa_{ff,R} \approx 40\,\text{cm}^2\,\text{g}^{-1}(1+X)(1-Z)\left(\frac{\rho}{1\,\text{g}\,\text{cm}^{-3}}\right)\left(\frac{T}{10^6\,\text{K}}\right)^{-3.5}. \tag{4.32}$$

Notice that the Rosseland mean opacity for pure hydrogen ($X = 1$) is an order of magnitude smaller than the opacity at ν_{peak} (Equation 4.24). There are two reasons why the Rosseland mean is so small. First, the weighting function dU_ν/dT for the Rosseland mean opacity has its maximum at $h\nu = 3.83kT$, higher than the value $h\nu_{peak} = 2.82kT$ at the maximum of U_ν itself. Since free–free opacity goes as $\kappa_{ff,\nu} \propto \nu^{-3}$, a weighting function shifted to higher frequency drives down the mean value of κ. Second, the Rosseland mean opacity is a *harmonic* mean: it involves finding the weighted mean of $1/\kappa_\nu$, not that of κ_ν. At $1/e$ of its maximum, the weighting function dU_ν/dT stretches from $h\nu \approx 1.5kT$ to $7.5kT$; within this fairly broad range, the higher frequencies (with lower values of κ_ν) dominate the calculation of the harmonic mean. The upshot is that when free–free absorption dominates the opacity, radiative transport is provided by fairly high-energy photons; half the contribution to $1/\kappa_R$, and thus half the contribution to the energy flux $\vec{F}$, comes from the interquartile range $h\nu \approx 6 \to 10kT$.

Bound–free opacity, resulting from the loss of photons to photoionization, is actually more important than free–free opacity in all but the lowest-metallicity stars. Previously in this chapter, we made the assumption that the gas in a star's interior is fully ionized. This is a good approximation for hydrogen and helium, as long as $T \gg 60\,000\,\text{K}$. However, heavier elements are more resistant to complete ionization. A hydrogenic ion of element i (that is, an ion with a single bound electron) has an ionization energy $I = Z_i^2(13.60\,\text{eV})$, where Z_i is the atomic number of element i. Thus, removing the last bound electron from oxygen, with $Z_O = 8$, requires an energy $I = 870\,\text{eV}$; this means that half the oxygen will be fully ionized at $T_{1/2} \sim 0.1I/k \sim 10^6\,\text{K}$. Removing the last bound electron from iron, with $Z_{Fe} = 26$, requires $I = 9200\,\text{eV}$, corresponding to $T_{1/2} \sim 0.1I/k \sim 10^7\,\text{K}$. Even at the high temperatures present in stellar interiors, heavy elements like

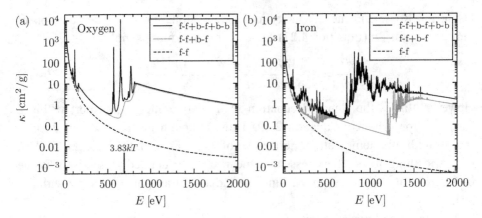

Figure 4.2 Contributions to opacity in gas of solar abundance with $T = 2.1 \times 10^6$ K and $\rho = 0.17\,\mathrm{g\,cm^{-3}}$. (a) Contribution from oxygen. (b) Contribution from iron. In each panel, dashed = free–free, gray = free–free + bound–free, black = free–free + bound–free + bound–bound. Vertical tick at $3.83kT = 698$ eV shows the location of the peak in dU_ν/dT.

iron will not be fully ionized. When ionization is only partial, the possibility of bound–free absorption exists. Accurately calculating the bound–free opacity κ_{bf} requires knowing the exact chemical composition of a star and knowing the photoionization cross section for each ion of each isotope as a function of frequency. The situation becomes still more complicated when including the non-negligible contribution from bound–bound transitions.

To illustrate the complexity of the problem, Figure 4.2 examines the opacity as a function of photon energy for gas of solar abundance, at a density and temperature chosen to match conditions just below the Sun's convection zone: $\rho = 0.17\,\mathrm{g\,cm^{-3}}$ and $T = 2.1 \times 10^6$ K. A gas consisting only of hydrogen and helium would be fully ionized at this density and temperature, and would have a Rosseland mean opacity (from free–free absorption and electron scattering) of $\kappa_R \approx 1\,\mathrm{cm^2\,g^{-1}}$. However, at solar abundance, the computed opacity is $\kappa_R = 20.8\,\mathrm{cm^2\,g^{-1}}$. Thus, metals dominate the opacity under these conditions. Unsurprisingly, a significant contribution comes from oxygen, the most abundant metal (Figure 4.2(a)). The opacity from free–free interactions involving oxygen ions (dashed line) is relatively small. However, at $T \sim 2 \times 10^6$ K, oxygen is not yet fully ionized. Thus, when bound–free opacity is added (gray line) there is a contribution above $h\nu = 870$ eV from photoionization of hydrogenic oxygen, above 740 eV from ionization of helium-like oxygen, and above 140 eV from ionization of lithium-like oxygen. Adding in bound–bound transitions (black line) boosts the opacity at energies that correspond to electronic transitions between bound states. Most prominent is the bound–bound opacity from the Lyman α transition in hydrogenic oxygen ($h\nu = 653$ eV). The corresponding opacity plot for iron (Figure 4.2(b)) is more complicated than for oxygen. When bound–free opacity is included (gray line), a jump in opacity can be seen at the ionization energy of neon-like Fe (1270 eV). However, a number of resonance features are seen, associated with autoionization of iron. Once bound–bound transitions are added (black line), iron is shown to provide high opacity in the range $h\nu \approx 800$–1200 eV, given $T \sim 2 \times 10^6$ K. Understanding the opacity of stars thus requires understanding the atomic physics of iron and other relatively abundant iron peak elements.

Once the Rosseland mean is taken, the bound–free and bound–bound opacities, like the free–free opacity, follow Kramers' law: $\kappa_R \propto \rho T^{-3.5}$. For a chemical composition similar to that of the Sun, a useful approximation is

$$\kappa_{\text{bf,R}} \approx 40\,000 \text{ cm}^2 \text{ g}^{-1} Z(1+X) \left(\frac{\rho}{1\,\text{g cm}^{-3}}\right) \left(\frac{T}{10^6\,\text{K}}\right)^{-3.5}, \tag{4.33}$$

where X is the hydrogen mass fraction and Z is the mass fraction of metals. Comparing Equation 4.33 for bound–free absorption to Equation 4.32 for free–free absorption, we find that

$$\frac{\kappa_{\text{bf,R}}}{\kappa_{\text{ff,R}}} \approx 1000 \frac{Z}{1-Z}, \tag{4.34}$$

and free–free absorption dominates over bound–free only in low-metallicity stars ($Z < 0.001 \approx 0.06 Z_{\odot,0}$).

Given the $T^{-3.5}$ dependence of Kramers' law upon temperature, electron scattering opacity (Equation 4.12) dominates over absorption opacity (Equations 4.32 and 4.33) in the high-temperature limit. The temperature at which electron scattering starts to take over is

$$T \approx 1 \times 10^7 \text{ K} \left(\frac{Z}{Z_{\odot,0}} + 0.06\right)^{0.29} \left(\frac{\rho}{1\,\text{g cm}^{-3}}\right)^{0.29}. \tag{4.35}$$

At the Sun's central density, $\rho_{c,\odot} = 153 \text{ g cm}^{-3}$, electron scattering opacity would dominate only at a temperature $T > 4 \times 10^7$ K, about three times the actual temperature at the Sun's center.

Determining the opacity of stellar interiors requires computing the frequency-dependent free–free, bound–free, and bound–bound cross sections of all the relevant ionization states of all the relevant elements in stellar gas. Figure 4.3, for instance, shows the computed opacity as a function of temperature for gas of solar abundance ($X = 0.706$, $Y = 0.277$, and $Z = 0.017$). In the limit of high temperature, the opacity approaches the electron scattering value $\kappa \approx 0.34 \text{ cm}^2 \text{ g}^{-1}$. At lower temperatures, the $\kappa \propto T^{-3.5}$ dependence predicted for bound–free and free–free absorption provides a good fit down to temperatures $T \sim 30\,000$ K.

In the temperature range 3000 K $< T <$ 8000 K, the major source of opacity in Sun-like stars is the H$^-$ ion. This negative ion is fragile; it takes an energy of just $I_- = 0.754$ eV to strip away the excess electron. On the one hand, this fragility

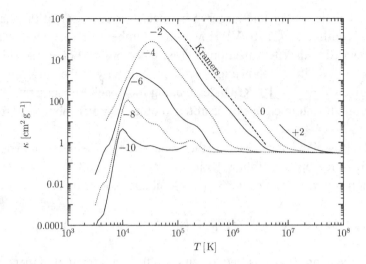

Figure 4.3 Rosseland mean opacity as a function of temperature for gas of solar abundance. Each curve is labeled with $\log \rho$, where ρ is in units of $\mathrm{g\,cm^{-3}}$. The dashed line shows the Kramers' law approximation (Equations 4.32 and 4.33) at a density of $\rho = 0.1\,\mathrm{g\,cm^{-3}}$. [Data from the Opacity Project]

means that photodetachment of H^- can absorb all photons with $\lambda < hc/I_- = 1.64\,\mu m$, providing bound–free opacity at near-infrared, visible, and ultraviolet wavelengths. On the other hand, the fragility of H^- means that very little H^- exists at temperatures $T > I_-/k \sim 8000\,K$, since it is destroyed by collisional detachment of the second electron. At the other end of the temperature scale, very little H^- exists at $T < 3000\,K$, because at such low temperatures, there are essentially no free electrons available to bond with neutral hydrogen atoms. Of the 20 most common metals in the solar photosphere, the one with the lowest first ionization energy is potassium, with $I_K = 4.3\,eV$. In the relatively low density of a stellar atmosphere, we expect potassium to be primarily in its singly ionized state at temperatures greater than $T_{1/2} \sim 0.1 I_K/k \sim 5000\,K$. However, when the temperature drops to $T \sim 3000\,K \sim 0.06 I_K/k$, even atoms far on the exponential tail of the Maxwellian energy distribution don't have enough kinetic energy to collisionally ionize potassium and produce free electrons.

In the temperature interval from 3000 K to 8000 K, which includes the Sun's photospheric temperature, there are enough H^- ions available to provide the main source of opacity in stars of solar abundance. The H^- opacity has the approximate temperature dependence

$$\kappa_{H^-} \propto Z\rho^{0.5}T^9. \tag{4.36}$$

In the temperature range $2300\,K < T < 3000\,K$, characteristic of the atmospheres of the coolest M dwarfs, molecules provide the main source of opacity. These molecules include not only H_2O but also scarcer oxides, such as TiO and VO, which have strong electronic absorption bands[2] at $\lambda \sim 1\,\mu m$. At $T < 2300\,K$, characteristic of the atmospheres of brown dwarfs, additional opacity is provided by molecules such as CH_4, NH_3, and H_2S, as well as by dust grains that start to condense out of the gas phase.

4.2 Convection

When the opacity κ at some radius r becomes very large, the radiative diffusion coefficient $D_\gamma \propto 1/\kappa$ becomes very small, and radiative transport becomes inefficient. Whenever the opacity becomes sufficiently large, **convection** takes over as the main energy transport mechanism inside stars. Convection is the transport of heat by bulk motions of hot gas; in practice, convection generally occurs through turbulent flow. Given the chaotic nature of turbulence, it can be difficult to describe analytically. However, we have to grapple with convection at some level, since it is known to occur in stars. For instance, the photosphere of the Sun shows granulation, as depicted in Figure 4.4. Each granule is a convection cell in the Sun's atmosphere. The brighter central region of the granule contains hot gas rising from the interior; the dark rim contains cooler gas falling back to the interior. Convection cells visible in the solar photosphere are typically $\ell \sim 1000\,km$ in width and have circulation speeds, found from Doppler measurements, of $v \sim 2\,km\,s^{-1}$.

The visible solar granulation represents only a thin layer of convective motion. Beneath the photosphere of the Sun lies a **convective zone** about $200\,000\,km$ thick where convection is the main mechanism for energy transport. The convective zone stretches from $r = 0.71\,R_\odot$ to the base of the photosphere at $r = 1\,R_\odot$. Although the convective zone represents 64% of the Sun's volume, it contains only 2.4% of the Sun's mass. Beneath the convective zone lies the **radiative zone** of the Sun, where radiative transport is dominant. However, the Sun's arrangement for energy transport doesn't apply to all stars. Dwarfs with $M_\star < 0.3\,M_\odot$ are convective throughout their interiors; main sequence stars with $M_\star > 1.2\,M_\odot$ have an outer radiative zone and an inner convective zone, opposite to the Sun's arrangement.

Start by considering a relatively small blob of gas inside a star. The blob has a temperature T, density ρ, and chemical composition X_i. If radiation pressure is negligible, the pressure in the blob will be

$$P = \frac{k}{m_{\text{AMU}}} \frac{\rho T}{\mu}, \qquad (4.37)$$

---

[2] Figure 1.8, for instance, shows the spectrum of an M5 dwarf ($T_{\text{eff}} \approx 3000\,K$) with TiO absorption bands seen at $\lambda \approx 0.68\,\mu m$ and $0.72\,\mu m$.

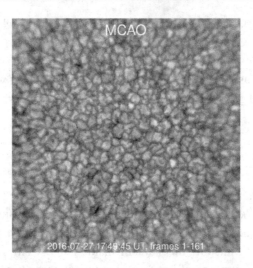

Figure 4.4 Solar granulation in a 53×53 arcsec field of view (corresponding to $19\,000 \times 19\,000$ km), imaged using multi-conjugate adaptive optics on the New Solar Telescope of the Big Bear Solar Observatory. [Schmidt *et al.* 2017]

where the mean molecular mass $\bar{\mu}$ can be computed when the chemical composition and ionization state are known. In the realm of thermodynamics, it is useful to know the specific internal energy u of the gas. In addition to the kinetic energy associated with random thermal motions of gas particles, the specific internal energy also includes the rotational and vibrational energy of molecules in the gas. (If the gas undergoes chemical or nuclear reactions, the specific internal energy also includes the potential energy associated with chemical or nuclear bonds.) For a monatomic gas, the specific internal energy u is

$$u_{\mathrm{mon}} = \frac{3}{2} \frac{kT}{m_{\mathrm{AMU}}\bar{\mu}} = \frac{3}{2}\frac{P}{\rho}. \qquad (4.38)$$

More generally, the specific internal energy can be written in the form

$$u = \frac{f}{2} \frac{kT}{m_{\mathrm{AMU}}\bar{\mu}} = \frac{f}{2}\frac{P}{\rho}, \qquad (4.39)$$

where f is the number of degrees of freedom for a gas particle. An atom has $f = 3$ degrees of freedom, from its translational motions in three dimensions. A diatomic molecule has an additional two degrees of freedom from its rotational motion, as long as the gas is hot enough for the rotational modes of the molecule to be excited; since the fundamental frequency of rotation for a diatomic molecule is typically $\hbar\omega_r \sim 10^{-3}$ eV, molecules will rotate at temperatures above $\hbar\omega_r/k \sim 10$ K. A diatomic molecule will also have an additional degree of freedom from its vibration, as long as the gas is hot enough for the vibrational modes of the molecule to be excited; since the fundamental vibrational frequency for a diatomic molecule is $\hbar\omega_v \sim 0.3$ eV, molecules won't vibrate unless the temperature approaches $\hbar\omega_v/k \sim 3000$ K. At room temperature, then, it is usually assumed that diatomic molecules have $f = 3 + 2 = 5$ degrees of freedom.

If the specific internal energy u of the blob changes by an amount du, it must be the result of heat flowing into (or out of) the blob and of PdV work being done by (or done to) the blob:

$$du = dq - Pd\left(\frac{1}{\rho}\right), \tag{4.40}$$

where dq is the amount of heat per unit mass flowing into the blob, and $1/\rho$ is a "specific volume"; that is, the volume in cubic centimeters occupied by one gram of gas. If there is no net heat flow ($dq = 0$), then the blob is undergoing an **adiabatic** process. Since the specific entropy change ds is related to the change in specific heat by the relation $dq = Tds$, an adiabatic process is one in which entropy is conserved.

For an adiabatic process,

$$du + Pd\left(\frac{1}{\rho}\right) = 0. \tag{4.41}$$

If we assume $u = (f/2)(P/\rho)$, then Equation 4.41 for an adiabatic process tells us that

$$\frac{f}{2}\frac{1}{\rho}dP - \frac{f}{2}\frac{P}{\rho^2}d\rho - \frac{P}{\rho^2}d\rho = 0, \tag{4.42}$$

leading to the power-law relation

$$P \propto \rho^\gamma, \tag{4.43}$$

where

$$\gamma = \frac{2+f}{f}. \tag{4.44}$$

The index γ, known as the **adiabatic index**, is $\gamma = 5/3 \approx 1.67$ for a monatomic gas with $f = 3$, and $\gamma = 7/5 = 1.4$ for a cool gas of diatomic molecules with $f = 5$. If the gas of diatomic molecules is hot enough for the molecular vibrational modes to be excited, then $f = 6$ and $\gamma = 4/3 \approx 1.33$. Note that as the gas particles attain more degrees of freedom, the adiabatic index $\gamma \to 1$.

One way in which the adiabatic index γ can be driven close to unity is to have a gas in the process of being ionized. A hydrogen atom in its ground state has an ionization energy $I = 13.6$ eV. Thus, a gas of hydrogen atoms has a specific ionization energy $u_{\text{ion}} = I/m_H$. As long as the hydrogen gas is at very low temperatures, this specific ionization energy is untapped. However, as the temperature approaches the half-ionization temperature $T_{1/2}$, the specific energy of ionization

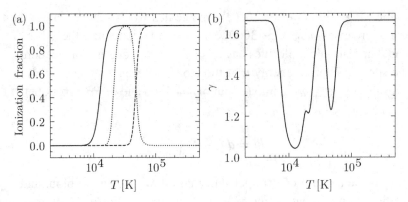

Figure 4.5 (a) Ionization fractions for a metal-free gas with $\rho = 10^{-7}\,\mathrm{g\,cm^{-3}}, X = 0.72$, and $Y = 0.28$. Solid line = fraction of H that is ionized; dotted line = fraction of He that is singly ionized; dashed line = fraction of He that is doubly ionized. (b) Resulting adiabatic index γ.

becomes accessible, and the specific energy of the hydrogen gas is

$$u = \frac{3}{2}\frac{P}{\rho} + \frac{I}{m_{\mathrm{H}}}x, \tag{4.45}$$

where $x = n_e/n_{\mathrm{H}}$ is the fractional ionization. Since this internal specific energy is greater than the value for a monatomic gas, this means that the effective number of degrees of freedom is greater than the $f = 3$ value expected for a fully neutral or fully ionized monatomic gas. By using the Saha equation (Equation 4.4) to write x in terms of ρ and P, it can be shown (with a certain amount of algebra) that the adiabatic index for partially ionized hydrogen is

$$\gamma(x) = \frac{5 + (5/2 + I/kT)^2 x(1-x)}{3 + [3/2 + (3/2 + I/kT)^2]x(1-x)}. \tag{4.46}$$

If we assume we are in the low-density limit, where $x = 1/2$ at $kT \approx 0.1I$, this leads to an adiabatic index of $\gamma \approx 1.2$ for a half-ionized gas.

For gases made of a mix of elements, the adiabatic index γ can be a complicated function of T, best computed numerically. For instance, Figure 4.5 shows the ionization behavior of gas with $X = 0.72$ and helium mass fraction $Y = 1 - X = 0.28$ at a density of $\rho = 10^{-7}\,\mathrm{g\,cm^{-3}}$; this relatively low density is found in the outer, cooler layers of a typical star. Figure 4.5(a) shows the ionization fraction of hydrogen as the solid line; at $T > 13\,000$ K, more than half the hydrogen in this low-density gas is ionized. The fraction of helium that is singly ionized is shown as the dotted line; in the temperature range $21\,000$ K $< T < 48\,000$ K, more than half the helium is singly ionized. The fraction of helium that is doubly ionized is shown as the dashed line; at $T > 48\,000$ K, the majority of the helium is doubly ionized. The resulting adiabatic index $\gamma(T)$ for the mix of hydrogen and helium is shown in the right panel. Even without metals present to complicate the situation, the plot of γ versus T shows multiple minima. The dip at $T \approx 13\,000$ K

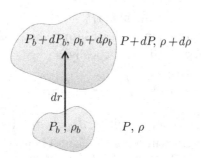

Figure 4.6 A blob of gas is moved upward in a star by a distance dr; will it continue to rise?

corresponds to ionization of hydrogen, the smaller dip at $T \approx 21\,000$ K corresponds to the first ionization of helium, and the dip at $T \approx 48\,000$ K corresponds to the second ionization of helium.

Let us return to our blob of gas, which has density ρ_b and pressure P_b. The gas in its immediate vicinity has density ρ and pressure P; since the blob is in equilibrium with its surroundings, $P_b = P$ and $\rho_b = \rho$. Now, however, we move the blob upward by some distance dr, to a level where the ambient density is $\rho + d\rho$ and the pressure is $P + dP$, as shown in Figure 4.6. We now want to ask if the blob will naturally continue to rise, triggering convective flow. (In the alternate scenario, the blob falls back downward and undergoes damped oscillations about its starting point, tamely ending up where it began.)

For a star in hydrostatic equilibrium,

$$dP = \frac{dP}{dr}dr < 0, \tag{4.47}$$

and the blob, once it has been moved upward, will expand in order to come back into pressure equilibrium with its new surroundings. Its new pressure, after expansion, will be $P_b + dP_b$, where $dP_b = dP$. This expansion takes a finite length of time; $t_{\text{exp}} \sim \ell_b/c_s$, where ℓ_b is the radius of the blob, and c_s is the sound speed within the blob. Once the blob is moved upward, it will generally not be at the same temperature as the surrounding gas. The time it takes for heat to flow out of the blob is $t_{\text{heat}} \sim \ell_b^2/D_\gamma$, where D_γ is the radiative diffusion coefficient. We then find that the expansion of the blob is adiabatic when $t_{\text{heat}} \gg t_{\text{exp}}$, or

$$\ell_b \gg \frac{D_\gamma}{c_s} \sim 2000 \,\text{cm} \left(\frac{T_b}{10^6 \,\text{K}}\right)^{-1/2} \left(\frac{\lambda_{\text{mfp}}}{2 \,\text{cm}}\right), \tag{4.48}$$

where T_b is the initial blob temperature.

Let's suppose that our blob is big enough that its expansion is adiabatic, with $P_b \propto \rho_b^\gamma$. This implies that the change in the blob's density as it expands is

$$\frac{d\rho_b}{\rho_b} = \frac{1}{\gamma}\frac{dP_b}{P_b}. \tag{4.49}$$

Given the initial conditions for the blob ($P_b = P$, $\rho_b = \rho$) and the requirement that it end in pressure equilibrium with its surroundings ($dP_b = dP$), the change in the blob's density as it expands adiabatically is

$$d\rho_b = \frac{\rho_b}{\gamma} \frac{dP_b}{P_b} = \frac{\rho}{\gamma} \frac{dP}{P}. \tag{4.50}$$

If the blob is denser than its surroundings, $d\rho_b > d\rho$, it will sink back down to where it started, and the gas will be *stable* against convective flow. If the blob is less dense than its surroundings, $d\rho_b < d\rho$, it will continue to rise, and the gas will be *unstable* against convective flow. Making use of Equation 4.50, we can rewrite the criterion for instability, $d\rho_b < d\rho$, as

$$\frac{dP}{P} < \gamma \frac{d\rho}{\rho} \qquad \text{[unstable]}. \tag{4.51}$$

This criterion for the onset of convection is called the **Schwarzschild criterion**, after Karl Schwarzschild, who published his derivation in 1906.

It is frequently convenient to write the Schwarzschild instability criterion of Equation 4.51 in terms of the temperature $T(r)$ rather than the mass density $\rho(r)$ within a star. The ideal gas law tells us that $P = \rho kT/m_{\mathrm{AMU}}\bar{\mu}$. If there are no changes in the ionization of the blob as it expands, then $\bar{\mu}$ is constant, and

$$\frac{dP}{P} = \frac{d\rho}{\rho} + \frac{dT}{T}. \tag{4.52}$$

Using Equation 4.52 to substitute for $d\rho/\rho$, we can rewrite Equation 4.51 as

$$\frac{P}{T} \frac{dT}{dP} > \frac{\gamma - 1}{\gamma} \qquad \text{[unstable]}. \tag{4.53}$$

In deriving Equation 4.53, keep in mind that we are assuming $dP < 0$ and $\gamma > 1$; hence, $dT < 0$ (the gas cools at it rises) and $dT/dP > 0$.

If you want to plunge into the literature on stellar convection, it's helpful to know some of the symbols involved. For instance, stellar astronomers find it useful to talk about the adiabatic temperature gradient ∇_{ad} ("del ad"). This is defined as

$$\nabla_{\mathrm{ad}} \equiv \left(\frac{\partial \ln T}{\partial \ln P} \right)_{\mathrm{ad}}, \tag{4.54}$$

where the subscript indicates that the changes in T and P take place adiabatically. For an ideal gas with an adiabatic index γ,

$$\nabla_{\mathrm{ad}} = \frac{\gamma - 1}{\gamma}. \tag{4.55}$$

Thus, ∇_{ad} is a number smaller than one. If $\gamma \approx 1.67$, as expected for a fully ionized or fully neutral monatomic gas, than $\nabla_{\mathrm{ad}} \approx 0.4$. If $\gamma \approx 1.1$, as Figure 4.5 tells us to expect at $T \sim 15\,000$ K, then $\nabla_{\mathrm{ad}} \approx 0.1$.

The Schwarzschild instability criterion (Equation 4.53) can now be written simply as

$$\nabla_{\text{act}} > \nabla_{\text{ad}} \quad \text{[unstable]}, \tag{4.56}$$

where

$$\nabla_{\text{act}} \equiv \frac{P}{T}\frac{dT}{dP} = \frac{d\ln T}{d\ln P} \tag{4.57}$$

("actual del") is the *actual* gradient in the star that you are investigating.

When radiation is the primary means of transporting energy through a star, the temperature gradient is, from Equation 2.49,

$$\frac{dT}{dr} = -\frac{3\kappa\rho L}{16\pi ac T^3 r^2}. \tag{4.58}$$

For a star in hydrostatic equilibrium, the pressure gradient is, from Equation 2.47,

$$\frac{dP}{dr} = -\frac{GM\rho}{r^2}. \tag{4.59}$$

A star with radiative heat transport must then have

$$\frac{dT}{dP} = \frac{3\kappa L}{16\pi ac T^3 GM}. \tag{4.60}$$

Thus, we may define ∇_{rad} ("del rad") so that it is equal to $d\ln T/d\ln P$ for a star in which heat transport is *entirely* radiative:

$$\nabla_{\text{rad}} \equiv \frac{3\kappa PL}{16\pi ac T^4 GM}. \tag{4.61}$$

If pressure is supplied entirely by the ideal gas law, then

$$\nabla_{\text{rad}} = \frac{3k}{16\pi ac m_{\text{AMU}}} \frac{\kappa L\rho}{\bar{\mu} T^3 GM}. \tag{4.62}$$

Earlier, we stated that convection sets in when the opacity becomes "sufficiently large." Now we have a way of quantifying "sufficiently large." As long as $\nabla_{\text{rad}} < \nabla_{\text{ad}}$, the star is not convective. Thus, we find that in a radiative zone, $\nabla_{\text{act}} = \nabla_{\text{rad}} < \nabla_{\text{ad}}$. When convection does not occur, the temperature gradient is given by the radiative equation:

$$\frac{dT}{dr} = \nabla_{\text{rad}}\frac{T}{P}\frac{dP}{dr} = -\frac{3\kappa\rho L}{16\pi ac T^3 r^2} \quad \text{[radiative]}. \tag{4.63}$$

However, it can happen that ∇_{rad} rises above ∇_{ad} at some radius in a star. This can happen if ∇_{rad} increases (through an increase in κ, for instance) or if ∇_{ad} decreases (through a decrease in γ, perhaps because of ionization). When $\nabla_{\text{rad}} > \nabla_{\text{ad}}$, convection begins. Convection will produce a value of ∇_{act} that is larger than ∇_{ad}; that's the instability criterion that produces convection, after all. However, it is frequently found that ∇_{act} is only slightly larger than ∇_{ad} in the convective zones of stars. Thus, we find that in a convective zone, $\nabla_{\text{ad}} \approx \nabla_{\text{act}} < \nabla_{\text{rad}}$. In this case, we can make the approximation that the temperature gradient in a convective zone is

$$\frac{dT}{dr} \approx \nabla_{ad}\frac{T}{P}\frac{dP}{dr} \approx -\frac{\gamma - 1}{\gamma}\frac{m_{AMU}\bar{\mu}}{k}\frac{GM}{r^2} \qquad \text{[convective]} \qquad (4.64)$$

when the pressure is described by an ideal gas law.

4.3 Mixing Length Theory

While estimating the temperature gradient dT/dr in a convective zone, we glossed over vast amounts of physics in stating $\nabla_{act} \approx \nabla_{ad}$. In truth, the physics of a convective zone is complicated, involving turbulent motions of compressible gas. In the age before fast, cheap computers, astronomers relied on relatively simple analytic theories such as **mixing length theory**. Even now, use of mixing length theory can help us understand the physics of convection. The basic assumption behind mixing length theory is that when a hot blob of gas satisfies the Schwarzschild criterion for instability, it moves upward through a distance ℓ before blending in with its surroundings.[3]

In standard mixing length theory, the blob of gas rises because its density ρ_b is slightly less than the density ρ of the surrounding gas; that is, $\Delta\rho \equiv \rho_b - \rho < 0$ and $|\Delta\rho| \ll \rho$. If the speed v with which the blob rises is small compared to the sound speed c_s of the gas, then it remains in pressure equilibrium with its surroundings: $P_b = P$. If the mean molecular mass $\bar{\mu}_b$ of the blob is the same as that of the surrounding gas, then the ideal gas law implies that the temperature T_b of the blob is slightly higher than the temperature T of the surrounding gas. That is, if we define $\Delta T \equiv T_b - T$, then $\Delta T/T = -\Delta\rho/\rho \ll 1$.

Consider a hot blob at a distance r from the star's center. Its buoyancy produces an acceleration

$$\frac{dv}{dt} = -\frac{\Delta\rho}{\rho}g, \qquad (4.65)$$

where $g = GM(r)/r^2$. If the mixing length ℓ is small compared to r, then the blob experiences a nearly constant value of g during its lifetime. Starting at rest, by the time the blob has traveled a distance ℓ upward, it reaches a speed

$$v \sim \left(-\frac{\Delta\rho}{\rho}g\ell\right)^{1/2} \sim \left(\frac{\Delta T}{T}g\ell\right)^{1/2}. \qquad (4.66)$$

(Since the fractional temperature excess, $\Delta T/T$, changes as the gas moves upward, Equation 4.66 is implicitly adopting an average value for $\Delta T/T$ over

---

[3] The length scale ℓ is the eponymous "mixing length" of mixing length theory.

the blob's lifetime.) In terms of the pressure scale height of the star,

$$h \equiv \left| \frac{1}{P} \frac{dP}{dr} \right|^{-1} = \frac{P}{\rho} \frac{r^2}{GM} \approx \frac{c_s^2}{g}, \tag{4.67}$$

the upward speed of the blob shortly before its demise (Equation 4.66) can be written as

$$v \sim \left(\frac{\Delta T}{T} \frac{\ell}{h} \right)^{1/2} c_s. \tag{4.68}$$

Numerical simulations of convection lead us to expect a mixing length that is comparable to the pressure scale height ($\ell \sim h$). Thus, the assumption that $\Delta T/T \ll 1$ implies subsonic convection ($v < c_s$).

Since the subsonically moving blob keeps in pressure equilibrium with its surroundings, its specific heat capacity is the specific heat capacity at constant pressure,

$$c_{\rm P} = \frac{\gamma}{\gamma - 1} \frac{k}{m_{\rm AMU} \bar{\mu}}. \tag{4.69}$$

Its specific internal energy, in excess of its surroundings, is then

$$\Delta u = c_{\rm P} \Delta T. \tag{4.70}$$

The upward flux of energy carried by the warm blob is then, combining Equations 4.68 and 4.70,

$$F_{\rm conv} = \rho_b (\Delta u)\, v \approx \rho (\Delta u)\, v \sim (\rho c_{\rm P} \Delta T) \left(\frac{\Delta T}{T} \frac{\ell}{h} \right)^{1/2} c_s \tag{4.71}$$

$$\sim (\rho c_{\rm P} T) \left(\frac{\Delta T}{T} \right)^{3/2} \alpha_\ell^{1/2} c_s, \tag{4.72}$$

where $\alpha_\ell \equiv \ell/h$ is the **mixing length parameter**, which we expect to be of order unity. Equation 4.72 tells us that blobs with larger values of $\Delta T/T$ carry a larger flux of energy, in part because they have a higher specific internal energy, and in part because they rise more rapidly. However, can we make a reasonable estimate of what we expect $\Delta T/T$ to be for a rising blob of gas?

We can start by computing what $\Delta T/T$ must be within the convective zone of the Sun in order to be in agreement with the Sun's observed properties. In general, given the Sun's mass $M_\odot$ and radius $R_\odot$, we expect, from Equation 3.6, that the pressure in its central regions will be

$$P_c \sim \frac{GM_\odot^2}{R_\odot^4}, \tag{4.73}$$

with a numerical factor in front that depends on the Sun's density profile. If we assume that all numerical factors are of order unity (thus committing ourselves to an order-of-magnitude calculation), we can take a typical pressure inside the Sun to be

$$P \sim G\rho_\odot^2 R_\odot^2, \tag{4.74}$$

where $\rho_\odot$ is the mean density of the Sun. Assuming an ideal gas law (Equation 3.32), this implies a typical temperature of

$$T \sim \frac{m_{\mathrm{AMU}}\bar{\mu}}{k}\frac{P}{\rho_\odot} \sim \frac{m_{\mathrm{AMU}}\bar{\mu}}{k}G\rho_\odot R_\odot^2. \tag{4.75}$$

The internal energy density of the gas will then be

$$U = \rho c_\mathrm{P} T \sim G\rho_\odot^2 R_\odot^2, \tag{4.76}$$

and the sound speed will be

$$c_s = \left(\gamma\frac{kT}{m_{\mathrm{AMU}}\bar{\mu}}\right)^{1/2} \sim (G\rho_\odot)^{1/2}R_\odot. \tag{4.77}$$

Substituting these values back into Equation 4.72, the convective energy flux in the Sun will be of order

$$F_{\mathrm{conv}} \sim G^{3/2}\rho_\odot^{5/2}R_\odot^3 \left(\frac{\Delta T}{T}\right)^{3/2}\alpha_\ell^{1/2}. \tag{4.78}$$

In terms of the freefall time $t_{\mathrm{ff},\odot}$ and gravitational potential energy $W_\odot$ of the Sun (Equations 2.7 and 2.19), the convective flux can be written as

$$F_{\mathrm{conv}} \sim \frac{-W_\odot}{t_{\mathrm{ff},\odot}R_\odot^2}\left(\frac{\Delta T}{T}\right)^{3/2}\alpha_\ell^{1/2}. \tag{4.79}$$

Integrated over a spherical surface just below the Sun's photosphere, the complete convective luminosity must be

$$L_{\mathrm{conv}} \sim R_\odot^2 F_{\mathrm{conv}} \sim \frac{-W_\odot}{t_{\mathrm{ff},\odot}}\left(\frac{\Delta T}{T}\right)^{3/2}\alpha_\ell^{1/2}. \tag{4.80}$$

(The numerical factor in front will depend on the spacing between the upward moving blobs.) Outside the central energy-generating core of the Sun, the actual luminosity is $L_{\mathrm{act}} \approx L_\odot$. By setting the actual luminosity equal to the convective luminosity of Equation 4.80, we find that we require

$$\left(\frac{\Delta T}{T}\right)^{3/2}\alpha_\ell^{1/2} \sim \frac{L_\odot t_{\mathrm{ff},\odot}}{-W_\odot} \sim \frac{t_{\mathrm{ff},\odot}}{t_{\mathrm{KH},\odot}}. \tag{4.81}$$

The freefall time for the Sun, as we saw in Section 2.3, is very much shorter than the Kelvin–Helmholtz time, with $t_{\mathrm{ff}}/t_{\mathrm{KH}} \approx 1.8 \times 10^{-12}$. We expect from simulations that the mixing length parameter α_ℓ will be of order unity. Thus, we expect

$$\frac{\Delta T}{T} \sim 10^{-8}\alpha_\ell^{-1/3} \tag{4.82}$$

for the roughly estimated solar properties that we have used. This implies, from Equation 4.68, that blobs move upward with a speed

$$v \sim 10^{-4}\alpha_\ell^{1/3}c_s \sim 20\,\mathrm{m\,s^{-1}}\alpha_\ell^{1/3}, \tag{4.83}$$

given a typical solar sound speed $c_s \sim (G\rho_\odot)^{1/2}R_\odot \sim 200\,\mathrm{km\,s^{-1}}$. This result tells us that the solar granulation seen in Figure 4.4 (the only solar convection we observe directly) differs in its properties from convection deeper in the Sun. Solar granulation has an observed temperature contrast $\Delta T/T \sim 0.03$ between the center of a bright granule and the dark lane surrounding the granule; the speed of rising and falling gas in solar granulation is $v \sim 2\,\mathrm{km\,s^{-1}}$, about 20% of the local sound speed. However, this unusually swift convection doesn't produce an overwhelmingly large convective flux F_{conv}, since the internal energy density $U = \rho c_P T$ is small at the base of the photosphere, where $\rho \approx 2 \times 10^{-7}\rho_\odot$ and $T \approx 4 \times 10^{-4}T_{c,\odot}$.

The value $\Delta T/T \sim 10^{-8}$ found deep in the Sun's convection zone implies that ∇_{act} is only slightly larger than ∇_{ad}. To see why this is true, consider a hot blob moving upward at a distance r from the star's center. The ambient gas surrounding the blob has a temperature gradient that can be written as

$$\frac{dT}{dr} = \frac{d\ln T}{d\ln P}\frac{T}{P}\frac{dP}{dr} = -\nabla_{\mathrm{act}}\frac{T}{h}, \tag{4.84}$$

where ∇_{act} is the actual value of $d\ln T/d\ln P$ in the star (Equation 4.57), and h is the pressure scale height of the ambient gas. As the blob moves upward, its temperature T_b decreases:

$$\frac{dT_b}{dr} = \frac{d\ln T_b}{d\ln P_b}\frac{T_b}{P_b}\frac{dP_b}{dr}. \tag{4.85}$$

The subsonically rising blob remains in pressure equilibrium with the surrounding gas ($P_b = P$). Thus, we can write

$$\frac{dT_b}{dr} = -\frac{d\ln T_b}{d\ln P}\frac{T_b}{h}. \tag{4.86}$$

As we found in Equation 4.48, the expansion of a sufficiently large blob is adiabatic, with

$$\frac{d\ln T_b}{d\ln P} = \nabla_{\mathrm{ad}}, \tag{4.87}$$

implying that the temperature change of the blob as it rises is given by the equation

$$\frac{dT_b}{dr} = -\nabla_{\mathrm{ad}}\frac{T_b}{h} \approx -\nabla_{\mathrm{ad}}\frac{T}{h}. \tag{4.88}$$

Combining Equations 4.84 and 4.88, we see that the difference in temperature between the blob and the ambient gas changes at the rate

$$\frac{d(\Delta T)}{dr} = (\nabla_{\mathrm{act}} - \nabla_{\mathrm{ad}})\frac{T}{h}. \tag{4.89}$$

If the blob starts out with only an infinitesimal difference in temperature from its surroundings, by the time it travels upward by a distance ℓ, it will have attained a fractional temperature excess of

$$\frac{\Delta T}{T} \sim (\nabla_{\text{act}} - \nabla_{\text{ad}})\frac{\ell}{h} \sim (\nabla_{\text{act}} - \nabla_{\text{ad}})\alpha_\ell. \qquad (4.90)$$

Thus, the value of $\Delta T/T$ given in Equation 4.82 implies

$$\nabla_{\text{act}} - \nabla_{\text{ad}} \sim 10^{-8}\alpha_\ell^{-4/3}. \qquad (4.91)$$

Assuming reasonable properties for convection in a Sun-like star, we therefore expect convective transport to be very nearly adiabatic, with ∇_{act} exceeding ∇_{ad} by only one part in 100 million.

4.4 Convective Overshoot

As mentioned in Section 4.2, main sequence stars with $M_\star < 0.3\,M_\odot$ are completely convective, but more massive main sequence stars are divided between a convective zone and a radiative zone. At the simplest level of approximation, the boundary between a convective zone and a radiative zone can be defined as the spherical surface on which $\nabla_{\text{ad}} = \nabla_{\text{rad}}$. On the side of the surface where $\nabla_{\text{rad}} < \nabla_{\text{ad}}$, it is assumed that radiation does all the energy transport; on the other side of the surface, it is assumed that convection does all the transport. This simplest level of approximation, however, ignores some physical effects that are important in the structure and evolution of stars.

To see why this strict division between radiative and convective transport is incomplete, consider a main sequence star with $M_\star > 1.2\,M_\odot$. Such a star has an outer radiative zone surrounding an inner convective zone, often called a **convective core**. In the star, the surface on which $\nabla_{\text{ad}} = \nabla_{\text{rad}}$ is at a distance R_{cr} from the star's center. At $r < R_{\text{cr}}$, $\nabla_{\text{ad}} < \nabla_{\text{rad}}$, and convection dominates the energy transport. However, consider a hot blob that starts rising upward from a distance $\ell/2$ below the boundary at $r = R_{\text{cr}}$. At the moment when it reaches the boundary, it is traveling upward at a non-zero speed v. Thus, it will take a finite length of time to decelerate once it crosses the boundary. The penetration of hot gas blobs into the overlying radiative zone is called **convective overshoot**. A similar phenomenon occurs in main sequence stars with $0.3\,M_\odot < M_\star < 1.2\,M_\odot$, which have a convective zone on top of a radiative zone. In these intermediate mass stars, cooler gas, as it sinks down to replace the buoyant, rising hot blobs, penetrates into the radiative zone below. (Cool gas sinking into an underlying radiative zone is occasionally called convective "undershoot," to distinguish it from the "overshoot" associated with hot gas rising into an overlying radiative zone. However, we will follow the more usual convention and refer to both situations as "convective overshoot.")

Because of convective overshoot, the boundary between a convective zone and a radiative zone is not an infinitesimally thin surface, but a layer of finite thickness. To see why decelerating a hot blob to a standstill needs a significant braking length, consider a massive main sequence star with a convective core. If a rising hot blob of gas reaches the boundary at $r = R_{cr}$, it has a positive velocity v. It also has a small but positive temperature excess ($\Delta T/T = -\Delta\rho/\rho \ll 1$); this means it has an upward buoyancy acceleration (Equation 4.65). Thus, the blob is speeding past the boundary with its metaphorical foot on the accelerator. However, as the blob moves upward, its temperature excess obeys the relation given in Equation 4.89:

$$\frac{d(\Delta T)}{dr} = (\nabla_{act} - \nabla_{ad})\frac{T}{h}. \tag{4.92}$$

Before the blob crosses the boundary, $\nabla_{act} - \nabla_{ad}$ is a small but positive number, and ΔT increases as the blob rises. After the blob crosses the boundary, $\nabla_{act} = \nabla_{rad} < \nabla_{ad}$, and ΔT decreases as the blob rises. When ΔT reaches zero, the blob is neutrally buoyant, but still has an upward velocity. It is only when the blob rises further, and becomes cooler than its surroundings ($\Delta T < 0$), that the metaphorical foot is switched to the brake, and the upward speed of the blob decreases until it reaches $v = 0$. Semi-empirical studies find that the thickness of the overshoot layer is typically $\ell_{over} \sim 0.2h$ for main sequence stars with $M_\star > 2\,M_\odot$.

The presence of an overshoot layer blurs the boundary between the convective and radiative zones of a star, which can influence its evolution. Consider, for instance, a massive main sequence star with a convective core. Without overshoot, the star can fuse only the material within $r = R_{cr}$. However, as helium-enriched material enters the overshoot layer, hydrogen-rich material from the radiative envelope moves downward to replace it. This mixing of hydrogen from the radiative envelope into the convective core increases the amount of hydrogen available for fusion by $\sim$50% over what it would be in the absence of convective overshoot.

Exercises

4.1 Consider two ionization states of the same isotope: species r has lost r electrons, while species $r + 1$ has lost $r + 1$ electrons. The ionization energy required to go from r to $r + 1$ is I_r. In ionization equilibrium at temperature T, the Saha equation for these two ionization states is

$$n_e n_{r+1} = n_r \frac{g_e G_{r+1}}{G_r} f(I_r, T), \tag{4.93}$$

where $g_e = 2$ is the statistical weight of the electron, G_r and G_{r+1} are the partition functions of the two states, and

$$f(I_r, T) = \left(\frac{2\pi m_e kT}{h^2} \right)^{3/2} \exp \left(-\frac{I_r}{kT} \right). \tag{4.94}$$

In a mixture of hydrogen and helium (with no metals), the number density of hydrogen is $n_H = n_p + n_{HI}$ and the number density of helium is $n_{He} = n_{HeIII} + n_{HeII} + n_{HeI}$.

(a) Write down all the relevant Saha equations for a hydrogen/helium mixture. (You do not need to solve the combined set of equations, but you should write them all down.) You can keep the partition functions G_r and the function $f(I_r, T)$ in their general form.

(b) Consider two boxes filled with gas. The two gases have the same number density n of atomic nuclei, but box #1 contains pure helium ($n_{He} = n$), while box #2 contains equal numbers of hydrogen and helium nuclei ($n_H = n_{He} = n/2$). We increase the temperature T of the boxes until the hydrogen in box #2 is almost fully ionized but the helium is still mostly neutral ($n_{HeI} > n_{HeII}$). At this temperature, do you expect the helium to be more highly ionized in box #1 or box #2? Now increase the temperature T until the helium is mostly doubly ionized ($n_{HeIII} > n_{HeII}$). At this higher temperature, do you expect the helium to be more highly ionized in box #1 or box #2? Justify your answer.

4.2 The Sun's convective zone stretches from $r = 0.71 \, R_\odot$ to $r = 1 \, R_\odot$. Suppose that within the convection zone, $\nabla_{act} = \nabla_{ad}$ exactly, and that the pressure P is described by the ideal gas law. Since the mass of the convection zone is small, we can approximate the gravitational acceleration within the convection zone as $g(r) = GM_\odot/r^2$. Given these assumptions, write down an expression for the temperature $T(r)$ within the convection zone. What is the temperature at the base of the convection zone (in kelvin)? Justify any additional assumptions you make (about abundances, ionization states, or any other properties of the convection zone).

5

Stars as Fusion Reactors

*It is held that the formation of helium from hydrogen would not be
appreciably accelerated at stellar temperatures, and must therefore be
ruled out as a source of stellar energy. But the helium which we handle
must have been put together at some time and some place. We do not argue
with the critic who urges that the stars are not hot enough for this process;
we tell him to go and find a* **hotter place**.

Arthur Stanley Eddington (1882–1944)
The Internal Constitution of the Stars, Chapter 11 [1926]

In the year 1913, the geologist Arthur Holmes estimated that the oldest rocks of the Earth's crust were $\sim$1.5 Gyr old, using the radiometric techniques that he had helped to develop. Since this was about 50 times the Kelvin–Helmholtz time for the Sun, a non-gravitational source of energy was obviously required to keep the Sun shining over the age of the solar system. A hint of what that energy source could be was provided by the physicist Francis Aston in 1920. Aston found that the mass of four hydrogen atoms, taken together, was $\sim$1% greater than the mass of a single helium atom. Thus, as Arthur Eddington pointed out, if four protons could be fused together to form a helium nucleus, the excess mass, when converted to energy, would keep the Sun shining for a time

$$t \sim \frac{0.01 \mathrm{M}_\odot c^2}{\mathrm{L}_\odot} \sim 150\,\mathrm{Gyr}, \tag{5.1}$$

if the Sun started as pure hydrogen and ended as pure helium. The difficulty, as Eddington's critics correctly pointed out, is that at the Sun's central temperature of $\sim$$10^7$ K, the thermal energy of protons is much smaller than the energy of the Coulomb barrier between them. Although Eddington, in his polite Quaker fashion, merely advised the critics to go to ... a hotter place, it was George Gamow, in the late 1920s, who used the new concepts of quantum mechanics to show how protons could quantum tunnel through the Coulomb barrier.

5.1 Quantum Tunneling and Fusion

Consider an atomic nucleus with atomic number Z and mass number A; that is, it contains Z protons and $A - Z$ neutrons. If the mass of the nucleus is m_{nuc}, we can define a mass deficit of

$$\Delta m = m_{\text{nuc}} - Z m_p - (A - Z) m_n, \tag{5.2}$$

where $m_p = 1.6726 \times 10^{-24}$ g is the mass of a free proton and $m_n = 1.6749 \times 10^{-24}$ g is the mass of a free neutron. The mass difference between a proton and a neutron is $m_n - m_p = 2.53 m_e$, where $m_e = 9.1094 \times 10^{-28}$ g is the mass of an electron. A free neutron can thus spontaneously decay via the reaction

$$n \rightarrow p + e^- + \overline{\nu}_e, \tag{5.3}$$

with a lifetime,[1] for an unbound neutron, of $\tau_n = 882$ s.

For a bound atomic nucleus, the mass deficit Δm is a negative number; this can be translated into a binding energy per nucleon,

$$B/A \equiv -\frac{\Delta m c^2}{A}. \tag{5.4}$$

The binding energy per nucleon is shown in Figure 5.1 for all stable nuclei. Deuterium (^{2}H, or D) is the most loosely bound of the stable nuclei, with $B/A = 1.1123$ MeV. The most tightly bound nuclei are those near the "iron peak" of the binding energy curve. There is no universally accepted definition of what constitutes an iron peak element. The broadest definition stretches from Sc ($Z = 21$) to Ge ($Z = 32$); the narrowest definition, which includes only nuclei with $B/A > 8.76$ MeV, stretches from Cr ($Z = 24$) to Ni ($Z = 28$). The most tightly bound nucleus of all is ^{62}Ni, with $B/A = 8.7946$ MeV, followed closely by ^{58}Fe, with $B/A = 8.7923$ MeV, and ^{56}Fe, with $B/A = 8.7904$ MeV. From an energetic viewpoint, the baryonic universe is striving to become ^{62}Ni. If we look at the standard solar abundances, however, nickel provides just 80 parts per million of the mass, and ^{62}Ni provides less than 4% of the nickel mass. Only three parts per million of the baryonic universe has reached fusion's final goal. Stars are slackers! (To be more charitable toward the activity of stars, two parts per thousand of the baryonic universe consists of isotopes near the iron peak, where B/A differs very little from one nucleus to another.)

In Figure 5.1, ^{4}He is seen to be particularly strongly bound for such a light element, with $B/A = 7.0739$ MeV. Light elements that are "multiples" of ^{4}He, such as ^{12}C, ^{16}O, and ^{20}Ne, are also more tightly bound than their immediate neighbors in the binding energy chart. The unusual stability of nuclei with $2N$ protons and $2N$ neutrons extends as far as $N = 10$; the ^{40}Ca nucleus is the heaviest

---

[1] An unstable particle, unlike a star, has a "lifetime" that is probabilistic in nature. If an unstable particle is in existence at time $t = 0$, the probability that it is still undecayed at a time $t > 0$ is $P = \exp(-t/\tau)$, where τ is the particle lifetime.

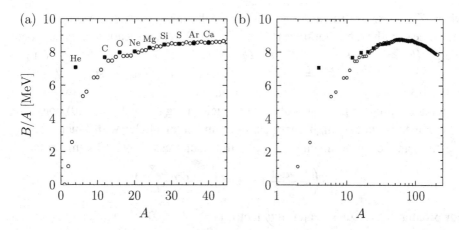

Figure 5.1 (a) Binding energy per nucleon (B/A) versus mass number (A) for light stable nuclei; nuclei with $2N$ protons and $2N$ neutrons are shown as filled squares. (b) B/A versus A (logarithmic scale) for all stable nuclei; note the "iron peak" maximum at $A \sim$ 60. [Data from AME2020]

stable nucleus with equal numbers of protons and neutrons. Even ^{100}Sn, with 50 protons and 50 neutrons, is relatively long-lived, despite being far into the mass range where neutron-rich nuclei are favored (the most abundant isotopes of tin are ^{118}Sn and ^{120}Sn). In view of the general stability of the tightly bound multiples of ^{4}He, the instability of ^{8}Be, which has four protons and four neutrons, is initially puzzling. The lifetime of beryllium-8 is $\tau_8 \approx 1.2 \times 10^{-16}$ s, which can be compared to $\tau_{100} \approx 1.6$ s for its massive cousin ^{100}Sn. The instability of ^{8}Be results from the fact that its binding energy per nucleon, $B/A = 7.0624$ MeV, is 0.16% smaller than that of ^{4}He. In splitting through the reaction

$$^8\text{Be} \rightarrow 2\,^4\text{He}, \tag{5.5}$$

the ^{8}Be nucleus can release 92 keV of energy without violating any conservation laws. For ^{12}C, with a binding energy of $B/A = 7.6801$ MeV, splitting into ^{4}He plus ^{8}Be (or any other combination of nuclei) requires an input of energy, rather than resulting in a release of energy.

Since only $\sim$0.2% of the baryons in the universe are in iron peak elements, there must be obstacles that prevent nuclear fusion in stars from rapidly reaching the goal of an iron/nickel alloy. One obstacle to fusion is the **Coulomb barrier**; atomic nuclei are positively charged, and thus repel each other. The range of the strong nuclear force that holds nuclei together is $r_{\text{sn}} \approx 1.4$ fm $\approx 1.4 \times 10^{-13}$ cm. Since $e^2 = 1.440$ MeV fm, two nuclei, in order to fuse, must overcome a Coulomb barrier with energy

$$E_{\text{coul}} = \frac{Z_j Z_k e^2}{r_{\text{sn}}} \approx 1.0 \,\text{MeV}\, (Z_j Z_k), \tag{5.6}$$

where Z_j and Z_k are the atomic numbers of the nuclei. At the center of a main sequence star, the temperature is typically $T_c \sim 10^7$ K. This leads to a mean kinetic energy $\langle E \rangle = (3/2)kT_c = 1.29\,\text{keV}\,T_7$, where $T_7 \equiv T_c/10^7$ K.[2] Even for a pair of protons, with $Z_j = Z_k = 1$, the mean kinetic energy is too small by a factor $\sim 0.0013\,T_7$ for the protons to clamber over the Coulomb barrier between them. Although a Maxwellian distribution of particle energy (Equation 3.40) does have an exponential tail to high energies, the number of nuclei with kinetic energy greater than the Coulomb energy E_{coul} will be suppressed by a Boltzmann factor

$$\exp\left(-\frac{E_{\text{coul}}}{kT_c}\right) = \exp\left(-1160\frac{Z_j Z_k}{T_7}\right). \tag{5.7}$$

For protons at the Sun's center, this factor is $e^{-740} \sim 10^{-320}$. Fusion of hydrogen in the Sun is thus impossible from a classical physics viewpoint.

The key to nuclear fusion at relatively low temperatures ($T_7 \sim 1$) is the quantum tunneling of nuclei through the Coulomb barriers that separate them. At the temperatures typical of stellar cores, the mean kinetic energy $\langle E \rangle \sim 1\,\text{keV}$ is much smaller than the rest energy of an atomic nucleus. This means that the motions of nuclei will be non-relativistic; a nucleus with mass m_j and kinetic energy $\langle E \rangle$ will then have a speed $v = (2\langle E \rangle/m_j)^{1/2}$ and momentum $p = m_j v = (2m_j \langle E \rangle)^{1/2}$. In this non-relativistic limit, a typical de Broglie wavelength for atomic nuclei at a temperature T is

$$\lambda_{\text{dB}} \equiv \frac{h}{p} \approx \frac{h}{(2A_j m_{\text{AMU}} \langle E \rangle)^{1/2}} \approx \frac{8 \times 10^{-11}\,\text{cm}}{(T_7 A_j)^{1/2}}. \tag{5.8}$$

At the temperatures found within stars, the de Broglie wavelength for atomic nuclei short of the iron peak is much longer than the range $r_{\text{sn}} \sim 10^{-13}$ cm of the strong nuclear force. Thus, when we consider the possible fusion of two nuclei, we cannot treat them as classical particles. We must take quantum effects into account.

To examine the circumstances under which fusion occurs, consider a pair of atomic nuclei. One nucleus has atomic number Z_j and mass $m_j \approx A_j m_{\text{AMU}}$. The other nucleus has atomic number Z_k and mass $m_k \approx A_k m_{\text{AMU}}$. The reduced mass of the two nuclei is

$$m_r = \frac{m_j m_k}{m_j + m_k} \approx m_{\text{AMU}} \frac{A_j A_k}{A_j + A_k}. \tag{5.9}$$

The two nuclei are moving toward each other with a relative speed $v \ll c$. If the velocities of the two nuclei are drawn from a Maxwellian distribution at temperature T, the root mean square expectation value of v is

$$\langle v^2 \rangle^{1/2} = \left(\frac{3kT}{m_r}\right)^{1/2} = 1.67 \times 10^{-3} c \left(\frac{A_j + A_k}{A_j A_k}\right)^{1/2} T_7^{1/2}. \tag{5.10}$$

---

[2] In the study of stellar cores, where 10^7 K is a useful scaling temperature, $T_7 = T/10^7$ K is a frequently used temperature parameter. At the Sun's center, for instance, $T_7 = 1.57$.

However, we are open-minded, and acknowledge the possibility that the actual value of v might be significantly larger or smaller than this root mean square value. In the center-of-mass frame, each nucleus has a momentum $p = m_r v$, and the total kinetic energy of the two nuclei is $E = m_r v^2/2$.

For the two nuclei to quantum tunnel through the Coulomb barrier, their separation must be comparable to or less than the de Broglie wavelength $\lambda_{dB} = h(2m_r E)^{-1/2}$, so that their wave functions overlap. The cross section for such a "de Broglie encounter" is

$$\pi \lambda_{dB}^2 \approx 2.6 \times 10^{-20} \, \text{cm}^2 \left(\frac{A_j + A_k}{A_j A_k} \right) \left(\frac{E}{1 \, \text{keV}} \right)^{-1}. \tag{5.11}$$

However, coming within a de Broglie wavelength of another atomic nucleus is far from a guarantee that quantum tunneling will occur. George Gamow showed that the probability of tunneling, given a separation $r \leq \lambda_{dB} \propto E^{-1/2}$, is

$$P_{\text{tun}} \approx \exp\left(-\sqrt{\frac{E_G}{E}} \right). \tag{5.12}$$

The characteristic energy E_G, called the **Gamow energy**, is given by the relation

$$E_G = \pi^2 \alpha^2 Z_j^2 Z_k^2 (2m_r c^2) \approx 0.979 \, \text{MeV} \, Z_j^2 Z_k^2 \left(\frac{A_j A_k}{A_j + A_k} \right), \tag{5.13}$$

where $\alpha \approx 1/137$ is the fine structure constant. For a pair of protons, $Z_j = Z_k = A_j = A_k = 1$, and the Gamow energy is $E_G = 0.49 \, \text{MeV}$. For a pair of ^{4}He nuclei, however, the Gamow energy jumps to $E_G = 31.3 \, \text{MeV}$; this is comparable to the Gamow energy for a proton encountering a ^{12}C nucleus ($E_G = 32.5 \, \text{MeV}$). At a given temperature T, and hence a given mean kinetic energy $\langle E \rangle$, encounters between the smallest nuclei (those with lowest A and Z) have the highest tunneling probability. However, at the typical thermal energy $E \approx 2 \, \text{keV}$ of the Sun's core, even proton–proton pairs (voted "most likely to tunnel") have a tunneling probability of just $e^{-15.7} \sim 10^{-7}$.

Even a successful quantum tunneling event is not necessarily a guarantee that nuclear fusion will occur. If two protons, for instance, tunnel through the Coulomb barrier between them, the probability of their fusing is still very small: $P_{\text{fus}} \ll 1$. The process by which two protons fuse to form a deuteron (D) can be written as

$$p + p \rightleftharpoons {}^2\text{He}, \tag{5.14}$$

$$^2\text{He} \rightarrow \text{D} + e^+ + \nu_e. \tag{5.15}$$

The wildly unstable isotope ^{2}He, also known as a "diproton," decays back to a pair of protons on a very short timescale, comparable to the light crossing time for a diproton, $\tau_\times \sim 10^{-23}$ s. The alternative decay path to a deuteron (Equation 5.15) is a beta-plus decay, involving the emission of a positron and an electron neutrino. All known beta-plus decays have a lifetime $\tau_+ \geq 0.01$ s. Thus, formation of a deuteron is strongly disfavored, with $P_{fus} \sim \tau_\times/\tau_+ \leq 10^{-21}$. The exact value of P_{fus} for proton–proton fusion has not been determined experimentally at $E <$ 10 keV; most lab fusion experiments are at much higher energies. However, we know from the Sun's continuing existence as a fusion reactor that the probability of a single de Broglie encounter resulting in fusion is $P_{tun}P_{fus} \sim 10^{-32}$. Given a tunneling probability $P_{tun} \sim 10^{-7}$ at the Sun's central temperature, we expect $P_{fus} \sim 10^{-25}$ for proton–proton fusion. For other reactions, however, the value of P_{fus} is not so minuscule. For example, consider the fusion of a deuteron with a proton to form light helium:

$$p + D \rightarrow {}^3He + \gamma. \tag{5.16}$$

Although the Coulomb barrier for this reaction is as high as for a proton–proton reaction, once the deuteron and proton tunnel through the barrier, their value of P_{fus} is very much higher, since there is no need for a weak nuclear reaction to convert a proton to a neutron.

To summarize, for two nuclei to fuse, they must first come within about a de Broglie wavelength of each other. Then they must quantum tunnel through the Coulomb barrier between them. Then they must fuse into a larger nucleus via the strong nuclear force; sometimes, as with proton–proton fusion to form a deuteron, this step also requires the (improbable) conversion of a proton to a neutron through the weak nuclear force. The total fusion cross section for two nuclei of species j and k is then

$$\sigma_{jk}(E) \approx \pi \lambda_{dB}^2 P_{tun} P_{fus} \approx \frac{\pi h^2 P_{fus}}{2m_r} \frac{1}{E} \exp\left(-\sqrt{\frac{E_G}{E}}\right), \tag{5.17}$$

making use of Equations 5.11 and 5.12. It is customary to write the fusion cross section in the form

$$\sigma_{jk}(E) = S_{jk}(E) \frac{1}{E} \exp\left(-\sqrt{\frac{E_G}{E}}\right), \tag{5.18}$$

where the **astrophysical S-factor** $S_{jk}(E)$ contains the energy dependence of P_{fus} (the fusion probability once a quantum tunneling event has occurred).[3] Since S_{jk} has dimensionality of energy times area, it is frequently quoted in units of "keV barn," where 1 keV barn $= 1.60 \times 10^{-33}$ erg cm$^2 = 6.06 \times 10^{-5} h^2/m_{AMU}$. Since $S_{jk}(E)$ can have resonances at values of E matching a nuclear excited state, its functional form can be complicated, and is often determined experimentally rather than computed theoretically. Figure 5.2, for instance, shows the experimentally determined astrophysical S-factor for the reaction

$$ {}^{14}N + p \rightarrow {}^{15}O + \gamma. \tag{5.19}$$

---

[3] The S-factor, defined in this way, was introduced by Edwin Salpeter in 1952. (Since Salpeter was a notably modest man, it is unlikely that he chose "S" to stand for "Salpeter.")

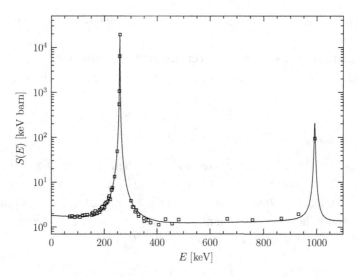

Figure 5.2 Astrophysical S-factor for the reaction $^{14}\text{N} + p \rightarrow {}^{15}\text{O} + \gamma$. Data are from laboratory experiments. [Adapted from Xu *et al.* 2013, Figure 79]

Notice the strong resonance at $E = 260\,\text{keV}$, the weaker resonance at $E = 990\,\text{keV}$, and the lack of experimental data at $E < 70\,\text{keV}$. Most fusion reactions in stars occur in the low-energy limit, $E \ll E_\text{G}$. Unless a reaction is known to happen at a resonance energy, it is often safe to use the approximation $S_{jk}(E) = S_{jk}(0)$, with the value of $S_{jk}(0)$ being found by extrapolating to the limit $E = 0$. For instance, as we show in Section 5.4, fusion of ^{14}N with a proton takes place within the Sun at an energy $E \sim 30\,\text{keV}$. At this relatively low energy, the extrapolated data of Figure 5.2 predict an astrophysical S-factor of $S_{1,14}(0) = 1.8\,\text{keV barn}$.[4]

5.2 Energy Generation and Composition Change

Instead of considering an isolated pair of nuclei attempting to fuse, let's now plunge into the jostling crowd of nuclei at the center of a star. Nuclei of species j, with atomic number Z_j and mass number A_j, have a number density n_j. Nuclei of species k, with atomic number Z_k and mass number A_k, have a number density n_k. The temperature of the gas is T, and the total density n_tot is low enough that the Maxwellian distribution is appropriate. The normalized number density of particles, $n(E)/n_\text{tot}$, is then (Equation 3.40)

$$F_\text{M}(E) = \frac{2}{\sqrt{\pi}} \frac{1}{kT} \left(\frac{E}{kT}\right)^{1/2} \exp\left(-\frac{E}{kT}\right). \tag{5.20}$$

---

[4] Extrapolation is often regarded as a risky daredevil stunt; however, in the case of ^{14}N fusion, nuclear physics theory leads us to *not* expect any resonances lurking at $E < 70\,\text{keV}$.

The result of fusing species j and k is species ℓ, which has atomic number Z_ℓ and mass number A_ℓ.

If all encounters between nuclei occurred with the same kinetic energy E, we could write the rate of increase of n_ℓ, the number density of fusion products, as

$$\frac{dn_\ell}{dt} = \sigma_{jk} v \frac{n_j n_k}{1 + \delta_{jk}}, \tag{5.21}$$

where δ_{jk} is the Kronecker delta, equal to one when $j = k$ and zero otherwise. (This keeps the accounting correct when species j and species k represent the same type of nucleus.) However, since there is a range of kinetic energies E, we need to compute the **rate coefficient** $K_{jk} \equiv \langle \sigma_{jk} v \rangle$, where the averaging is done over the Maxwellian distribution of Equation 5.20. That is,

$$K_{jk} = \int_0^\infty \sigma_{jk}(E)\, v(E)\, F_{\mathrm{M}}(E)\, dE. \tag{5.22}$$

Using the fusion cross section σ_{jk} from Equation 5.18, and assuming a constant astrophysical S-factor $S_{jk}(0)$, we find the rate coefficient is

$$K_{jk} = \left(\frac{8}{\pi}\right)^{1/2} \frac{S_{jk}(0)}{m_r^{1/2}(kT)^{3/2}} \int_0^\infty f_{\mathrm{GP}}(E)dE, \tag{5.23}$$

where the integrand $f_{\mathrm{GP}}(E)$ is

$$f_{\mathrm{GP}}(E) \equiv \exp\left[-\sqrt{\frac{E_{\mathrm{G}}}{E}} - \frac{E}{kT}\right]. \tag{5.24}$$

The function $f_{\mathrm{GP}}(E)$, called the **Gamow peak**, is the product of the tunneling probability, which is small when $E^{1/2} \ll E_{\mathrm{G}}^{1/2}$, and the Boltzmann factor, which is small when $E \gg kT$.

The Gamow peak for proton–proton fusion is shown in Figure 5.3. The dashed line shows the tunneling probability for the proton–proton reaction ($E_{\mathrm{G}} = 490$ keV), while the dotted line shows the Boltzmann factor at the Sun's central temperature ($kT_{c,\odot} = 1.35$ keV). The solid line shows the product $f_{\mathrm{GP}}(E)$ of the tunneling probability and the Boltzmann factor. Unfortunately, the area under the Gamow peak, $\int f_{\mathrm{GP}}dE$, does not have a simple analytic form. However, we can simply describe some useful properties of the Gamow peak. For instance, the maximum of $f_{\mathrm{GP}}(E)$, representing the energy at which a successful tunneling is most likely to occur at temperature T, is at

$$E_{\mathrm{peak}} = 2^{-2/3}(kT)^{2/3}E_{\mathrm{G}}^{1/3} \tag{5.25}$$

$$= 5.66\,\mathrm{keV}(Z_j Z_k)^{2/3}\left(\frac{A_j A_k}{A_j + A_k}\right)^{1/3} T_7^{2/3}. \tag{5.26}$$

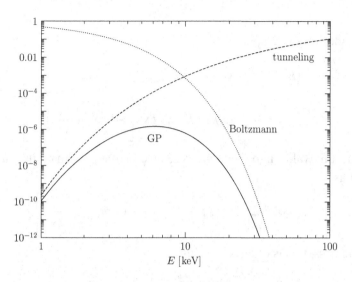

Figure 5.3 The Gamow peak (solid line) is the product of the Boltzmann factor (dotted line) and the tunneling probability (dashed line). Values are computed for proton–proton fusion at $T = 1.57 \times 10^7$ K.

For proton–proton fusion in the Sun's center, this yields $E_{peak} \approx 6.1$ keV, about three times the mean proton kinetic energy, but only 0.6% of the Coulomb barrier energy. At the energy E_{peak}, the value of f_{GP} is found by substitution to be

$$f_{GP}(E_{peak}) = \exp\left[-\sqrt{\frac{E_G}{E_{peak}}} - \frac{E_{peak}}{kT}\right] = \exp\left[-1.890\left(\frac{E_G}{kT}\right)^{1/3}\right]. \quad (5.27)$$

For proton–proton fusion in the Sun's center, this has the numerical value $f_{GP}(E_{peak}) \approx 1.4 \times 10^{-6}$. We can characterize the width of the Gamow peak by taking its width at $1/e$ of its maximum. This particular definition yields a width of

$$\Delta E \approx \frac{4}{\sqrt{3}}(E_{peak}kT)^{1/2} \approx 1.83(kT)^{5/6}E_G^{1/6}, \quad (5.28)$$

far more strongly dependent on the gas temperature T than on the Gamow energy E_G of the fusion reaction in question. For proton–proton fusion at the Sun's center, this width has the value $\Delta E \approx 6.6$ keV. Note that the dimensionless width of the Gamow peak is

$$\frac{\Delta E}{E_{peak}} \approx 2.91\left(\frac{kT}{E_G}\right)^{1/6}. \quad (5.29)$$

For proton–proton fusion in the Sun, this is $\Delta E/E_{peak} \approx 1.1$; for all fusion reactions in stars, it is a number of order unity that is quite insensitive to the temperature T.

We can now make the rough approximation that the area under the Gamow peak is height times width, or

$$\int_0^\infty f_{GP}(E)dE \approx f_{GP}(E_{peak})\Delta E \tag{5.30}$$

$$\approx 1.83(kT)^{5/6}E_G^{1/6}\exp\left[-1.890\left(\frac{E_G}{kT}\right)^{1/3}\right]. \tag{5.31}$$

Substituting this value into Equation 5.23, we can approximate the fusion rate coefficient as

$$K_{jk} \approx 2.9\frac{S_{jk}(0)E_G^{1/6}}{m_r^{1/2}}(kT)^{-2/3}\exp\left[-1.890\left(\frac{E_G}{kT}\right)^{1/3}\right]. \tag{5.32}$$

For compactness, this is often written as

$$K_{jk} = K_{jk,0}\tau_{jk}^2 e^{-\tau_{jk}}, \tag{5.33}$$

where τ_{jk} is a dimensionless parameter that includes all the temperature dependence, and is defined as

$$\tau_{jk} \equiv 1.890\left(\frac{E_G}{kT}\right)^{1/3} = 19.72\left(Z_j^2 Z_k^2 \frac{A_j A_k}{A_j + A_k}\right)^{1/3} T_7^{-1/3}. \tag{5.34}$$

The normalization for the rate coefficient in Equation 5.33 is

$$K_{jk,0} \approx \frac{25}{Z_j Z_k}\frac{A_j + A_k}{A_j A_k}\frac{S_{jk}(0)}{m_{AMU}c} \tag{5.35}$$

$$\approx 8 \times 10^{-19}\,\mathrm{cm^3\,s^{-1}}\frac{1}{Z_j Z_k}\frac{A_j + A_k}{A_j A_k}\left(\frac{S_{jk}(0)}{1\,\mathrm{keV\,barn}}\right). \tag{5.36}$$

(Given the coarseness of our "height times width" approximation of the area under the Gamow peak, the numerical factor of 25 in Equation 5.35 should be taken as a very approximate number.)

In terms of the rate coefficient K_{jk}, we can now write the volumetric rate of production for the fusion product, which we have labeled species ℓ. Starting with Equation 5.21, we can write

$$\frac{dn_\ell}{dt} = K_{jk}\frac{n_j n_k}{1 + \delta_{jk}}. \tag{5.37}$$

The energy released by a single fusion event is Q_{jk}. The specific energy generation rate ϵ, in erg s^{-1} g^{-1}, is then

$$\epsilon = \frac{Q_{jk}}{\rho}\frac{dn_\ell}{dt} = Q_{jk}K_{jk}\frac{n_j n_k}{(1 + \delta_{jk})\rho}, \tag{5.38}$$

where ρ is the local mass density. We can also write this in terms of the mass fraction, which for species j is $X_j \approx n_j m_{AMU}A_j/\rho$ and for species k is $X_k \approx$

$n_k m_{\mathrm{AMU}} A_k / \rho$. Using mass fractions, the specific energy generation rate can be written as

$$\epsilon = Q_{jk} K_{jk} \frac{X_j X_k}{1 + \delta_{jk}} \frac{\rho}{A_j A_k m_{\mathrm{AMU}}^2}. \tag{5.39}$$

Using Equation 5.33 for the rate coefficient K_{jk}, we can write the specific energy generation rate as a function of ρ and T:

$$\epsilon(\rho, T) \propto X_j X_k \rho \tau_{jk}^2 e^{-\tau_{jk}}, \tag{5.40}$$

where $\tau_{jk} \propto T^{-1/3}$. In quantifying the sensitivity of energy generation to changes in temperature, it is useful to write $\epsilon(\rho, T) \propto \rho T^\nu$, where the temperature exponent is

$$\nu \equiv \frac{\partial \ln \epsilon}{\partial \ln T} = -\frac{1}{3} \frac{\partial \ln \epsilon}{\partial \ln \tau_{jk}}. \tag{5.41}$$

Using the relation for ϵ given in Equation 5.40, the dependence on temperature is

$$\nu = -\frac{2}{3} + \frac{\tau_{jk}}{3}. \tag{5.42}$$

In the absence of resonances, a reaction's sensitivity to temperature always declines as T increases. For instance, proton–proton fusion has $\tau_{11} = 15.65 T_7^{-1/3}$, and thus $\nu = -0.67 + 5.22 T_7^{-1/3}$. At the Sun's central temperature, this yields $\nu = 3.8$; doubling the temperature would drop the dependence to $\nu = 2.9$, while halving the temperature would raise the dependence to $\nu = 5.0$. In addition, at a fixed temperature T, reactions with a larger Gamow energy are more sensitive to temperature. As an example, the fusion of a proton with a ^{14}N nucleus has $\tau_{1,14} = 70.5 T_7^{-1/3}$ and thus $\nu = -0.67 + 23.5 T_7^{-1/3}$; at the Sun's central temperature, this yields $\nu = 19.6$, much steeper than the $\nu = 3.8$ temperature dependence for proton–proton fusion.

In addition to computing the specific energy generation rate ϵ, we can also compute the rate of change of the amount of each species present. Since each fusion reaction converts one j nucleus and one k nucleus into a single ℓ nucleus, the rate of change in the number density of each species is

$$\frac{dn_j}{dt} = \frac{dn_k}{dt} = -K_{jk} n_j n_k, \qquad \frac{dn_\ell}{dt} = +K_{jk} \frac{n_j n_k}{1 + \delta_{jk}}. \tag{5.43}$$

The rate of change can also be written in terms of the mass fractions X_j, X_k, and X_ℓ:

$$\frac{dX_j}{dt} = -A_j R_{jk} X_j X_k, \tag{5.44}$$

$$\frac{dX_k}{dt} = -A_k R_{jk} X_j X_k, \tag{5.45}$$

$$\frac{dX_\ell}{dt} = +A_\ell R_{jk} \frac{X_j X_k}{1 + \delta_{jk}}, \tag{5.46}$$

where the rate R_{jk}, with units of s^{-1}, is defined as

$$R_{jk}(\rho, T) \equiv \frac{\rho K_{jk}}{m_{AMU} A_j A_k}. \tag{5.47}$$

In terms of the rate R_{jk}, the specific energy generation rate (Equation 5.39) can be written as

$$\epsilon_{jk} = Q_{jk} R_{jk} \frac{X_j X_k}{m_{AMU}(1 + \delta_{jk})}. \tag{5.48}$$

In general, species ℓ can be formed by multiple different fusion reactions, not just the fusion of species j with species k. In addition, species ℓ might itself fuse with other species to make still heavier isotopes. Thus, the differential equation for dX_ℓ/dt in real stellar conditions will have as many terms on the right-hand side as there are distinct fusion reactions in which species ℓ can be involved. Each term has its own temperature-dependent fusion rate R_{jk}, which depends in turn on its astrophysical S-factor S_{jk}. If all the relevant fusion rates are known, then the network of differential equations for mass fractions provides a powerful framework for the study of composition change within a star.

5.3 The pp Chain

Main sequence stars, including the Sun, are powered by the fusion of ^{1}H into ^{4}He. Consider a collection of four protons, with four associated electrons to preserve charge neutrality, that are converted to a single ^{4}He nucleus, with two associated electrons. (The two lost electrons annihilate with positrons produced in deuterium synthesis (Equation 5.15) and form gamma rays.) The initial rest energy of four protons and four electrons is $4(m_p + m_e)c^2 = 3755.132$ MeV; the final rest energy of a helium nucleus and two electrons is $(m_\alpha + 2m_e)c^2 = 3728.401$ MeV. Thus, fusing hydrogen with helium produces 26.731 MeV per helium nucleus formed, or 6.683 MeV per initial proton. The energy produced represents 0.712% of the initial mass of hydrogen; in other words, the efficiency of "hydrogen burning" is $\eta_{nuc} = 0.007\,12$. If the Sun's luminosity, $1\,L_\odot = 2.389 \times 10^{39}$ MeV s^{-1}, is provided entirely by hydrogen burning, this means that 3.6×10^{38} protons must be fused every second. This number of protons represents 6.0×10^8 metric tons of hydrogen per second.[5]

Stars can fuse hydrogen into helium by two distinct processes. In his famous 1939 paper, "Energy Production in Stars," Hans Bethe showed that ^{4}He could be made either by the **pp chain**, a process that can occur even in a pure hydrogen

---

[5] The mass fused per second, 600 million tons, is equivalent to the mass of all the cows on Earth, which in turn is slightly greater than the mass of all human beings.

plasma, or by the **CNO bi-cycle**, a process that requires carbon, nitrogen, and oxygen as catalysts. Let's start by considering the pp chain, the process which dominates in the Sun. The first step in the pp chain is the fusion of two protons to form a deuteron (Equations 5.14 and 5.15). This fusion reaction can be summarized as

$$p + p \rightarrow D + e^+ + \nu_e. \tag{5.49}$$

This reaction releases $Q_{11} = 1.442\,\text{MeV}$ per fusion event, including the gamma ray energy released when the positron annihilates with an electron. The proton–proton reaction has a relatively low Coulomb barrier and τ value ($\tau_{11} = 15.65 T_7^{-1/3}$). However, since it requires converting a proton to a neutron via the weak nuclear interaction, its astrophysical S-factor is tiny: only $S_{11} \approx 4.0 \times 10^{-22}\,\text{keV barn}$. This emphasizes an important rule for gauging the likely pathways for nuclear reactions: if a weak interaction is required, the cross section is very small and the reaction is strongly disfavored. The proton–proton reaction is important only because there is no other fusion pathway available in a pure mix of ^{1}H and ^{4}He at $T_7 \sim 1$: attempting to fuse ^{4}He with ^{4}He produces unstable ^{8}Be, while attempting to fuse ^{1}H with ^{4}He produces highly unstable ^{5}Li.

Once deuterium (D) is formed, it can undergo fusion reactions. It is more likely to fuse with ^{1}H than with ^{4}He (or any metals that may be present) because of the lower Coulomb barrier. The next step in the pp chain is therefore the proton–deuteron reaction:

$$p + D \rightarrow {}^3\text{He} + \gamma. \tag{5.50}$$

This reaction releases $Q_{12} = 5.493\,\text{MeV}$ per fusion event. The τ value for the proton–deuteron reaction is similar to that of the proton–proton reaction, with $\tau_{12} = (4/3)^{1/3}\tau_{11} = 17.23 T_7^{-1/3}$. However, its astrophysical S-factor is more than 17 orders of magnitude larger, with $S_{12} = 2.1 \times 10^{-4}\,\text{keV barn}$.

Consider a star in which the proton–proton reaction and the proton–deuteron reaction are the only fusion reactions occurring. The mass fractions of ^{1}H, ^{2}H, and ^{3}He (which we will call X_1, X_2, and Y_3 respectively) change with time according to the abundance equations

$$\frac{dX_1}{dt} = 1\left[-R_{11}X_1^2 - R_{12}X_1X_2\right], \tag{5.51}$$

$$\frac{dX_2}{dt} = 2\left[R_{11}\frac{X_1^2}{2} - R_{12}X_1X_2\right], \tag{5.52}$$

$$\frac{dY_3}{dt} = 3\left[R_{12}X_1X_2\right]. \tag{5.53}$$

Note that the sum $X_1 + X_2 + Y_3$ remains constant, given our assumption of only two possible fusion reactions. Looking at each species individually, the mass fraction of ^{1}H can only decline, while the mass fraction of ^{3}He can only increase. However, the mass fraction of ^{2}H can reach an equilibrium value, in which its rate of creation in the proton–proton reaction exactly balances its rate of destruction in the proton–deuteron reaction. From Equation 5.52, the equilibrium abundance of deuterium is

$$X_{2,\text{eq}} = \frac{R_{11}}{2R_{12}} X_1. \tag{5.54}$$

From the definition of the rates R_{11} and R_{12}, we can rewrite the equilibrium abundance in terms of the τ parameter and the astrophysical S-factor:

$$X_{2,\text{eq}} = \frac{4S_{11}}{3S_{12}} \left(\frac{\tau_{11}}{\tau_{12}} \right)^2 e^{(\tau_{12}-\tau_{11})} X_1. \tag{5.55}$$

Since τ_{11} and τ_{12} have the same temperature dependence, and since $\tau_{12} - \tau_{11}$ is small when $T_7 \sim 1$, the equilibrium abundance of deuterium is not extremely dependent on temperature, with

$$X_{2,\text{eq}} \approx 2.1 \times 10^{-18} \exp\left(1.575 T_7^{-1/3} \right) X_1. \tag{5.56}$$

This drops from $X_{2,\text{eq}} \approx 1.5 \times 10^{-17} X_1$ at $T_7 = 0.5$ to $X_{2,\text{eq}} \approx 7 \times 10^{-18} X_1$ at $T_7 = 2$. At plausible temperatures for a stellar core, the equilibrium deuterium abundance is tiny, thanks to the fact that $S_{11} \ll S_{12}$. Deuterium is destroyed in stellar interiors far more efficiently than it is produced. The primordial mixture of isotopes that came out of Big Bang Nucleosynthesis had $X_2 = 3.7 \times 10^{-5}$, much higher than the equilibrium value in a stellar interior. Thus, when a star is first formed, there is a non-equilibrium stage when its initial supply of deuterium is burned. This deuterium-burning stage is important for the pre-main sequence evolution of stars, as discussed in Section 7.5.[6]

The proton–proton and proton–deuteron reactions represent the first two links in the pp chain. However, how does a star convert ^{3}He into the more abundant, and more tightly bound, isotope ^{4}He? A logical next reaction might be

$$p + {}^3\text{He} \rightarrow {}^4\text{He} + e^+ + \nu_e, \tag{5.57}$$

known as the "hep" reaction. However, the hep reaction has the same problem as the proton–proton reaction: a weak interaction is required, implying a very low cross section. Although the hep reaction is known to occur, it is a very rare side branch of the pp chain.

Inside a star, once ^{3}He forms, it is overwhelmingly likely to fuse with another helium nucleus. It can do this following one of three branches, individually known

---

[6] The protosolar abundance of deuterium was $X_2 = 2.8 \times 10^{-5}$; comparing this to the higher primordial value, we deduce that ~25% of the gas from which the Sun formed was "recycled" material that had earlier undergone deuterium burning within a protostar.

as the pp I chain, pp II chain, and pp III chain. At the Sun's center, the most likely branch is the pp I chain, which requires just one more step to create ^{4}He:

$$^3\text{He} + {}^3\text{He} \rightarrow {}^4\text{He} + 2p \qquad [\text{pp I}]. \qquad (5.58)$$

This reaction releases $Q_{33} = 12.859$ MeV per fusion event. The τ value for the ^{3}He–^{3}He reaction is $\tau_{33} = 56.89T_7^{-1/3}$; this is larger than the value for proton–proton or proton–deuteron reactions because of the higher Coulomb barrier between helium nuclei. However, once the Coulomb barrier is tunneled through, the astrophysical S-factor is large, with $S_{33} \approx 5200$ keV barn.

For the pp I chain, we can now write down the full set of abundance change equations (using Y_4 as the mass fraction of ^{4}He):

$$\frac{dX_1}{dt} = 1\left[-R_{11}X_1^2 - R_{12}X_1X_2 + 2R_{33}\frac{Y_3^2}{2}\right], \qquad (5.59)$$

$$\frac{dX_2}{dt} = 2\left[R_{11}\frac{X_1^2}{2} - R_{12}X_1X_2\right], \qquad (5.60)$$

$$\frac{dY_3}{dt} = 3\left[R_{12}X_1X_2 - R_{33}Y_3^2\right], \qquad (5.61)$$

$$\frac{dY_4}{dt} = 4\left[R_{33}\frac{Y_3^2}{2}\right]. \qquad (5.62)$$

Note that the equilibrium abundance of deuterium is unchanged by adding the ^{3}He–^{3}He reaction: it is still

$$X_{2,\text{eq}} = \frac{R_{11}}{2R_{12}}X_1. \qquad (5.63)$$

Now, however, there is also an equilibrium abundance of ^{3}He. From Equation 5.61, its value is

$$Y_{3,\text{eq}} = \left(\frac{R_{12}}{R_{33}}X_1X_{2,\text{eq}}\right)^{1/2} = \left(\frac{R_{11}}{2R_{33}}\right)^{1/2}X_1, \qquad (5.64)$$

independent of the rate R_{12} of the proton–deuteron reaction that actually makes ^{3}He. Expressed in terms of τ_{11} and S_{11} for the proton–proton reaction and τ_{33} and S_{33} for the ^{3}He–^{3}He reaction, the equilibrium abundance of ^{3}He is

$$Y_{3,\text{eq}} = \left(\frac{2S_{11}}{3S_{33}}\right)^{1/2}\frac{\tau_{11}}{\tau_{33}}e^{-(\tau_{11}-\tau_{33})/2}X_1 \qquad (5.65)$$

$$\approx 5.6 \times 10^{-13}\exp\left(20.62T_7^{-1/3}\right)X_1. \qquad (5.66)$$

Since τ_{33} is significantly greater than τ_{11}, the equilibrium abundance of ^{3}He is highly temperature dependent, plummeting from $Y_{3,\text{eq}} \sim 0.1X_1$ at $T_7 = 0.5$ to $Y_{3,\text{eq}} \sim 10^{-5}X_1$ at $T_7 = 2$.

Because of the strong dependence of ^{3}He abundance on temperature, the ^{3}He abundance in the Sun has a distinctive radial profile, shown in Figure 5.4. Near

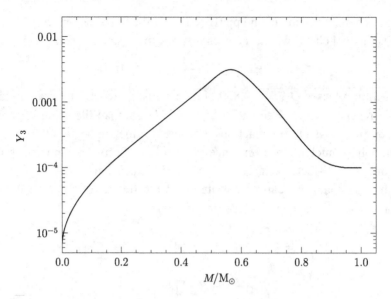

Figure 5.4 The mass fraction of ^{3}He within a standard non-rotating solar model, plotted as a function of enclosed mass M.

the Sun's surface ($M \approx 1 \, M_\odot$), the ^{3}He mass fraction is $Y_3 = 10^{-4}$; this includes both the ^{3}He present in the interstellar gas from which the Sun formed and the ^{3}He formed during the protosolar deuterium-burning stage. Moving inward from the surface, the temperature rises from $T_7 \ll 1$ at the surface to $T_7 = 1.57$ at the Sun's center. As we travel inward to higher temperatures, first the proton–proton reaction and the resulting proton–deuteron reaction start up, causing a non-equilibrium rise in the ^{3}He mass fraction. At the present day, the ^{3}He mass fraction has a maximum value of $Y_3 = 3.2 \times 10^{-3}$ at $M = 0.56 \, M_\odot$, where the temperature is $T_7 = 0.72$. As the temperature increases further, the ^{3}He–^{3}He reaction kicks in, permitting the ^{3}He mass fraction to settle at its equilibrium value. Since the equilibrium abundance $Y_{3,\mathrm{eq}}$ decreases with temperature (Equation 5.66), the mass fraction of ^{3}He then decreases with decreasing radius, eventually dropping to a level at the Sun's core that is lower than the protosolar value. The nucleosynthesis of ^{3}He is therefore not straightforward: some regions of a star produce ^{3}He while others destroy it. Whether a star ends up increasing or decreasing our galaxy's ^{3}He supply depends on which portions of the star's interior are returned to the interstellar medium.

Given a knowledge of the reaction rates R_{jk}, we can use Equation 5.47 to compute the specific energy generation rate for the pp I chain:

$$\epsilon_{\mathrm{pp\,I}} = \frac{1}{m_{\mathrm{AMU}}} \left[Q_{11} \frac{X_1^2 R_{11}}{2} + Q_{12} X_1 X_2 R_{12} + Q_{33} \frac{Y_3^2 R_{33}}{2} \right]. \quad (5.67)$$

By using the equilibrium abundances for deuterium (Equation 5.63) and ^{3}He (Equation 5.64), this can be simplified to

$$\epsilon_{pp\,I} = \frac{1}{m_{AMU}}\left[Q_{11}\frac{X_1^2 R_{11}}{2} + Q_{12}\frac{X_1^2 R_{11}}{2} + Q_{33}\frac{X_1^2 R_{11}}{4}\right] \qquad (5.68)$$

$$= \frac{2Q_{11} + 2Q_{12} + Q_{33}}{4m_{AMU}}X_1^2 R_{11}, \qquad (5.69)$$

independent of the rates of the proton–deuteron and ^{3}He–^{3}He reactions. Numerically, the energy generation rate can be written as

$$\epsilon_{pp\,I} \approx 6.4 \times 10^{18}\ \mathrm{erg\ g}^{-1}X_1^2 R_{11}, \qquad (5.70)$$

where the factor $6.4 \times 10^{18}\ \mathrm{erg\ g}^{-1} = 0.007\,12c^2$ represents the efficiency with which the initial hydrogen mass is converted to energy.

In the Sun's core, the pp I chain is the dominant method of producing ^{4}He. However, in the presence of ^{4}He, there are two other ways of "burning" ^{3}He to ^{4}He. Each of these two branches starts with the production of unstable ^{7}Be. In the pp II chain, the beryllium decays by electron capture:

$$^3\mathrm{He} + {}^4\mathrm{He} \rightarrow {}^7\mathrm{Be} + \gamma, \qquad (5.71)$$

$$^7\mathrm{Be} + e^- \rightarrow {}^7\mathrm{Li} + \nu_e, \qquad (5.72)$$

$$^7\mathrm{Li} + p \rightarrow 2\,{}^4\mathrm{He} \qquad [\mathrm{pp\ II}]. \qquad (5.73)$$

In the ionized interior of the Sun, the electrons captured by ^{7}Be nuclei are primarily free electrons. At the temperature and free electron density of the Sun's center, the lifetime of ^{7}Be is a few months; the ^{7}Li that is formed by electron capture survives for $\tau \sim 15\,\mathrm{min}$ before reacting with a proton to form two ^{4}He nuclei. The months-long lifetime of a ^{7}Be nucleus is sufficiently great that there is a finite chance that it will fuse with a proton before capturing an electron. This fusion leads to the pp III chain:

$$^3\mathrm{He} + {}^4\mathrm{He} \rightarrow {}^7\mathrm{Be} + \gamma, \qquad (5.74)$$

$$^7\mathrm{Be} + p \rightarrow {}^8\mathrm{B} + \gamma, \qquad (5.75)$$

$$^8\mathrm{B} \rightarrow {}^8\mathrm{Be} + e^+ + \nu_e, \qquad (5.76)$$

$$^8\mathrm{Be} \rightarrow 2\,{}^4\mathrm{He} \qquad [\mathrm{pp\ III}]. \qquad (5.77)$$

The lifetime of boron-8 before its beta-plus decay is $\tau \approx 1.1\,\mathrm{s}$, and the lifetime of beryllium-8 is $\tau \approx 1.2 \times 10^{-16}\,\mathrm{s}$. Thus, both the pp II and pp III chains run rapidly to completion once the ^{3}He and ^{4}He nuclei quantum tunnel through the Coulomb barrier between them.

In the Sun, 83% of the fusion energy comes from the pp I chain, 15% from the pp II chain, and only 0.02% from the pp III chain. (Yes, we are aware that

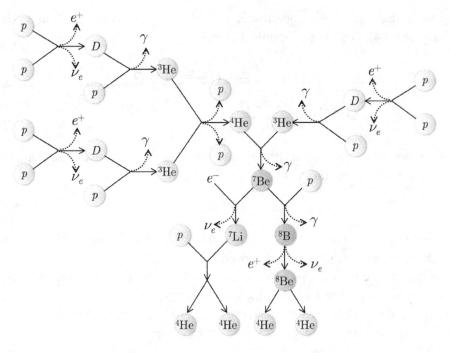

Figure 5.5 Reactions of the pp chain, including the three main branches, pp I, pp II, and pp III.

these add up to less than 100%; the rest of the Sun's fusion energy comes from the CNO bi-cycle, to be discussed in the next section.) The nuclear reactions of the pp chain, including all three main branches, are summarized in Figure 5.5.

Many of the nuclear reactions involved in the pp chain result in the production of an electron neutrino. The total flux of solar neutrinos at 1 au from the Sun is $\Phi = 6.5 \times 10^{10}$ cm^{-2} s^{-1}; since solar neutrinos are energetic enough to travel at very nearly the speed of light, this implies a density $n = \Phi/c = 2.2$ cm^{-3} of solar neutrinos in and around your body.[7] Figure 5.6 shows the spectrum of solar neutrinos as a function of neutrino energy. About 92% of these neutrinos are produced by proton–proton fusion, the essential first step in every branch of the pp chain. These neutrinos, labeled as "pp" in Figure 5.6, have a maximum energy of $E = 0.420$ MeV. Electron capture by ^{7}Be, a step in the pp II chain, produces about 7% of the solar neutrino flux. There is an 89.6% chance that the ^{7}Be goes straight to the ground state of ^{7}Li, producing a neutrino with $E = 0.862$ MeV. However, 10.4% of the time, electron capture by ^{7}Be leads to an excited state of ^{7}Li, emitting a neutrino with $E = 0.384$ MeV; the ^{7}Li nucleus then decays to its ground state, emitting a gamma-ray photon with $E = 0.478$ MeV. The third

---

[7] The number density of the cosmic neutrino background is $n_{C\nu B} = 336$ cm^{-3}; however, these neutrinos, redshifted relics of the dense early universe, are much lower in energy than the solar neutrinos.

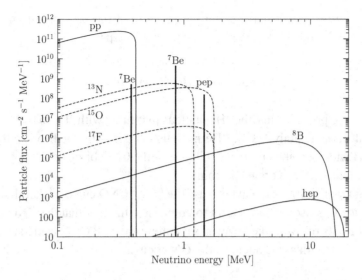

Figure 5.6 Particle flux of solar neutrinos at 1 au from the Sun. The ^{7}Be and pep emission lines are in units of $cm^{-2}\,s^{-1}$. [BP04+ solar model]

narrow emission line in the neutrino spectrum, labeled "pep" in Figure 5.6, comes from the reaction

$$p + e^- + p \rightarrow D + \nu_e, \tag{5.78}$$

called the "pep" reaction. Since this is an implausible three-body reaction, it does not contribute significantly to the overall deuterium production in the Sun. However, the neutrinos produced by the pep reaction have $E = 1.445\,MeV$; this is more energetic than neutrinos produced by the proton–proton reaction.

The remaining 1% of the solar neutrino flux is produced mainly by the CNO bi-cycle, examined in the next section. However, the neutrinos produced by the decay of ^{8}B in the pp III chain (Equation 5.76) and by the weak hep reaction (Equation 5.57) are of great interest to neutrino astronomers, despite their puny contribution to the overall solar neutrino flux. The decay of ^{8}B produces only 80 parts per million of the particle flux of neutrinos, while the hep reaction produces less than one part per million. However, these two reactions produce neutrinos with energy as high as $E \sim 15\,MeV$, permitting the study of an energy range not attained by other solar neutrinos (Figure 5.6).

Each branch of the pp chain starts with four protons and ends with one ^{4}He nucleus, releasing $Q_{He} = 26.731\,MeV$ per ^{4}He nucleus formed. The division of this energy between neutrinos and other particles depends on which branch you take; the usual quote that 2.3% of the energy goes to neutrinos is a weighted average over pp I, pp II, and pp III. For each branch of the pp chain, the rate-limiting step is always the initial step: fusing two protons to create a deuteron. Thus, for the pp II and pp III chains, as well as for the pp I chain, the specific energy generation rate ϵ depends on the rate R_{11} of the proton–proton reaction (Equation 5.69), and not on the rate of subsequent steps.

5.4 The CNO Bi-cycle

In the pp II and pp III chains, the ^{4}He nucleus that fuses with ^{3}He to make ^{7}Be can be thought of as a catalyst. One ^{4}He nucleus goes into the pp II or pp III chain, and two ^{4}He nuclei come out. In a similar vein, the CNO bi-cycle employs heavier nuclei as catalysts for hydrogen burning.

Although many astronomical texts use the term "CNO cycle" rather than "CNO bi-cycle," we will use the latter term to emphasize the fact that the CNO bi-cycle actually consists of two linked cycles, separately called the CN cycle and the NO cycle.[8] Starting with a ^{12}C nucleus, the **CN cycle** is

$$^{12}\text{C} + p \rightarrow \, ^{13}\text{N} + \gamma, \tag{5.79}$$

$$^{13}\text{N} \rightarrow \, ^{13}\text{C} + e^+ + \nu_e \quad [\tau = 14.4\,\text{min}], \tag{5.80}$$

$$^{13}\text{C} + p \rightarrow \, ^{14}\text{N} + \gamma, \tag{5.81}$$

$$^{14}\text{N} + p \rightarrow \, ^{15}\text{O} + \gamma, \tag{5.82}$$

$$^{15}\text{O} \rightarrow \, ^{15}\text{N} + e^+ + \nu_e \quad [\tau = 2.9\,\text{min}], \tag{5.83}$$

$$^{15}\text{N} + p \rightarrow \, ^{12}\text{C} + \, ^4\text{He}. \tag{5.84}$$

After the complete CN cycle, we are left with a ^{12}C nucleus, ready to start the CN merry-go-round again. The CN cycle is given its name because the catalytic nuclei are usually in the form of stable ^{12}C, ^{13}C, ^{14}N, or ^{15}N; the ^{15}O nuclei involved have a short lifetime of $\tau \sim 3$ min.

However, the interaction of ^{15}N with a proton does not necessarily result in ^{12}C $+\,^4$He, as given in Equation 5.84. It can result in fusion to form a nucleus of ^{16}O. If the ^{15}N nucleus takes this route, it diverts into the **NO cycle**. Starting with a nucleus of ^{14}N (we have to break into the cycle somewhere), the NO cycle is

$$^{14}\text{N} + p \rightarrow \, ^{15}\text{O} + \gamma, \tag{5.85}$$

$$^{15}\text{O} \rightarrow \, ^{15}\text{N} + e^+ + \nu_e \quad [\tau = 2.9\,\text{min}], \tag{5.86}$$

$$^{15}\text{N} + p \rightarrow \, ^{16}\text{O} + \gamma, \tag{5.87}$$

$$^{16}\text{O} + p \rightarrow \, ^{17}\text{F} + \gamma, \tag{5.88}$$

$$^{17}\text{F} \rightarrow \, ^{17}\text{O} + e^+ + \nu_e \quad [\tau = 1.55\,\text{min}], \tag{5.89}$$

$$^{17}\text{O} + p \rightarrow \, ^{14}\text{N} + \, ^4\text{He}. \tag{5.90}$$

After the complete NO cycle, we are left with a ^{14}N nucleus, ready to start the NO merry-go-round again, or perhaps jump aboard the CN cycle, as demonstrated in

---

[8] The CNO bi-cycle is also known as the "Bethe–Weizsäcker cycle," since the physicist Carl Friedrich von Weizsäcker had proposed, independently of Hans Bethe, a cycle that used carbon, nitrogen, and oxygen as catalysts for the fusion of hydrogen.

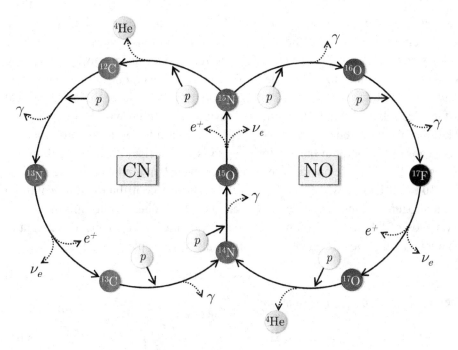

Figure 5.7 The CN cycle (left) and NO cycle (right), summarized in cartoon form.

Figure 5.7. The NO cycle is given its name because the catalytic nuclei are usually in the form of stable ^{14}N, ^{15}N, ^{16}O, or ^{17}O; the ^{17}F nuclei involved have a short lifetime of $\tau \sim 1.5$ min.

Fusion of a proton with a carbon nucleus ($Z = 6$) has a Gamow energy of $E_G \approx 66.5 E_G(pp) \approx 32.5$ MeV and a peak tunneling rate at an energy of

$$E_{\text{peak}}(C + p) \approx 18.2 \, \text{keV} \, T_7^{2/3}. \tag{5.91}$$

Similarly, fusion of a proton with a nitrogen nucleus ($Z = 7$) has a Gamow energy of $E_G \approx 91.5 E_G(pp) \approx 44.8$ MeV and a peak tunneling rate at

$$E_{\text{peak}}(N + p) \approx 20.2 \, \text{keV} \, T_7^{2/3}. \tag{5.92}$$

Finally, fusion of a proton with an oxygen nucleus ($Z = 8$) has $E_G \approx 120 E_G(pp) \approx 59.0$ MeV and a peak tunneling rate at

$$E_{\text{peak}}(O + p) \approx 22.2 \, \text{keV} \, T_7^{2/3}. \tag{5.93}$$

Compare these results to $E_{\text{peak}} \approx 4.5 \, \text{keV} \, T_7^{2/3}$ for proton–proton fusion, as well as to the mean particle thermal energy $\langle E \rangle = 1.29 \, \text{keV} \, T_7$; tunneling through the Coulomb barriers of C, N, and O occurs only for nuclei and protons far out on the exponential tail of the Maxwellian distribution.

The high Coulomb barriers around carbon, nitrogen, and oxygen nuclei mean that quantum tunneling rates in the CNO bi-cycle are small. The area under the

Gamow peak is $f_{peak}\Delta E \sim 4 \times 10^{-23}$ keV for proton–carbon fusion, 10^{-25} keV for proton–nitrogen fusion, and 4×10^{-28} keV for proton–oxygen fusion. This is contrasted with $f_{peak}\Delta E \sim 10^{-5}$ keV for proton–proton fusion, always assuming a temperature similar to that of the Sun's center. However, once tunneling occurs between a proton and a carbon, nitrogen, or oxygen nucleus, the astrophysical S-factor is relatively high: for instance, fusion of a proton with ^{14}N has $S_{1,14} \sim$ 2 keV barn, contrasted with $S_{11} \sim 4 \times 10^{-22}$ keV barn for proton–proton fusion.

Because the Coulomb barrier for oxygen is higher than those for carbon and nitrogen, the CN cycle operates solo (as a uni-cycle?) at temperatures below $T \sim 3 \times 10^7$ K. It is only when $T_7 > 3$ that the NO cycle contributes significantly to the hydrogen burning. When the CN cycle operates alone, its rate-limiting step is the fusion of ^{14}N with a proton to form ^{15}O (Equation 5.82). Thus, if we want to characterize the specific energy generation rate by a simple power law,

$$\epsilon_{CN} \propto \rho T^\nu, \tag{5.94}$$

then in our relation for ν (Equation 5.42),

$$\nu = -\frac{2}{3} + \frac{\tau_{jk}}{3} = -\frac{2}{3} + 0.63 \left(\frac{E_G}{kT}\right)^{1/3}, \tag{5.95}$$

we must use the value of E_G appropriate for the fusion of nitrogen with a proton. Using $E_G = 44.8$ MeV and $T = 2 \times 10^7$ K, this yields $\nu \approx 18$. Doubling the temperature drops the dependence to $\nu \approx 14$; halving the temperature results in $\nu \approx 23$.

Figure 5.8 illustrates the specific energy generation rates from the pp chain (solid line) and the CNO bi-cycle (dashed line). Because of the steep dependence of ϵ_{CN} on temperature, hydrogen burning is dominated by the pp chain at temperatures below $T \approx 1.8 \times 10^7$ K and by the CN cycle and CNO bi-cycle at higher temperatures. Although the Sun's central temperature of $T_{c,\odot} = 1.57 \times 10^7$ K is not far below the crossover temperature, the extreme temperature sensitivity of the CN cycle means that the CN fusion rate drops quickly away from the Sun's center. Only $\sim$1.6% of the Sun's ^{4}He manufacture is done by the CN cycle.

At the Sun's central temperature, the CN cycle runs efficiently, leading to equilibrium mass fractions of ^{12}C, ^{13}C, ^{14}N, and ^{15}N. As mentioned earlier, the fusion of a proton with ^{14}N is the rate-limiting step in the CN cycle. Its reaction rate $R_{1,14}$ is relatively low because a proton's reaction with ^{14}N has a smaller astrophysical S-factor than its reaction with ^{15}N, and a higher Coulomb barrier than its reaction with ^{12}C or ^{13}C. The resulting slow reaction rate leads to a pile-up of ^{14}N nuclei waiting to undergo fusion. Figure 5.9, for instance, illustrates the effect of the CN cycle on the relative abundances of ^{12}C and ^{14}N within the Sun. Near the Sun's surface ($M \approx 1$ M$_\odot$), the mass fractions of ^{14}N and ^{12}C have the relative values $X_{14} \approx 0.30 X_{12}$, representing their abundances in the interstellar gas from which the Sun formed. As the Sun's temperature increases with depth, the CN cycle

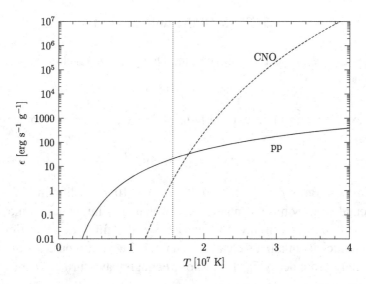

Figure 5.8 Energy generation rate versus temperature for the pp chain (solid line) and CNO bi-cycle (dashed line). The plot assumes $\rho = 160 \, \mathrm{g \, cm^{-3}}$, $X = 0.34$, and $X_{\mathrm{CNO}} = 0.01$ for C, N, and O combined. These are values appropriate for the Sun's center, whose temperature is shown as the vertical dotted line.

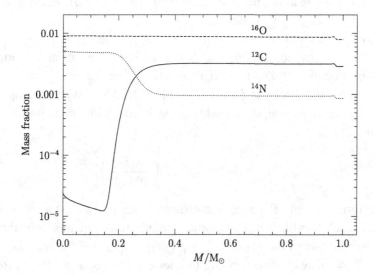

Figure 5.9 The mass fraction of ^{12}C, ^{14}N, and ^{16}O within a standard non-rotating solar model, plotted as a function of enclosed mass M.

starts up and the amount of ^{14}N rises. At the Sun's center, the nitrogen mass fraction is $X_{14} \approx 230 X_{12}$. Notice that since the NO cycle doesn't run efficiently at the Sun's central temperature, the mass fraction of ^{16}O is nearly constant throughout the Sun's interior.

5.5 Triple Alpha and Beyond

Nuclear fusion can proceed beyond helium. One possible reaction is

$$^4\text{He} + p \rightarrow\ ^5\text{Li}, \tag{5.96}$$

but ^5Li is extremely unstable, with a lifetime of $\tau_5 \sim 5 \times 10^{-22}$ s. Another possible reaction is

$$^4\text{He} +\ ^4\text{He} \rightarrow\ ^8\text{Be}. \tag{5.97}$$

The lifetime of ^8Be is $\tau_8 \approx 1.2 \times 10^{-16}$ s $\sim 2 \times 10^5 \tau_5$. A lifetime of $\sim 10^{-16}$ s doesn't sound long by most standards; however, a beryllium nucleus moving at a speed $v \sim 10^8$ cm s^{-1} can travel a distance $d \sim 10^5$ fm during that time. Thus, a beryllium nucleus in a hot stellar core can travel tens of thousands of times its own diameter before decaying. This is long enough that a further fusion step,

$$^8\text{Be} +\ ^4\text{He} \rightarrow\ ^{12}\text{C}, \tag{5.98}$$

has a significant probability of occurring, once temperatures and densities are sufficiently high. The process for making ^{12}C described in Equations 5.97 and 5.98 is known as the **triple alpha process**, since it converts three ^4He nuclei (or "alpha particles") into a single ^{12}C nucleus.

How high must temperatures and densities be for the triple alpha process to occur? Let's start by considering the fusion of two ^4He nuclei to form ^8Be, as given in Equation 5.97. The characteristic tunneling energy for encounters between helium nuclei is $E_G = 64 E_G(pp) \approx 31.3$ MeV. When averaged over a Maxwellian distribution, this yields a peak tunneling rate at the energy (Equation 5.26)

$$E_{\text{peak}}(\text{He} + \text{He}) \approx 92 \text{ keV} \left(\frac{kT}{10 \text{ keV}} \right)^{2/3}. \tag{5.99}$$

Since the formation of ^8Be is an endothermic reaction, with $\Delta mc^2 = 92$ keV required for each beryllium nucleus formed, we see from Equation 5.99 that there is a resonance between the peak tunneling energy E_{peak} and the reaction energy Δmc^2 at $kT \approx 10$ keV, corresponding to a temperature of $T \approx 10^8$ K. Because of this resonance, ^8Be production sets in rather suddenly at $T \approx 10^8$ K. Even at $T \approx 10^8$ K, the equilibrium state of the two-way reaction

$$^4\text{He} +\ ^4\text{He} \rightleftharpoons\ ^8\text{Be} \tag{5.100}$$

strongly favors helium over beryllium. At $T = 10^8$ K, the ^8Be to ^4He ratio is, from the nuclear physics equivalent of the Saha equation,

$$\frac{n(^8\text{Be})}{n(^4\text{He})} \approx 4 \times 10^{-10} \left[\frac{n(^4\text{He})}{10^{28} \text{ cm}^{-3}} \right]. \tag{5.101}$$

Even at high densities ($n \sim 10^{28}\,\mathrm{cm}^{-3}$ corresponds to $\rho \sim 70\,000\,\mathrm{g\,cm}^{-3}$), the fraction of ^{8}Be is small.

Around the year 1953, Fred Hoyle realized that the triple alpha process would not produce significant amounts of ^{12}C unless the fusion of ^{4}He with ^{8}Be went to an excited state of the ^{12}C nucleus:

$$^8\mathrm{Be} + {}^4\mathrm{He} \rightarrow {}^{12}\mathrm{C}^*. \tag{5.102}$$

Why does the existence of an excited state boost production of ^{12}C? Consider that production of ^{8}Be is enhanced at $kT \approx 10\,\mathrm{keV}$ because at that temperature, there is a resonance between the energy of peak tunneling, $E_{\mathrm{peak}}(\mathrm{He+He}) \approx 92\,\mathrm{keV}$, and the energy required for the formation of ^{8}Be from two helium nuclei, $\Delta mc^2 \approx 92\,\mathrm{keV}$. For encounters between ^{8}Be and ^{4}He, the characteristic tunneling energy is $E_G \approx 341 E_G(pp) \approx 168\,\mathrm{MeV}$. This means that the Gamow peak for such encounters has its maximum at the energy (Equation 5.26)

$$E_{\mathrm{peak}}(\mathrm{Be+He}) \approx 160\,\mathrm{keV} \left(\frac{kT}{10\,\mathrm{keV}}\right)^{2/3}. \tag{5.103}$$

Thus, if the triple alpha process takes place at the temperature $kT \approx 10\,\mathrm{keV}$ for which ^{8}Be production has a resonance, there will be a similar resonance in the production of ^{12}C if it is made in an endothermic reaction requiring an energy of $\Delta mc^2 \approx 160\,\mathrm{keV}$. However, the rest energy of ^{12}C in its nuclear ground state is $7.37\,\mathrm{MeV}$ lower than the rest energy of ^{8}Be + ^{4}He, instead of being $0.16\,\mathrm{MeV}$ higher. Thus, we conclude that enhancing the production of ^{12}C requires that it have a nuclear excited state at an energy $\Delta E \approx 7.37 + 0.16 \approx 7.53\,\mathrm{MeV}$ above its ground state.[9]

After Hoyle pointed out the astrophysical significance of such an excitation, Willy Fowler and his nuclear physics group found an excited state of ^{12}C at an energy of $\Delta E = 7.65\,\mathrm{MeV}$. Usually, the excited carbon nucleus ^{12}C* decays back to ^{8}Be and ^{4}He. Sometimes, however, it de-excites by emitting a pair of photons before it can fall apart. The complete triple alpha process can then be written more fully as

$$^4\mathrm{He} + {}^4\mathrm{He} \rightleftharpoons {}^8\mathrm{Be}, \tag{5.104}$$

$$^8\mathrm{Be} + {}^4\mathrm{He} \rightleftharpoons {}^{12}\mathrm{C}^*, \tag{5.105}$$

$$^{12}\mathrm{C}^* \rightarrow {}^{12}\mathrm{C} + 2\gamma. \tag{5.106}$$

Since the number density of ^{8}Be is $n(^8\mathrm{Be}) \propto n(^4\mathrm{He})^2 \propto Y^2\rho^2$, from Equation 5.101, the rate of production of ^{12}C is then

$$\frac{dn(^{12}\mathrm{C})}{dt} \propto n(^8\mathrm{Be})n(^4\mathrm{He}) \propto n(^4\mathrm{He})^3 \propto Y^3\rho^3. \tag{5.107}$$

---

[9] Hoyle actually predicted an excited state at $\Delta E = 7.68\,\mathrm{MeV}$, using a more sophisticated argument than our back-of-envelope calculation.

Converting this to a specific energy generation rate, we find

$$\epsilon_{3\alpha} = \frac{Q_{3\alpha}}{\rho} \frac{dn(^{12}C)}{dt} \propto Y^3 \rho^2, \qquad (5.108)$$

where $Q_{3\alpha} \approx 7.28\,\text{MeV}$ is the energy released in going from three helium nuclei to one carbon nucleus. Computing the temperature dependence of the triple alpha process is tricky, because of the resonances involved. However, the usual temperature dependence cited for $T \approx 10^8\,\text{K}$ is

$$\epsilon_{3\alpha} \propto Y^3 \rho^2 T^{40}. \qquad (5.109)$$

Once enough carbon is produced, the triple alpha process also goes on to make oxygen by the reaction

$$^{12}C + {}^4He \rightarrow {}^{16}O + \gamma, \qquad (5.110)$$

with the additional release of 7.16 MeV of energy per oxygen nucleus formed. The ratio of ^{16}O to ^{12}C produced by the triple alpha process is dependent on density and temperature. At the temperature ($T \sim 10^8\,\text{K}$) and density typical for helium burning within stars, roughly equal numbers of ^{16}O and ^{12}C nuclei are made by the time the helium is exhausted.

At a temperature $T \sim 6 \times 10^8\,\text{K}$, **carbon burning** begins. At this temperature, the fusion of two ^{12}C nuclei has a resonance with an excited state of ^{24}Mg:

$$^{12}C + {}^{12}C \rightarrow {}^{24}Mg^*. \qquad (5.111)$$

The excited state of ^{24}Mg can decay by multiple pathways, of which the most probable are

$$^{24}Mg^* \rightarrow {}^{23}Na + p, \qquad (5.112)$$
$$^{24}Mg^* \rightarrow {}^{20}Ne + {}^4He. \qquad (5.113)$$

The free protons produced by decay to ^{23}Na (Equation 5.112) trigger additional reactions such as

$$^{23}Na + p \rightarrow {}^{24}Mg + \gamma, \qquad (5.114)$$
$$^{23}Na + p \rightarrow {}^{20}Ne + {}^4He. \qquad (5.115)$$

The net yield of carbon burning is dependent on temperature and density; at typical conditions for carbon burning in massive stars, the main product is ^{20}Ne, with smaller amounts of ^{23}Na, ^{24}Mg, and other isotopes.

At a temperature $T \sim 1.2 \times 10^9\,\text{K}$, **neon burning** begins. At this temperature ($kT \sim 0.1\,\text{MeV}$), photons in the high-energy tail of the blackbody distribution become energetic enough to photodisintegrate neon:

$$^{20}Ne + \gamma \rightarrow {}^{16}O + {}^4He. \qquad (5.116)$$

The newly liberated alpha particles are able to fuse with neon through the reaction

$$^{20}\text{Ne} + {}^4\text{He} \rightarrow {}^{24}\text{Mg} + \gamma. \tag{5.117}$$

The effect of neon burning is thus to convert ^{20}Ne into a mix of a heavier element (^{24}Mg) and a lighter element (^{16}O).

At a temperature $T \sim 1.5 \times 10^9$ K, **oxygen burning** begins. At this temperature, the fusion of two ^{16}O nuclei has a resonance with an excited state of ^{32}S:

$$^{16}\text{O} + {}^{16}\text{O} \rightarrow {}^{32}\text{S}^*. \tag{5.118}$$

The excited state of ^{32}S can decay by multiple pathways, of which the most probable are

$$^{32}\text{S}^* \rightarrow {}^{31}\text{P} + p, \tag{5.119}$$

$$^{32}\text{S}^* \rightarrow {}^{28}\text{Si} + {}^4\text{He}. \tag{5.120}$$

The free protons produced by the decay to ^{31}P (Equation 5.119) trigger additional reactions such as

$$^{31}\text{P} + p \rightarrow {}^{32}\text{S} + \gamma, \tag{5.121}$$

$$^{31}\text{P} + p \rightarrow {}^{28}\text{Si} + {}^4\text{He}. \tag{5.122}$$

The net yield of oxygen burning is dependent on temperature and density; at typical conditions for oxygen burning in massive stars, the main product is ^{28}Si, with smaller amounts of ^{31}P, ^{32}S, and other isotopes.

At a temperature $T \sim 3 \times 10^9$ K, **silicon burning** begins. At this high temperature ($kT \sim 0.3$ MeV), photons in the high-energy tail of the blackbody distribution become energetic enough to photodisintegrate even resilient silicon:

$$^{28}\text{Si} + \gamma \rightarrow {}^{24}\text{Mg} + {}^4\text{He}. \tag{5.123}$$

The liberated alpha particles are able to fuse with the nuclei produced in the previous oxygen burning, in reactions such as

$$^{28}\text{Si} + {}^4\text{He} \rightarrow {}^{32}\text{S} + \gamma, \tag{5.124}$$

$$^{32}\text{S} + {}^4\text{He} \rightarrow {}^{36}\text{Ar} + \gamma. \tag{5.125}$$

This chain of alpha particle reactions extends to ^{40}Ca, ^{44}Ti, ^{48}Cr, ^{52}Fe, and ^{56}Ni. Although the more massive nuclei along this chain are unstable to beta-plus decay, at $T \sim 3 \times 10^9$ K they are more likely to be photodisintegrated than to suffer spontaneous decay. Once silicon burning begins, therefore, a network of many different fusion and photodisintegration reactions is possible. The competing reactions are almost (but not quite) in statistical equilibrium. However, there is a slow "leakage" to the tightly bound iron peak elements. Once fusion reaches the iron peak, no more energy can be squeezed out.

In stellar cores, once the initial step of hydrogen burning is complete, only a relatively small amount of energy can be extracted by the subsequent stages of fusion. As we computed in Section 5.3, fusing ^{1}H into ^{4}He releases energy with an efficiency $\eta = 7.12 \times 10^{-3}$, corresponding to 6.68 MeV per nucleon. Stars with $M_\star > 0.4\,M_\odot$ develop cores hot enough for helium burning to start. Fusing ^{4}He into a mix of ^{12}C and ^{16}O releases energy with an efficiency $\eta = 0.83 \times 10^{-3}$, corresponding to 0.78 MeV per nucleon.[10] The most massive stars, with initial mass $M_\star > 10\,M_\odot$, develop cores hot enough to burn all the way to iron. If an equal mix of ^{12}C and ^{16}O is fused to ^{56}Fe (or any other isotope near the iron peak), the energy release has an efficiency $\eta = 0.90 \times 10^{-3}$, corresponding to 0.84 MeV per nucleon. In other words, the ^{4}He nucleus is so tightly bound that the 6.68 MeV per nucleon released by hydrogen burning accounts for 80% of the total 8.30 MeV per nucleon that is released going from hydrogen all the way to the iron peak.

Exercises

5.1   A star with mass $M_\star = 0.1\,M_\odot$ has a central temperature $T_c = 4 \times 10^6$ K and a central ^{1}H mass fraction $X_1 = 0.70$. Fusion in the star's core occurs purely through the pp I chain. Assume that the deuterium mass fraction X_2 and light helium mass fraction Y_3 have reached their equilibrium values.
   (a) What are the values of $X_{2,eq}$ and $Y_{3,eq}$ at the center of this star?
   (b) What is the energy generation rate $\epsilon_{pp\,I}$ at the center of this star? Give an answer in units of erg s^{-1} g^{-1}.
   (c) What is the timescale for hydrogen depletion at the center of this star, $t = X_1/(dX_1/dt)$? Give an answer in gigayears.

5.2   Since the equilibrium abundances of deuterium and light helium are dependent on temperature T and hydrogen mass fraction X_1, we expect their abundances to change with time as the Sun evolves.
   (a) When it first started hydrogen burning in its core, the young Sun had a central temperature $T_c = 1.27 \times 10^7$ K and a central hydrogen mass fraction $X_1 = 0.71$. What were the equilibrium abundances of deuterium (X_2) and of light helium (Y_3) at the Sun's center?
   (b) Currently, the Sun has a central temperature $T_c = 1.57 \times 10^7$ K; hydrogen burning has reduced the central hydrogen mass fraction to $X_1 = 0.34$. What are the current equilibrium abundances of deuterium and light helium at the Sun's center?

[10] The quoted efficiency $\eta = 0.83 \times 10^{-3}$ assumes equal numbers of C and O nuclei are produced; the efficiency of the triple alpha process varies from $\eta = 0.65 \times 10^{-3}$ if it stops at ^{12}C to $\eta = 0.97 \times 10^{-3}$ if it goes all the way to ^{16}O.

Main Sequence Stars

I like the stars. It's the illusion of permanence, I think.
I mean, they're always flaring up and caving in and going out.
But from here, I can pretend . . . I can pretend that things last.

Neil Gaiman (1960–)
The Sandman: Brief Lives [1994]

The main sequence was discovered by Henry Norris Russell and Ejnar Hertz-sprung as an empirical correlation between absolute visual magnitude and spectral type. From a theorist's point of view, it is a correlation between a star's luminosity L and the effective temperature T_{eff} of its photosphere. However, the correlation between luminosity and temperature (and between luminosity and mass, and between radius and mass) is the outward visible sign of an inward physical state. In a 1925 paper, Russell pointed out that for main sequence stars to be in hydrostatic equilibrium, they all had to have a central temperature $T_c \sim 3 \times 10^7$ K; Russell then concluded,

> We must next suppose that as a temperature of some thirty million degrees is approached, a process comes into play which leads to the actual annihilation of the main mass of the stellar material, with a correspondingly great liberation of energy.[1]

In 1938, Ernst Öpik pointed out that if the process that comes into play is ther-monuclear fusion, then main sequence stars are powered by fusion in their cores, and red giants are stars that have exhausted their central fuel. In Öpik's words,

> giants must contain superdense cores, probably formed by the collapse of the central portions after the exhaustion of the original source of energy (exhaustion of hydrogen in the central region. . .).

---

[1] Russell's temperature estimate was actually too high, because he thought the Sun was made primarily of ionized metals (with $\bar{\mu} \sim 2$). It took him a few years to accept Cecilia Payne's 1925 result that the Sun is mostly hydrogen and helium (and thus has a lower $\bar{\mu}$).

We have already quoted, in Section 1.3, the proverbial wisdom that the main sequence is a *mass* sequence. In detail, however, the observed properties of stars also depend on their metallicity. On a Hertzsprung–Russell diagram like that of Figure 1.10, metal-poor main sequence stars are seen to lie below solar-metallicity stars with the same T_{eff}. Because of this relation, these metal-poor stars were called "subdwarfs" by Gerard Kuiper when he noted their existence in 1939. Subdwarfs are often assigned to a separate luminosity class VI. The nearest subdwarf to the Sun is Kapteyn's star, at $d = 3.93$ pc; it has spectral class M1VI. Kapteyn's star, with metallicity $Z \approx 0.1Z_{\odot}$, has the same temperature ($T_{\text{eff}} \approx 3600$ K) as the solar-metallicity M dwarf Lalande 21185. However, the luminosity of Kapteyn's star is barely half that of Lalande 21185. The displacement of subdwarfs from dwarfs on an H–R diagram results from the fact that the lower metallicity of subdwarfs leads to a lower opacity. More specifically, consider the outer layers of a star where the temperature lies in the range $3000\,\text{K} < T < 8000\,\text{K}$. At these temperatures, the opacity of solar-metallicity dwarfs is provided by the H^{-} ion. However, a metal-poor subdwarf, thanks to its lack of electron-donating metals, has fewer H^{-} ions; the resulting lower opacity enables photons to escape from a deeper, hotter layer of the subdwarf. Thus, a low-mass subdwarf has the same photospheric temperature as a higher-mass, higher-luminosity dwarf; to return to our example stars, Kapteyn's star has $M_{\star} = 0.28\,M_{\odot}$, while Lalande 21185 has $M_{\star} = 0.39\,M_{\odot}$.

To summarize the dependence of stellar properties on metallicity as well as on mass, astronomers cite the **Vogt–Russell theorem**, which states "the structure of a static, spherical star is uniquely determined by its total mass $M_{\star}$ and the chemical composition $X_i(M)$ in its interior." The Vogt–Russell theorem is not a theorem in the strictest mathematical sense; indeed, there are disproofs in the form of pairs of mathematically modeled stars with the same $M_{\star}$ and $X_i(M)$ but with different interior structures. However, the Vogt–Russell theorem is empirically useful for main sequence stars. That is, although you can make a variety of model stars for a given $M_{\star}$ and $X_i(M)$, the processes that create real main sequence stars always produce the same interior structure for a given $M_{\star}$ and $X_i(M)$.

Stars are made primarily of hydrogen and helium; since hydrogen is abundant, and can be converted to helium at relatively low temperatures, we expect it to be the first major source of fusion energy in stars. Hydrogen burning is also a relatively efficient source of energy, with an efficiency $\eta = 0.007\,12$ that is greater than that of any other fusion reaction occurring in stars. We can therefore identify the main sequence as the locus of hydrogen-burning stars on a Hertzsprung–Russell diagram. This intuitive reasoning is confirmed by detailed numerical models of stellar evolution. To link the observable properties of main sequence stars to the physical processes in their interiors, we now turn to the equations of stellar structure.

6.1 Internal Structure

The first three equations of stellar structure, describing mass conservation, hydrostatic equilibrium, and thermal equilibrium, are straightforward:

$$\frac{dM}{dr} = 4\pi\rho r^2, \tag{6.1}$$

$$\frac{dP}{dr} = -\frac{GM\rho}{r^2}, \tag{6.2}$$

$$\frac{dL}{dr} = 4\pi\rho r^2 \epsilon. \tag{6.3}$$

The form of the fourth equation, which tells us the temperature gradient inside the star, depends on whether energy transport is radiative or convective. To determine which type of transport is dominant, compute

$$\nabla_{rad} = \frac{3\kappa PL}{16\pi acT^4 GM} \tag{6.4}$$

(from Equation 4.61) and compare it to

$$\nabla_{ad} = \frac{\gamma - 1}{\gamma}. \tag{6.5}$$

If $\nabla_{rad} < \nabla_{ad}$, then energy transfer is radiative, with $\nabla_{act} = \nabla_{rad}$ and

$$\frac{dT}{dr} = -\frac{3\kappa\rho L}{16\pi acT^3 r^2}. \tag{6.6}$$

If $\nabla_{rad} > \nabla_{ad}$, then we must use mixing length theory to infer ∇_{act}; in the limit of high density, $\nabla_{act} \approx \nabla_{ad}$ and

$$\frac{dT}{dr} \approx -\frac{\gamma - 1}{\gamma}\frac{T}{P}\frac{GM\rho}{r^2}. \tag{6.7}$$

In addition to the four differential equations of stellar structure, we need to know properties that involve microphysics: the equation of state, $P(X_i, \rho, T)$, the specific energy generation rate $\epsilon(X_i, \rho, T)$, and the opacity $\kappa(X_i, \rho, T)$. Moreover, to compute ∇_{rad}, we need to know the adiabatic index $\gamma(X_i, \rho, T)$; since γ depends on the ionization state of the gas, we can't simply assume that $\gamma = 5/3$ everywhere in a star's interior. For detailed numerical simulations of stellar structure and evolution, full physical calculations of these properties are included. However, in many circumstances, simple analytic approximations are useful. For instance, within main sequence stars, the pressure is usually well described by the ideal gas law, with a significant contribution from radiative pressure in the hottest stars:

$$P(X_i, \rho, T) = P_{ideal} + P_{rad} = \frac{k}{m_{AMU}}\frac{\rho T}{\bar{\mu}} + \frac{1}{3}aT^4. \tag{6.8}$$

We have seen in Section 4.1 that the opacity can also be simply approximated at times. For instance, $\kappa_e \propto (1 + X)$ for Thomson electron scattering, $\kappa_{ff} \propto (1 + X)(1 - Z)\rho T^{-3.5}$ for free–free absorption, and $\kappa_{bf} \propto (1 + X)Z\rho T^{-3.5}$ for bound–free absorption. Similarly, in Sections 5.3 and 5.4, we found that the specific energy generation rate for hydrogen burning at the Sun's center can be approximated as $\epsilon_{pp} \propto \rho T^4$ for the pp chain and $\epsilon_{CNO} \propto \rho T^{18}$ for the CNO bi-cycle.

Mathematically, the equations of stellar structure are four equations in four unknowns; the mass $M_\star$ and abundance profile $X_i(M)$ are set as initial conditions. Setting the boundary conditions for the equations can be a little tricky. If we are modeling a star of specified mass $M_\star$ and composition $X_i(M)$, the central boundary conditions, at $M = 0$, are $r = 0$ and $L = 0$. The boundary conditions at the star's surface are more complicated, and involve matching the photospheric pressure P and temperature T to values found from modeling stellar atmospheres. This set of boundary conditions, some at the center and some at the surface, formally closes our set of equations for stellar structure, making it into a two-point boundary value problem. Note that we cannot specify in advance the pressure P_c and temperature T_c at the star's center; these are outcomes of the solution, and not boundary conditions. Similarly, we cannot specify in advance the luminosity $L_\star$ and radius $R_\star$ at the star's photosphere; these too are outcomes of the solution.

To follow a star's evolution on the main sequence, the four equations of stellar structure must be supplemented by equations for composition change. There are two important drivers for the change in $X_i(M)$ within stars: nuclear reactions and internal mixing. Consider, for example, hydrogen burning by the pp I chain. If deuterium and light helium have their equilibrium abundance, the rate at which the mass fraction of ^{4}He increases and the mass fraction of ^{1}H decreases (Equations 5.59 and 5.62) is

$$\frac{dY_4}{dt} = -\frac{dX_1}{dt} = R_{11}X_1^2, \tag{6.9}$$

where $R_{11}(\rho, T)$ is the fusion rate for the proton–proton reaction.[2] In the absence of mixing mechanisms, nuclear reaction products remain where they are created. In slowly rotating stars like the Sun, the sole important mixing process is convection, which occurs on a timescale much shorter than the nuclear burning time t_{nuc}. In rapidly rotating stars, rotational mixing can also be important. The changes in the abundance profile $X_i(M)$ brought about by nuclear fusion, convection, and rotational mixing are the primary drivers of stellar evolution on the main sequence.

---

[2] At the present day, $R_{11} \sim 0.8\,\text{Gyr}^{-1}$ for proton–proton fusion at the Sun's center. However, the rate was lower in the past, when the Sun's central temperature was lower.

If we specify $M_\star$ and $X_i(M)$, then the resulting solution of the four standard equations of stellar structure constitutes a snapshot in time of an evolving star. This solution, of course, will be only as good as the approximations that went into it. The standard equations omit all explicitly time-dependent terms; this ignores the structural impact of the gravitational potential energy term, crucial for the early stages of stellar evolution. The standard equations also omit the influence of binary companions, mass transfer, rotation, and magnetism. However, the simplified standard equations of stellar structure allow us to define a **theoretical** main sequence:

- The main sequence is the locus of time-independent solutions of the four stellar structure equations for stars with hydrogen at their center.
- The zero age main sequence (ZAMS) is the locus of time-independent solutions of the stellar structure equations for chemically homogeneous stars containing hydrogen.

In other words, when stars start their existence on the main sequence, they are assumed to be well-mixed, with the same chemical composition throughout. It is only once fusion begins that significant composition gradients can be built up in stars.

Given the existence of cheap, fast computers, solving the four standard equations of stellar structure can be done numerically. However, it is still useful to look at analytic approximations to stellar structure. It's time, therefore, to look at two useful tools in our analytic toolbox: polytropes and homology.

6.2 Polytropes

Let's start, as we often do when looking at stars, with the equation of hydrostatic equilibrium (Equation 6.2). This equation can be rewritten as

$$M = -\frac{r^2}{G\rho}\frac{dP}{dr}. \tag{6.10}$$

Taking the derivative of both sides, we find

$$\frac{dM}{dr} = -\frac{1}{G}\frac{d}{dr}\left(\frac{r^2}{\rho}\frac{dP}{dr}\right). \tag{6.11}$$

Substituting for dM/dr from the equation of mass conservation (Equation 6.1), we end with

$$-4\pi G\rho = \frac{1}{r^2}\frac{d}{dr}\left(\frac{r^2}{\rho}\frac{dP}{dr}\right). \tag{6.12}$$

This equation, which relates pressure to mass density, assumes mass conservation and hydrostatic equilibrium; it is oblivious to how energy is generated and transported within a star. Thus, it applies equally well to planets, brown dwarfs, white dwarfs, and self-gravitating spherical cows.

Equation 6.12 is a single equation in two unknowns, P and ρ. This means, for instance, that if we know a star's density profile $\rho(r)$, we can compute the pressure profile $P(r)$ required to keep it from collapsing. However, even if we don't know $\rho(r)$ *a priori*, we can still find solutions for Equation 6.12 if we have an additional equation that gives the relation between P and ρ. Fortunately, there are physically interesting cases where the pressure can be written in the simple form

$$P(\rho) = K_n \rho^{1+1/n}, \tag{6.13}$$

where K_n and n are constant throughout the star, or other spherical body, in question. The power law given in Equation 6.13 is called a **polytropic** relation between pressure and density. The word "polytropic" is derived from the ancient Greek *polutropon*, literally meaning "turning many ways," and figuratively meaning "versatile." A polytropic relation between pressure and density is indeed versatile: by choosing different values of the **polytropic index** n and **polytropic constant** K_n, we can provide useful approximations to different types of star. The polytropic index n can be thought of as describing the stiffness of the gas. A "stiff" material, in this context, is one for which a large change in P produces only a small change in ρ; conversely, a "soft" material is one for which a small change in P produces a large change in ρ. Thus, as $n \to 0$, a gas becomes increasingly stiff; as $n \to \infty$, it becomes increasingly soft.[3] The polytropic constant K_n is constant within an individual polytropic star; however, two stars with the same value of n may have different values of K_n.

The polytropic relation of Equation 6.13, with $P \propto \rho^{1+1/n}$, is identical in mathematical form to the equation of state, $P \propto \rho^\gamma$, that we derived for a gas undergoing adiabatic processes (Equation 4.43). However, these two equations have different physical interpretations. The polytropic relation tells us that for a system in equilibrium, the pressure P and density ρ at any location are related by the equation $P = K_n \rho^{1+1/n}$, where K_n is constant throughout the system. The adiabatic equation of state, by contrast, tells us what happens when the density ρ at some location is *perturbed* from its equilibrium value by an amount $d\rho$; if the perturbation is adiabatic, the pressure will be changed by an amount dP whose value is given by the relation

$$\frac{\rho}{P}\frac{dP}{d\rho} = \frac{d\ln P}{d\ln \rho} = \gamma, \tag{6.14}$$

where γ is the adiabatic index of the gas. Because of this difference in definition, it is not necessary for $1+1/n$ in the polytropic relation to be equal to the adiabatic index γ for the gas of which the polytropic system is made.

A useful application of polytropes to stellar structure was made by Arthur Eddington in the 1920s. At that time, energy generation and transport within stars were imperfectly understood. However, Eddington realized that a polytropic

---

[3] Negative values of n are not physically forbidden. However, they aren't useful for describing stars, so we'll assume $n \geq 0$.

model, based purely on the first two equations of stellar structure, could give a useful approximation for the distribution of pressure and density inside a star like the Sun. To follow in the path of Eddington, assume that at each radius a non-zero fraction f_γ of the total pressure is provided by radiation, with

$$P_{\text{rad}} = \frac{1}{3}aT^4 = f_\gamma P. \tag{6.15}$$

The remainder of the pressure is provided by the ideal gas law, with

$$P_{\text{ideal}} = \frac{k}{m_{\text{AMU}}} \frac{\rho T}{\bar{\mu}} = (1 - f_\gamma)P. \tag{6.16}$$

If we combine Equations 6.15 and 6.16 to eliminate the temperature T, we find the total pressure P as a function of the gas density ρ:

$$P = \left(\frac{k}{m_{\text{AMU}}}\right)^{4/3} \left[\frac{3f_\gamma}{a(1 - f_\gamma)^4}\right]^{1/3} \bar{\mu}^{-4/3} \rho^{4/3}. \tag{6.17}$$

Thus, if $\bar{\mu}$ and f_γ are constant with radius, a star can be treated as a polytrope with polytropic index $n = 3$ and polytropic constant

$$K_3 = \left(\frac{k}{m_{\text{AMU}}}\right)^{4/3} \left[\frac{3f_\gamma}{a\bar{\mu}^4(1 - f_\gamma)^4}\right]^{1/3} \tag{6.18}$$

$$= 1.658\hbar c \left[\frac{f_\gamma}{m_{\text{AMU}}^4 \bar{\mu}^4(1 - f_\gamma)^4}\right]^{1/3}, \tag{6.19}$$

where we have used Equation 3.69 to eliminate the radiation density constant a.

In general, we do not expect the values of $\bar{\mu}$ and f_γ, and hence of K_3, to be constant throughout a star. When Eddington attempted to model the Sun's interior, he simply assumed $K_3 = $ constant in order to make a simple model. However, if we are considering a young star, where fusion hasn't had a chance to increase $\bar{\mu}$ significantly in the core, the assumption of constant $\bar{\mu}$ in the star's ionized interior isn't a bad one. In addition, if $f_\gamma \ll 1$ (that is, if radiation provides only a small contribution to the total pressure), then $K_3 \propto f_\gamma^{1/3}$, and modest variations in f_γ won't strongly affect the value of K_3. Modern numerical models of the Sun reveal that it has $f_\gamma = (4 \to 12) \times 10^{-4}$ everywhere but in its outer convection zone. This factor of three variation in f_γ implies that K_3 varies by a factor of $3^{1/3} \approx 1.4$; in retrospect, Eddington's assumption of constant K_3 was bad only at the $\pm 20\%$ level, not at the "many orders of magnitude" level. The Sun can therefore be fitted moderately well with an $n = 3$ polytrope that has $f_\gamma = 7 \times 10^{-4}$. This implies

$$P \approx 4.5 \times 10^{14}\,\text{dyn cm}^{-2} \left(\frac{\rho}{1\,\text{g cm}^{-3}}\right)^{4/3}, \tag{6.20}$$

assuming $\bar{\mu}_\odot \approx 0.6$. In the Sun's interior, the adiabatic index is $\gamma \approx 5/3$; thus, this model for the Sun represents a case where $1 + 1/n \neq \gamma$.

It is definitely possible to have polytropes in which $1 + 1/n = \gamma$. For instance, consider a star with completely convective energy transport. As we saw in Section 4.3, convection in stellar interiors is almost, but not quite, adiabatic. In such a case, the adiabatic equation of state, $P \propto \rho^{\gamma}$, provides a good description of the relation between P and ρ in the stellar interior, and we can adopt $n = 1/(\gamma - 1)$ for a fully convective star. In regions of the star's interior where partial ionization doesn't affect γ, we can assume $\gamma \approx 5/3$ and $n \approx 1.5$.

The polytropic indices $n = 1.5$ and $n = 3$ are of particular interest to astronomers. A polytrope with $n = 1.5$, in addition to describing stars with convective energy transport, also describes white dwarfs supported by non-relativistic degeneracy pressure (Equation 3.52). A softer polytrope with $n = 3$, in addition to describing stars with radiative energy transport, also describes white dwarfs supported by relativistic degeneracy pressure (Equation 3.59). Once we decide on an appropriate value for n, we can solve to find the internal density, pressure, and temperature of a polytropic star. Substituting the general polytropic relation (Equation 6.13) into Equation 6.12, we find a second-order differential equation for the density ρ:

$$- 4\pi G\rho = K_n \left(1 + \frac{1}{n}\right) \frac{1}{r^2} \frac{d}{dr}\left(r^2 \rho^{1/n-1} \frac{d\rho}{dr}\right). \qquad (6.21)$$

To simplify this equation, define a dimensionless variable related to the density,

$$\theta \equiv \left(\frac{\rho}{\rho_c}\right)^{1/n}, \qquad (6.22)$$

where ρ_c is the central density of the polytropic sphere. With the substitution $\rho = \rho_c \theta^n$, Equation 6.21 becomes

$$- \theta^n = \left[\frac{(n+1)K_n}{4\pi G\rho_c^{1-1/n}}\right] \frac{1}{r^2} \frac{d}{dr}\left(r^2 \frac{d\theta}{dr}\right). \qquad (6.23)$$

The quantity in square brackets in Equation 6.23 has the dimensionality of length squared. Thus, we can define a characteristic length

$$\alpha_n \equiv \left[\frac{(n+1)K_n}{4\pi G\rho_c^{1-1/n}}\right]^{1/2} = \left[\frac{n+1}{4\pi G\rho_c} \frac{P_c}{\rho_c}\right]^{1/2}, \qquad (6.24)$$

which is proportional to the central sound speed times the central freefall time. We can now define a new dimensionless radius $\xi \equiv r/\alpha_n$.

Why bother to define the dimensionless quantities θ and ξ? They enable us to write Equation 6.23 in the simple form

$$\frac{1}{\xi^2} \frac{d}{d\xi}\left(\xi^2 \frac{d\theta}{d\xi}\right) = -\theta^n. \qquad (6.25)$$

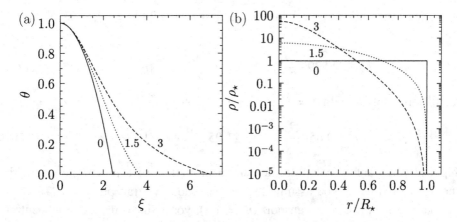

Figure 6.1 (a) Values of θ as a function of the dimensionless radius ξ for $n = 0$, 1.5, and 3. (b) Mass density ρ, in units of the mean density $\rho_\star$, as a function of the radius r, in units of the stellar radius $R_\star$.

Equation 6.25 is known as the **Lane–Emden equation**, after the astronomers J. Homer Lane and Robert Emden. Since the Lane–Emden equation is a second-order differential equation, we need two boundary conditions to solve it unambiguously. At the center, we have the requirement that $\theta = 1$ at $\xi = 0$. We can also impose the reasonable boundary condition $dP/dr = 0$ at the center, implying $d\theta/d\xi = 0$ at $\xi = 0$. With these boundary conditions, we can integrate the Lane–Emden equation using standard numerical techniques, yielding the results shown in Figure 6.1(a). All solutions of the Lane–Emden equation with $n < 5$ have θ going to zero at some finite value of the dimensionless radius ξ. This value (call it $\xi_\star$) can be converted to a physical radius for the star by the relation $R_\star = \alpha_n \xi_\star$.

The variable ξ is converted to a physical radius r by the relation $r = \alpha_n \xi$; the function $\theta(\xi)$ is converted to a physical density ρ by the relation $\rho = \rho_c \theta^n$. In Figure 6.1(b), the density ρ is shown as a function of the radius r. The plot shows that increasing the polytropic index n, and thus making the star softer, results in a more centrally concentrated star. For instance, a star with convective transport has $n = 1.5$. Such a star is relatively stiff; its density falls to half the central value at $r \approx 0.45R_\star$, and the value of the central density is $\rho_c = 5.991\rho_\star$, where $\rho_\star$ is the mean density of the star. By contrast, a star with $n = 3$ is relatively soft; its density falls to half its central value at $r \approx 0.18R_\star$, and the value of the central density is squeezed to $\rho_c = 54.18\rho_\star$.

The central pressure of a polytrope of given mass $M_\star$ and radius $R_\star$ is quite sensitive to the polytropic index n. A uniform density sphere, corresponding to the ultimate in stiffness ($n = 0$), has a central pressure (Equation 3.4)

$$P_c = 0.5 \frac{GM_\star}{R_\star} \rho_\star \approx 0.1194 \frac{GM_\star^2}{R_\star^4} \qquad [n = 0]. \qquad (6.26)$$

Numerical integration reveals that a softer sphere with $n = 1.5$ has a central pressure

$$P_c \approx 0.539 \frac{GM_\star}{R_\star} \rho_c \approx 0.771 \frac{GM_\star^2}{R_\star^4} \qquad [n = 1.5]. \qquad (6.27)$$

A still softer sphere with $n = 3$ has a central pressure

$$P_c \approx 0.854 \frac{GM_\star}{R_\star} \rho_c \approx 11.05 \frac{GM_\star^2}{R_\star^4} \qquad [n = 3]. \qquad (6.28)$$

Suppose you want to compute the central properties of a polytropic star with known $M_\star$ and $R_\star$. If you assume the star has radiative transport ($n = 3$) when it actually has convective transport ($n = 1.5$), you will overestimate its central density ρ_c by a factor of 9 and its central pressure P_c by a factor of 14. You will overestimate the central temperature $T_c \propto P_c / \rho_c$ by a factor of only 1.6; however, given the extreme sensitivity of the energy generation rate ϵ to temperature, this also represents a severe error.

6.3 Homology

Polytropes help us understand how the stiffness or softness of a star affects its internal structure. Adopting a polytropic relation yields a solution for $\rho(r)$, as plotted in Figure 6.1, and thus for $P(r)$, since the pressure goes simply as a power of the density. However, since polytropes don't directly address the question of how a star generates and transports energy, we need additional tools to understand how the luminosity $L(r)$ and temperature $T(r)$ of a star depend on the underlying physics. One valuable tool in this regard is the concept of **homology**. The word "homology" comes from the ancient Greek roots *homos* (meaning "same") and *logos* (meaning, among other things, "proportion" or "relation"). The first scientists to talk about homology were biologists. In evolutionary biology, homology refers to a similarity among different species that reveals they have a common ancestor. The pastern of a horse, for instance, is homologous with the two largest bones in a human's middle finger. In astronomy, homology refers to a similarity among different stars that reveals that they are shaped by the same underlying physical processes. In a homologous population of stars, the stars are related to each other by a simple change of scale.

To see how homology works, consider a series of zero age main sequence stars that have just begun hydrogen burning in their cores. We assume that all these stars have the same chemical composition and the same mean molecular mass $\bar{\mu}$, constant throughout each star. One important requirement for our homologous series of stars is that they all transport energy radiatively. In the case of radiative energy transport (Equation 6.6), the temperature gradient dT/dr is closely linked to the radiative energy flux $L/(4\pi r^2)$; this will ultimately allow us to infer the relation among mass, luminosity, and radius for our homologous stars. In the case of convective energy transport, by contrast, the temperature gradient dT/dr is almost completely decoupled from the flux, and the tools of homology are not applicable.

To be truly homologous, our family of stars must also have the same equation of state, opacity source, and source of nuclear energy. For example, stars with energy generation from the CNO bi-cycle will not be strictly homologous to stars powered by the pp chain. For homology to be a useful tool, like must be compared to like. To begin, we will start with the most general homology relations, then move to more specialized relations that rely on the nature of the microphysics to determine equations of state, opacity, and energy sources.

Let's start our study of homology by defining a dimensionless variable

$$x \equiv \frac{r}{R_\star}. \tag{6.29}$$

This will act as our radial coordinate; it ranges from $x = 0$ at a star's center to $x = 1$ at a star's photosphere. The mass inside the dimensionless radius x can be written as

$$M(r) = M_\star f_M(x), \tag{6.30}$$

where $M_\star$ is the star's total mass, and f_M is a dimensionless function that ranges from $f_M(0) = 0$ at the star's center to $f_M(1) = 1$ at the star's photosphere. Similarly, the enclosed luminosity L can be written as

$$L(r) = L_\star f_L(x), \tag{6.31}$$

where $L_\star$ is the star's total luminosity, and f_L varies from $f_L(0) = 0$ at the star's center to $f_L(1) = 1$ at the star's photosphere. The mass density ρ can be written as

$$\rho(r) = \rho_\star f_\rho(x), \tag{6.32}$$

where $\rho_\star$ is the mean density of the star, and f_ρ varies from $f_\rho(0) > 1$ at the star's center to $f_\rho(1) \ll 1$ at the star's photosphere.

Similarly, the pressure $P(r)$ and temperature $T(r)$ can be written as

$$P(r) = P_0 f_P(x) \tag{6.33}$$

and

$$T(r) = T_0 f_T(x), \tag{6.34}$$

where P_0 and T_0 are scaling coefficients with the dimensionality of a pressure and a temperature respectively, which have an amplitude that is convenient for the star in question. (The central pressure and temperature would be a useful choice for P_0 and T_0 were it not for the fact that we don't know them yet.)

Using our new dimensionless functions, we can write the mass conservation equation (Equation 6.1) as

$$\frac{M_\star}{R_\star}\frac{df_M}{dx} = 4\pi\rho_\star R_\star^2 f_\rho x^2. \tag{6.35}$$

Making the substitution

$$\rho_\star \equiv \frac{3M_\star}{4\pi R_\star^3}, \tag{6.36}$$

this reduces to the dimensionless form

$$\frac{df_M}{dx} = 3f_\rho x^2. \tag{6.37}$$

We can refer to a population of stars as **homologous** if they all have the same dimensionless density profile f_ρ, despite having different values of $M_\star$ and $R_\star$. Equation 6.37 then implies that they also have the same mass profile f_M. In a similar vein, an individual star is homologous if its dimensionless profile f_ρ remains unchanged when its mass $M_\star$ changes (by accretion or a strong stellar wind) or its radius $R_\star$ changes (by expansion or contraction). Equation 6.36 then gives us our first general homology relation: the mean density of a homologous star scales as $\rho_\star \propto M_\star R_\star^{-3}$. Since the star's central density is $\rho_c = \rho_\star f_\rho(0)$, and since $f_\rho(0)$ never changes for a homologous star, the central density also must scale as $\rho_c \propto M_\star R_\star^{-3}$.

Using our dimensionless functions, we can write the equation of hydrostatic equilibrium (Equation 6.2) as

$$\frac{P_0}{R_\star}\frac{df_P}{dx} = -\frac{GM_\star\rho_\star}{R_\star^2}\frac{f_M f_\rho}{x^2}. \tag{6.38}$$

This suggests a convenient choice for our scaling coefficient P_0. If we take

$$P_0 \equiv \frac{GM_\star\rho_\star}{R_\star} = \frac{3GM_\star^2}{4\pi R_\star^4}, \tag{6.39}$$

then the dimensionless form of the hydrostatic equilibrium equation is

$$\frac{df_P}{dx} = -\frac{f_M f_\rho}{x^2}. \tag{6.40}$$

Equation 6.40 tells us that a homologous star, with f_ρ and f_M unchanged over time, must also have a dimensionless pressure profile $f_P(x)$ that is unchanged. Equation 6.39 then gives us our second homology relation: the pressure scaling coefficient for a homologous star goes as $P_0 \propto M_\star^2 R_\star^{-4}$. Since the star's central pressure is $P_c = P_0 f_P(0)$, and since $f_P(0)$ never changes for a homologous star, the central pressure also must scale as

$$P_c \propto \frac{M_\star^2}{R_\star^4}. \tag{6.41}$$

Since stars maintain strict hydrostatic balance, we can use this second homology relation to derive an important result: the central pressure of a homologous star *must* respond to changes in radius. If $M_\star$ is constant but $R_\star$ is changed, then the central pressure must alter as $R_\star^{-4}$. Conversely, if the central pressure changes, the radius of a homologous star will alter as necessary to restore hydrostatic equilibrium.

To proceed further, we must choose an equation of state for our homologous stars. The ideal gas law ($P \propto \rho T/\bar{\mu}$) is an excellent approximation for most stars. Having made our choice of P_0, we can then make an obvious choice for the scaling temperature T_0 from the ideal gas law:

$$P_0 f_P = \frac{k}{m_{\mathrm{AMU}}\bar{\mu}} \rho_\star T_0 f_\rho f_T. \tag{6.42}$$

If we choose

$$T_0 \equiv \frac{m_{\mathrm{AMU}}\bar{\mu}}{k}\frac{P_0}{\rho_\star} = \frac{m_{\mathrm{AMU}}\bar{\mu}}{k}\frac{GM_\star}{R_\star}, \tag{6.43}$$

then the dimensionless ideal gas law is simply

$$f_P = f_\rho f_T, \tag{6.44}$$

implying that a homologous star must also have a dimensionless temperature profile f_T that is unchanged with time. Equation 6.43 then gives us our third homology relation: the temperature scaling coefficient for a homologous star goes as $T_0 \propto \bar{\mu}M_\star/R_\star$. Since the star's central temperature is $T_c = T_0 f_T(0)$, and since $f_T(0)$ never changes for a homologous star, the central temperature also must scale as

$$T_c \propto \frac{\bar{\mu}M_\star}{R_\star}. \tag{6.45}$$

This is the first of our homology relations that depends on the composition of a star. Since stars are made primarily of H and He, and since the interiors of stars are highly ionized, we can use Equation 3.33 to write the mean molecular mass as

$$\bar{\mu} \approx \frac{4}{8 - 5Y}, \tag{6.46}$$

where Y is the helium mass fraction. This implies that the central temperature of a homologous star (and by extension, its luminosity) depends strongly on its helium abundance. This dependence permits astronomers to infer the initial helium abundance of stars whose mass, luminosity, and age are known. A ZAMS star with no helium ($Y = 0$) would have an exceptionally low central temperature (and by extension, an exceptionally low luminosity) for a given mass and radius. Historically, the lack of anomalously dim stars gave an important piece of evidence for the Hot Big Bang model: the observed luminosity of even the most metal-poor stars (made from gas that hadn't been recycled from earlier generations of stars) requires an initial helium abundance well above zero. Big Bang Nucleosynthesis during the first 15 minutes of the universe provided a means of creating this primordial helium.

To proceed further in our study of homologous stars, we must make additional assumptions about energy generation and transport. The energy generation rate can be approximated as $\epsilon \propto \rho T^{\nu}$, while the opacity that regulates the rate of radiative energy transport can be written as a generalized Kramers' law: $\kappa \propto \rho^{\alpha} T^{-\beta}$. For the highest-mass stars, CNO energy generation and electron scattering opacity are good approximations ($\nu \sim 18$, $\alpha = \beta = 0$), while for lower-mass stars, pp energy generation combined with free–free and bound–free opacity provides a reasonable approximation ($\nu \sim 4$, $\alpha = 1$, $\beta = 3.5$).

Let's start by examining energy transport. If we write the opacity as a generalized Kramers' law,

$$\kappa = \kappa_0 \rho^{\alpha} T^{-\beta}, \tag{6.47}$$

then the equation of radiative energy transport (Equation 6.6) becomes

$$\frac{df_T}{dx} = -D \frac{f_\rho^{1+\alpha} f_L}{x^2 f_T^{3+\beta}}, \tag{6.48}$$

where

$$D \equiv \frac{3\kappa_0 \rho_\star^{1+\alpha} L_\star}{16\pi\, acR_\star T_0^{4+\beta}} \tag{6.49}$$

is a dimensionless constant. In a homologous family of stars, D will be the same for every star. Previously, we were able to fiddle the amplitudes of P_0 and T_0 to ensure that the dimensionless equation of hydrostatic equilibrium (Equation 6.40) and the dimensionless equation of state (Equation 6.44) were free of constants like D. However, we are unable to fiddle the amplitude of κ_0 to make $D = 1$ since κ_0 is fixed by the microphysics of opacity.

Since D and κ_0 are constant, Equation 6.49 gives us another homology relation:

$$L_\star \propto R_\star T_0^{4+\beta} \rho_\star^{-1-\alpha} \propto \bar{\mu}^{4+\beta} M_\star^{3-\alpha+\beta} R_\star^{3\alpha-\beta}, \tag{6.50}$$

making use of the homology relations for T_0 and $\rho_\star$. Notice that when the opacity is provided by electron scattering ($\alpha = \beta = 0$), this reduces to the elegantly simple relation

$$L_\star \propto \bar{\mu}^4 M_\star^3, \tag{6.51}$$

independent of the stellar radius $R_\star$. When the opacity is provided by free–free and bound–free absorption ($\alpha = 1$, $\beta = 3.5$), the equivalent relation is

$$L_\star \propto \bar{\mu}^{7.5} M_\star^{5.5} R_\star^{-0.5}. \tag{6.52}$$

Although luminosity has a weak dependence on radius in this case, the primary dependence is on the mean molecular mass $\bar{\mu}$ and the stellar mass $M_\star$. Thus, given a reasonable source of opacity, stars with radiative energy transport must have a strong dependence of luminosity on mass, regardless of their energy source. There must also be a strong dependence on $\bar{\mu}$, with the luminosity increasing as the mean molecular mass goes upward.

So far, all our homology relations have been independent of the energy generation rate: they would be the same if stars were powered by nuclear fusion or by immortal draft horses trotting on treadmills. However, energy generation does play an important role in stars, for two different reasons. First, the amount of fuel and the efficiency η of energy generation determine the star's lifetime. Second, since the energy generation rate ϵ from fusion is very sensitive to temperature, there is only a narrow range of T_c that generates a luminosity in agreement with the value of $L_\star$ required by energy transport (Equations 6.51 and 6.52). Given that $T_c \propto \bar{\mu} M_\star / R_\star$, this implies that for a given stellar mass and composition, there is only a narrow range of $R_\star$ that yields the required luminosity. Speaking broadly, energy transport determines a star's luminosity, while energy generation determines its lifetime and radius.

To illustrate the importance of energy generation, write the specific energy generation rate in the form

$$\epsilon = \epsilon_0 \rho T^\nu. \tag{6.53}$$

The equation of thermal equilibrium (Equation 6.3) then becomes

$$\frac{L_\star}{R_\star} \frac{df_L}{dx} = 4\pi \rho_\star^2 R_\star^2 T_0^\nu \epsilon_0 f_\rho^2 f_T^\nu x^2. \tag{6.54}$$

This can be rewritten as

$$\frac{df_L}{dx} = C f_\rho^2 f_T^\nu x^2, \tag{6.55}$$

where

$$C \equiv 3 \frac{\epsilon_0 M_\star \rho_\star T_0^\nu}{L_\star} \tag{6.56}$$

is a dimensionless constant. In a homologous family of stars, C will be the same for every star.

Since C and ϵ_0 are constants, Equation 6.56 gives us one last homology relation:

$$L_\star \propto M_\star \rho_\star T_0^\nu \propto \bar{\mu}^\nu M_\star^{2+\nu} R_\star^{-3-\nu}. \tag{6.57}$$

We now have two competing homology relations for the luminosity: Equation 6.50 from energy transport and Equation 6.57 from energy generation. These two equations can be reconciled only if

$$\log R_\star = K + \frac{\nu - 4 - \beta}{\nu + 3 + 3\alpha - \beta} \log \bar{\mu} + \frac{\nu - 1 + \alpha - \beta}{\nu + 3 + 3\alpha - \beta} \log M_\star. \tag{6.58}$$

In general, the slope of the mass–radius relation for homologous stars depends on the properties of the opacity ($\kappa \propto \rho^\alpha T^{-\beta}$) as well as the properties of the energy generation rate ($\epsilon \propto T^\nu$).

Let's look at the interesting limit when the energy generation rate becomes overwhelmingly sensitive to temperature ($\nu \to \infty$). In this case, the general mass–radius relation of Equation 6.58 reduces to a linear relation, regardless of opacity: $R_\star \propto \bar{\mu} M_\star$. This implies that the star's central temperature, $T_c \propto \bar{\mu} M_\star / R_\star$, is pinned to a unique value, regardless of stellar mass and composition. In the limit that $\nu \to \infty$, the star's luminosity has the dependence $L_\star \propto \bar{\mu}^{4+3\alpha} M_\star^{3+2\alpha}$.

As a more realistic scenario, consider a hot massive star producing energy by the CNO bi-cycle, with $\nu \approx 18$. Such a hot star will have electron scattering opacity, with $\alpha = \beta = 0$. In this case, Equation 6.58 predicts a mass–radius relation $R_\star \propto \bar{\mu}^{0.7} M_\star^{0.8}$. This is somewhat steeper than the $R_\star \propto M_\star^{0.5}$ relation actually observed for massive stars (Figure 1.13). Since $18 < \infty$, homologous stars powered by the CNO bi-cycle will have a mild dependence of central temperature on mass and composition, with $T_c \propto \bar{\mu} M_\star / R_\star \propto \bar{\mu}^{0.3} M_\star^{0.2}$. The central pressure, however, decreases quite strongly with mass: $P_c \propto M_\star^2 / R_\star^4 \propto \bar{\mu}^{-2.7} M_\star^{-1.4}$. As we found earlier, massive stars with electron scattering opacity have $L_\star \propto M_\star^3$, regardless of the value of ν; this is an excellent fit to the mass–luminosity relation actually observed for massive stars (Figure 1.12).

It is also possible to make homology relations for lower-mass stars producing energy by the pp chain, with $\nu \approx 4$, and with opacity dominated by free–free and bound–free absorption, implying $\alpha = 1$ and $\beta = 3.5$. In this case, Equation 6.58 predicts a very weak dependence of radius on mass, with $R_\star \propto \bar{\mu}^{-0.5} M_\star^{0.1}$. This is significantly shallower than the $R_\star \propto M_\star$ relation actually observed for lower-mass stars (Figure 1.13). The resulting central temperature from the homology relations is $T_c \propto \bar{\mu}^{1.5} M_\star^{0.9}$. The predicted central pressure in this case increases quite strongly with mass: $P_c \propto \bar{\mu}^{2.2} M_\star^{1.7}$. Finally, the mass–luminosity relation predicted for these lower-mass stars from the homology relations is $L_\star \propto \bar{\mu}^{0.2} M_\star^{5.5}$. If we exclude M dwarfs with $M_\star < 0.4\,M_\odot$, which are fully convective, the observed mass–luminosity relation for lower-mass stars on the main sequence is $L_\star \propto M_\star^{4.5}$, not as steep as the homology prediction.

6.4 Solar Models

Although main sequence stars are all powered by fusion of hydrogen into helium, they harbor enough variety that homologous models are inadequate to describe them perfectly. Fusion can occur by the pp chain and by the CNO bi-cycle. Opacity can occur by electron scattering or by free–free and bound–free absorption; in the star's cooler atmosphere, ions such as H^- and molecules such as H_2O, TiO, and VO can provide opacity. The internal pressure can be purely from an ideal gas law or with a significant contribution from radiation pressure or degeneracy pressure. Energy transport can be by convection or by radiation; many main sequence stars have both a convection zone and a radiation zone, with convective overshoot blurring the boundary between the zones. Thus, to have a full picture of the main sequence, we need a different model for each stellar mass, instead of a one-size-fits-all homologous model that scales simply with mass.

Unsurprisingly, the numerical stellar model that we will examine in most detail is the solar model, since the Sun is the star for which we have the most information available. Any model we make for the present-day Sun must match the Sun's observed mass ($1\,M_\odot = 1.9884 \times 10^{33}$ g), its luminosity ($1\,L_\odot = 3.828 \times 10^{33}$ erg s^{-1}), and its photospheric radius ($1\,R_\odot = 6.957 \times 10^{10}$ cm).

In Section 3.1, we made an amusingly naïve guess at the Sun's central pressure by assuming that the Sun's density was everywhere equal to its mean density $\rho_\odot = 1.41\,\mathrm{g\,cm}^{-3}$. A uniform-density mock Sun has an internal pressure found by integrating the equation of hydrostatic equilibrium:

$$P(r) = P_{c,\odot}\left(1 - r^2/R_\odot^2\right), \tag{6.59}$$

where the central pressure of the uniform-density Sun is (Equation 3.5)

$$P_{c,\odot} = \frac{2\pi}{3} G\rho_\odot^2 R_\odot^2 = 1.345 \times 10^{15}\,\mathrm{dyn\,cm}^{-2}. \tag{6.60}$$

The pressure profile of the uniform-density Sun, shown as the dashed line in Figure 6.2(a), has a large central region of nearly constant pressure. The pressure doesn't fall to half of its central value until a radius $r = R_\odot/\sqrt{2}$. Assuming the pressure in our uniform-density Sun is given by an ideal gas law, the internal temperature is

$$T(r) = \frac{m_{AMU}}{k}\frac{\bar{\mu}_\odot}{\rho_\odot}P(r) = T_{c,\odot}\left(1 - r^2/R_\odot^2\right), \tag{6.61}$$

where the central temperature of the uniform-density Sun is

$$T_{c,\odot} = \frac{2\pi}{3}\frac{Gm_{AMU}}{k}\bar{\mu}_\odot\rho_\odot R_\odot^2 \approx 7.0 \times 10^6\,\mathrm{K}, \tag{6.62}$$

assuming $\bar{\mu}_\odot \approx 0.61$. As shown by the dashed line in Figure 6.2(b), a mock uniform-density Sun has a nearly constant temperature in its central regions, as well as having a nearly constant pressure.

Of course, modeling a star like the Sun as a stiff, incompressible sphere is not a good match to reality. As we saw in Section 6.2, a better approximation for the Sun is a polytrope of index $n = 3$. Assuming this polytropic model, the Sun's central density is $\rho_c = 54.18\rho_\odot \approx 76.4\,\mathrm{g\,cm}^{-3}$, using the appropriate central-to-average density ratio for an $n = 3$ polytrope. When we model the Sun as a polytrope, the translation from density to pressure is given by Equation 6.17, which can be

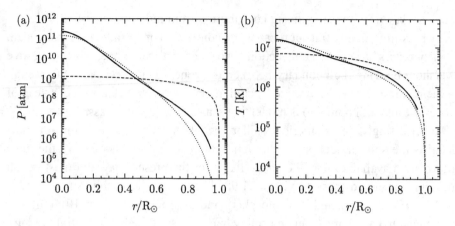

Figure 6.2 (a) Pressure as a function of radius for a uniform-density Sun (dashed line), a polytropic Sun (dotted line), and the BP04 standard solar model (heavy solid line). (b) Temperature as a function of radius for the same models.

rewritten as

$$P = P_* \left(\rho/\rho_\odot\right)^{4/3}, \tag{6.63}$$

where

$$P_* = 1.658 \hbar c \left[\frac{f_\gamma}{(1-f_\gamma)^4}\right]^{1/3} \left(\frac{\rho_\odot}{m_{\text{AMU}}\bar{\mu}_\odot}\right)^{4/3} \approx 7.2 \times 10^{14} \, \text{dyn cm}^{-2}, \tag{6.64}$$

assuming $\bar{\mu}_\odot \approx 0.61$ and $f_\gamma \approx 7 \times 10^{-4}$. The central pressure of the polytropic Sun is then

$$P_{c,\odot} = (54.18)^{4/3} P_* \approx 1.5 \times 10^{17} \, \text{dyn cm}^{-2}, \tag{6.65}$$

over 100 times the pressure at the center of a uniform-density Sun. The complete run of pressure with radius inside a polytropic Sun is shown as the dotted line in Figure 6.2(a). Although the central pressure of the soft polytropic Sun is ~100 times that of the stiff uniform-density Sun, at a radius $r \sim 0.8 \, \text{R}_\odot$ the pressure of the polytropic Sun is only 0.01 times that of the uniform-density Sun. If the polytropic Sun obeys an ideal gas law, its temperature–density relation is

$$T = \frac{m_{\text{AMU}}\bar{\mu}_\odot}{k} \frac{P}{\rho} = T_* \left(\rho/\rho_\odot\right)^{1/3}, \tag{6.66}$$

where

$$T_* = 1.658 \frac{\hbar c}{k} \left[\frac{f_\gamma}{(1-f_\gamma)^4}\right]^{4/3} \left(\frac{\rho_\odot}{m_{\text{AMU}}\bar{\mu}_\odot}\right)^{1/3} \approx 3.8 \times 10^6 \, \text{K}. \tag{6.67}$$

The central temperature of the polytropic Sun is then

$$T_{c,\odot} = (54.18)^{1/3} T_* \approx 1.4 \times 10^7 \, \text{K}, \tag{6.68}$$

which is twice the central temperature of a uniform-density Sun with the same radius and mass. The temperature within a polytropic Sun, as shown in Figure 6.2(b), is within a factor of two of the uniform-density temperature throughout the range $r = 0 \to 0.7\,R_\odot$. It's harder to botch temperature estimates within stars than it is to bungle the pressure estimates, since the temperature depends less strongly on the softness or stiffness assumed for the star.

Since the mid twentieth century, advances in computing have made possible increasingly sophisticated numerical models of the Sun. "Standard solar models" are models of the Sun which contain all the relevant physics, constrained by the best available observations.[4] One difficulty in producing a standard solar model is that the chemical composition of the Sun today is the result of 4.57 Gyr of solar history. To compute a model of the Sun today, we must typically start by modeling the Sun as it first appeared on the zero age main sequence. Dividing the ZAMS Sun into thin radial shells, the internal properties of the ZAMS Sun are computed using the differential equations of stellar structure with appropriate values for the opacity κ, specific energy generation rate ϵ, and adiabatic index γ; the ZAMS Sun is assumed to have uniform chemical composition. The mass of the ZAMS Sun is fixed at $M_{\text{ZAMS}} = 1\,M_\odot$, since its mass loss on the main sequence is negligible. The model Sun is then stepped forward in time, keeping track of changes in the composition profile, until at $t_\odot = 4.57\,\text{Gyr}$, the model (we hope) matches the present-day value of $L_\odot$ and $R_\odot$.

But what if our hopes are crushed, and the model doesn't correctly match $L_\odot$ and $R_\odot$? The two main free parameters in a solar model, which we can tweak to give a better match, are the convective mixing length parameter α_ℓ and the initial helium abundance $Y_{\odot,0}$. Neither is directly measured or strongly constrained by theory. In practice, the current solar radius $R_\odot$ is sensitive to the adopted α_ℓ, and the luminosity $L_\odot$ is highly sensitive to $Y_{\odot,0}$. Recent standard solar models have found good fits to the current Sun by adopting a mixing length parameter in the range $\alpha_\ell = 1.5\text{--}2$, while adopting an initial helium abundance from the narrower range $Y_{\odot,0} = 0.272\text{--}0.277$.

The solar mixing length is a phenomenological tool, not of direct physical interest. However, the initial solar helium abundance deduced from numerical models is a useful measure of helium enrichment between the time of Big Bang Nucleosynthesis and the time of the Sun's formation. Given a primordial value $Y_p = 0.247$, solar models imply a relatively modest helium enrichment $\Delta Y = Y_{\odot,0} - Y_p = 0.025\text{--}0.030$ for the protosolar gas. When compared to the metallicity enhancement $\Delta Z = Z_{\odot,0} - Z_p \approx Z_{\odot,0}$, the models imply a helium-to-metal enrichment ratio $\Delta Y/\Delta Z \sim 1.5$.

Once suitable values of α_ℓ and $Y_{\odot,0}$ are determined, the resulting solar model tells us how the Sun's properties change with time on the main sequence. As

---

[4] For the figures in this text, we primarily use the standard solar model described by Bahcall and Pinsonneault in 2004; this is called the BP04 model, for short.

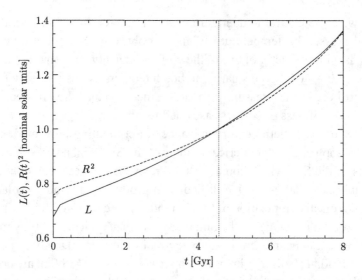

Figure 6.3 Luminosity (solid line) and surface area (dashed line) of the Sun as a function of time. Values are normalized to the nominal solar values at the present day ($t = 4.57\,\text{Gyr}$, shown as the vertical dotted line). Numbers are from the BP04 solar model.

an example, Figure 6.3 shows the computed luminosity $L(t)$ of the BP04 solar model as the solid line. The luminosity of the ZAMS Sun was $L = 0.68\,L_\odot$; at an age $t = 8\,\text{Gyr}$, the Sun's luminosity will be $L = 1.36\,L_\odot$. The dashed line in Figure 6.3 shows how the surface area of the Sun ($\propto R^2$) increases with time in the BP04 solar model. Until $t \approx 6\,\text{Gyr}$, the Sun's luminosity increases more rapidly than its surface area. Thus, the effective temperature of the Sun slightly increases, going from $T_\text{eff} = 5620\,\text{K}$ on the zero age main sequence to a maximum value of $T_\text{eff} = 5790\,\text{K}$ at $t \approx 6\,\text{Gyr}$.

The Sun's increasing luminosity results from the changes in chemical composition in the Sun's core. As Figure 6.4(a) shows, the Sun's composition today is no longer homogeneous. At the Sun's surface, the BP04 model has $X = 0.740$ and $Y = 0.243$. At the Sun's center, however, hydrogen burning has reduced the hydrogen mass fraction to $X = 0.340$ and increased the helium mass fraction to $Y = 0.640$. Note also in Figure 6.4 that the Sun is chemically homogeneous at $r > 0.71\,R_\odot$; this is the **convective zone**, where the turbulent convection that transports energy also acts to keep the Sun well-mixed.

Because of the conversion of hydrogen to helium, the mean molecular mass at the Sun's center today is significantly higher than at the Sun's surface.[5] As shown in Figure 6.4(b), the BP04 model for the present-day Sun has a mean molecular mass that goes from $\bar{\mu} = 0.60$ in the outer convective zone to $\bar{\mu} = 0.85$ in the Sun's core. The change in the mean molecular mass $\bar{\mu}_c(t)$ at the Sun's center has

---

[5] The preference for ^{14}N over ^{12}C in the CN cycle (see Figure 5.9) also increases $\bar{\mu}$ at the center; however, this is a minor effect.

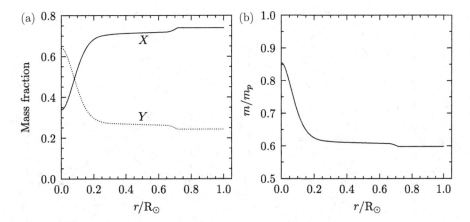

Figure 6.4 (a) The Sun's mass fraction of hydrogen (X) and helium (Y) as a function of radius. (b) Resulting mean molecular mass $\bar{\mu}$ in the Sun's interior. Numbers are from the BP04 solar model.

important implications for the central temperature $T_c(t)$ of the Sun. If we assume that the evolution of the Sun on the main sequence is roughly homologous, then we can use Equation 6.45 to write

$$T_c(t) \propto \frac{\bar{\mu}_c(t)M_\odot}{R(t)}. \tag{6.69}$$

Here, we have assumed a constant mass for the Sun, since its mass-energy loss on the main sequence is insignificant. The mean molecular mass in the core of the ZAMS Sun was $\sim$30% smaller than its current value. Although the radius of the ZAMS Sun was smaller than that of the present Sun, it was smaller by just 13%. Thus, the temperature at the Sun's core has been increasing with time, with

$$\frac{T_c(t_\odot)}{T_c(0)} \approx \left(\frac{\bar{\mu}_c(t_\odot)}{\bar{\mu}_c(0)} \right) \left(\frac{R(0)}{R_\odot} \right) \approx \left(\frac{0.85}{0.60} \right) \left(\frac{0.87}{1} \right) \approx 1.2. \tag{6.70}$$

As the Sun's central temperature increases with time, the CNO bi-cycle plays an increasingly important role in energy generation. In the BP04 model, the CNO bi-cycle produces 1.6% of the Sun's luminosity today, but by $t = 8\,\mathrm{Gyr}$ it will produce 22% of the Sun's (increased) luminosity.

The pressure inside the BP04 model at the present day ($t_\odot = 4.57\,\mathrm{Gyr}$) is shown as the solid line in Figure 6.2(a); temperature is shown as the solid line in Figure 6.2(b). Except in the convection zone at $r > 0.71\,\mathrm{R_\odot}$, the soft $n = 3$ polytrope, shown as the dotted line, gives a fair fit to the pressure and temperature. Because of the Sun's softness, its mass is centrally concentrated. For the BP04 model at the present day, half the Sun's mass is enclosed within a sphere of radius $r = 0.26\,\mathrm{R_\odot}$, representing 1.8% of the Sun's volume. Because of the density and temperature dependence of the nuclear energy generation rate ϵ, the Sun's luminosity is even more centrally concentrated than its mass. Half the Sun's

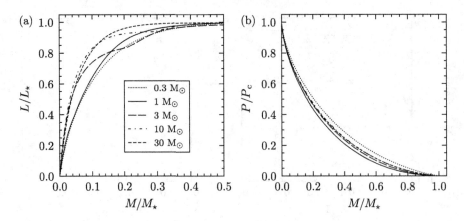

Figure 6.5 (a) Enclosed luminosity, normalized to $L_\star$, versus enclosed mass, normalized to $M_\star$. (b) Pressure, normalized to the central pressure P_c, versus enclosed mass, normalized to $M_\star$. Results are from numerical models of stars.

luminosity is produced in a sphere of radius $r = 0.11\,R_\odot$, representing just 0.13% of the Sun's volume and containing 8.8% of the Sun's mass.

6.5 Zero Age Main Sequence Models

The main sequence is a central concept in stellar theory, and a natural place to start our tour of the life cycle of stars. In this section, we will look at model stars with the same value of α_ℓ and Y_0 as the BP04 solar model, but with a range of stellar masses $M_\star$. For simplicity, we start with the assumption that the stars are on the zero age main sequence (ZAMS), and that their homogeneous chemical composition is the same as that of the ZAMS Sun. The individual values of $M_\star$ that we choose are selected to illustrate the full range of stellar properties. A low-mass star with $M_\star = 0.3\,M_\odot$ is fully convective. A $1\,M_\odot$ star (a solar twin) has a radiative core and convective envelope, while a $3\,M_\odot$ star has a convective core and radiative envelope. Going to higher mass, a $10\,M_\odot$ star is massive enough to eventually become a supernova, while a $30\,M_\odot$ star is luminous enough for radiation-driven mass loss to become important during its main sequence lifetime.

With a set of model stars that include all sources of pressure, opacity, and energy generation, we can test some of the simplifying assumptions that went into our homology relations (Section 6.3). For each numerically modeled star, we can compute the dimensionless functions $f_M = M/M_\star$, $f_L = L/L_\star$, and so forth, that were assumed to be the same for all stars in a homologous family. By comparing these dimensionless functions for models with different $M_\star$, we can see if they are in fact the same, or even roughly similar.

Figure 6.5(a), for instance, shows f_L versus f_M for numerical models of stars with different mass. In the central core of a star ($M/M_\star < 0.1$), the lower-mass

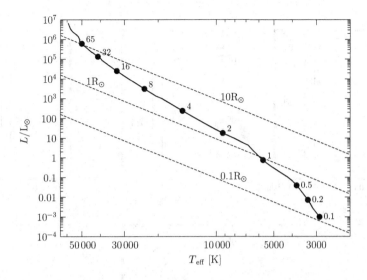

Figure 6.6 Zero age main sequence (solid line) for stars with solar metallicity. Dots are labeled with the stellar mass (in units of the nominal solar mass). Dashed lines indicate lines of constant stellar radius, at $R_\star = 0.1$, 1, and $10\,R_\odot$.

stars ($M_\star = 0.3$ and $1\,M_\odot$) are homologous to each other, while the higher-mass stars ($M_\star = 3$, 10, and $30\,M_\odot$) follow a different homology relation. This reflects the fact that the lower-mass stars are powered by the pp chain, while the higher-mass stars are powered by the CNO bi-cycle, with its steeper temperature dependence. Figure 6.5(b) shows the dimensionless pressure f_P, renormalized to one at the star's center, as a function of f_M. In general, the pressure relations are nearly homologous. Notice, however, that the pressure for the lowest-mass star ($M_\star = 0.3\,M_\odot$) falls the least steeply from its central value; this indicates that fully convective low-mass stars, with polytropic index $n = 1.5$, are stiffer than more massive stars where radiative energy transport is dominant.

Figure 6.6 shows the location on the "theorists' H–R diagram" ($\log L$ versus $\log T_{\rm eff}$) of modeled zero age main sequence stars with solar metallicity. Comparing the ZAMS to lines of constant stellar radius (the dashed lines in Figure 6.6), we find that stars start their lives on the main sequence with radii ranging from $R_\star \sim 10\,R_\odot$ for the most massive stars to $R_\star \sim 0.1\,R_\odot$ for the least massive stars. This is in agreement with the observed range of stellar radii for main sequence stars of assorted ages (Figure 1.13). The steep mass–luminosity relation seen in real stars is also naturally reproduced.

To provide more insight into how the observable properties of stars rely on their internal structure, Figure 6.7 shows different properties of stars as a function of the enclosed mass M (or, in the bottom two panels, of the local pressure P). The plot of pressure versus mass in Figure 6.7(a) shows the rather counterintuitive result that the higher-mass stars, with $M_\star = 10$ or $30\,M_\odot$, have a lower central

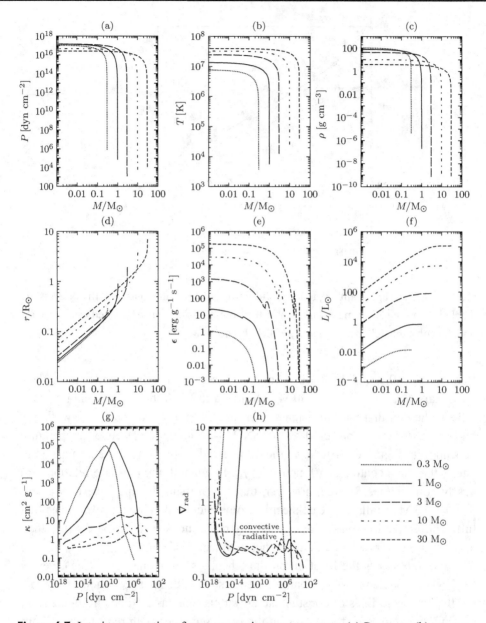

Figure 6.7 Interior properties of zero age main sequence stars. (a) Pressure, (b) temperature, (c) density, (d) radius, (e) energy generation rate, and (f) enclosed luminosity are plotted versus enclosed mass. (g) Opacity and (h) $\nabla_{\rm rad}$ are plotted versus pressure.

pressure than stars of lower mass. However, the lower-mass stars, with $M_\star = 0.3$, 1, and 3 $M_\odot$, all have nearly the same central pressure. This suggests that we can regard main sequence stars with $M_\star \le 3\,M_\odot$ (a mass range that includes the vast majority of main sequence stars) as varying their density and temperature in order to maintain a central pressure $P_c \sim 10^{17}\,\mathrm{dyn\,cm^{-2}}$. Excluding the highest-mass stars, the central temperature increases from $T_{c,7} \approx 0.78$ for a 0.3 $M_\odot$ ZAMS star to $T_{c,7} \approx 2.45$ for a 3 $M_\odot$ star. Conversely, the central density decreases from $\rho_c \approx 110$ g cm^{-3} for a 0.3 $M_\odot$ star to $\rho_c \approx 43$ g cm^{-3} for a 3 $M_\odot$ star.

Moving away from a star's center, the pressure profile is dictated by the law of hydrostatic equilibrium: this leads to a dramatic drop in pressure in the outermost layers of all the models, tied to a drop in temperature and density. Note that at a given enclosed mass M, a more massive star always has a higher temperature than a less massive star (Figure 6.7(b)). However, in the central regions, the temperature of a 30 $M_\odot$ star is only 20% higher than that of a 10 $M_\odot$ star, while a 1 $M_\odot$ star has a central temperature 75% higher than that of a 0.3 $M_\odot$ star. The smaller range in T_c for higher-mass stars is a result of the different temperature dependence of the CNO bi-cycle (dominant in high-mass stars) and the pp chain (dominant in low-mass stars). A higher stellar mass $M_\star$ implies a much higher luminosity $L_\star$; we found this was universally true for homologous stars, regardless of their opacity source. For the pp chain, with $\nu \approx 4$, a large increase in luminosity requires a fairly large increase in T_c; however, for the CNO bi-cycle, with $\nu \approx 18$, only a small uptick in T_c is needed to produce a large increase in luminosity.

The plots in Figure 6.7 are taken as far as the stellar photosphere, ignoring the transparent outer atmosphere of the star. Because of the steep pressure drop required by hydrostatic equilibrium, the photospheric pressure is smaller by 11 to 14 orders of magnitude than the central pressure. Since the photospheric temperature is smaller by only three orders of magnitude than the central pressure, the star's pressure gradient is driven largely by the density gradient between the center and the photosphere. In the ZAMS Sun, for instance, the density ranges from a central value of $\rho_c \approx 80$ g cm^{-3} (about four times the density of solid gold) to a photospheric value of $\rho \approx 2 \times 10^{-7}$ g cm^{-3} (the density of the Earth's atmosphere at an altitude of 60 km). Because of the steep drop in density at larger values of M, all stars have virtually no mass in their outer layers. This is seen in a plot of r versus M (Figure 6.7(d)) as a steep rise in r as M approaches $M_\star$.

The specific energy generation rate ϵ and enclosed luminosity L for the model stars are shown in Figures 6.7(e) and 6.7(f). These plots vividly display the extreme temperature sensitivity of energy generation. The central temperature of a 30 $M_\odot$ star is only 20% greater than that of a 10 $M_\odot$ star; however, its value of ϵ is greater by a factor of seven. The energy flux at a given value of M goes as L/r^2. Since the dependence of L on total stellar mass $M_\star$ (Figure 6.7(f)) is far greater than the dependence of r^2 on $M_\star$ (Figure 6.7(d)), the flux within high-mass stars is higher than in low-mass stars, both at a fixed M and at a fixed $M/M_\star$.

To evaluate energy transport within stars, we need to know the opacity κ, shown in Figure 6.7(g), and the resulting value of ∇_{rad}, shown in Figure 6.7(h). These two quantities are plotted as a function of pressure, to give more emphasis to the outer envelope of the star. For the 0.3 and 1 $M_\odot$ stars, the opacity has a maximum value of $\kappa \sim 10^5$ cm^2 g^{-1} when the pressure is $P \sim 10^{10}$ dyn cm^{-2}. This pressure occurs in a layer of the star's envelope with $T \approx 40\,000\,\text{K}$ and $\rho \sim 10^{-3}\,\text{g cm}^{-3}$. Within this highly opaque layer, the mean free path for photons is $\lambda_{\text{mfp}} \sim 0.01\,\text{cm}$ (about a hair's breadth); this is much shorter than the mean free path $\lambda_{\text{mfp}} \sim 2\,\text{cm}$ that we found for electron scattering in the Sun's denser, hotter interior. The high opacity and short mean free path in a low-mass star's envelope are the result of bound–free absorption, which is extremely effective at $T \approx 40\,000\,\text{K}$ when the density is $\rho \geq 10^{-4}\,\text{g cm}^{-3}$ (see Figure 4.3).

Because of the high opacity provided by bound–free absorption, the value of ∇_{rad} becomes enormous in the envelopes of low-mass stars; by truncating Figure 6.7(h) at $\nabla_{\text{rad}} = 10$, we obscure the fact that ∇_{rad} can rise as high as $\sim 10^5$ in low-mass stars. Any value of ∇_{rad} greater than 0.4, shown by the horizontal dashed line in Figure 6.7(h), leads to convection in a fully ionized gas. The upturn in ∇_{rad} at high P, corresponding to small values of M and r, results from the high flux of energy in the physically compact, energy-generating core of the star. Even given the relatively modest opacity in the core of a ZAMS star (Figure 6.7(g)), the flux is too high to be transported by photons, and convection kicks in. (Notice that a $1\,\text{M}_\odot$ ZAMS star has a small convective core. However, as the central opacity drops, due to the fusion-driven decrease in X and increase in T, convection at the center of Sun-like stars eventually gives way to radiative transport.)

So far, we have been using the loose terms "high-mass" and "low-mass." Like the distinction between an island and a continent, the boundary between high and low mass can seem arbitrary. However, after looking at the dependence of stellar properties on $M_\star$, we can point to some properties that provide a physically meaningful separation between high-mass stars and low-mass stars.

High-mass stars generate energy by the CNO bi-cycle and transport energy through a convective core and radiative envelope. Since opacity is relatively low in the hot cores of high-mass stars, the convection in their cores is driven by high flux. The CNO bi-cycle, with its extreme temperature sensitivity, concentrates energy generation within a small volume, necessarily resulting in a high flux of energy. The strong dependence of the central energy generation rate on $M_\star$ for high-mass stars (Figure 6.7(e)) leads to larger convective cores in stars with greater total mass $M_\star$.

Low-mass stars generate energy by the pp chain and transport energy through a radiative core and convective envelope; at the very lowest masses, the convective zone stretches all the way to the star's center. Since flux is relatively low in the envelopes of low-mass stars, the convection in their envelopes is driven by high opacity. For example, the zone within a star where H or He ionization takes place is generally convective; not only is the opacity high, as a result of bound–free absorption, but the decrease in the adiabatic index γ also drives down ∇_{ad}, thus lowering the threshold for Schwarzschild instability. Ionization of relatively common metals, such as oxygen, carbon, nitrogen, and the iron peak elements can also contribute to opacity and trigger convection.

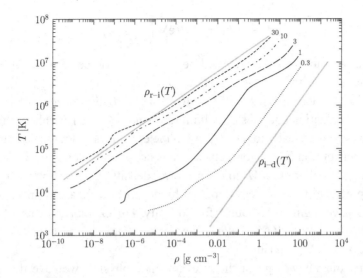

Figure 6.8 Temperature versus density within models of zero age main sequence stars; labels are mass in units of $M_\odot$. Heavy gray lines indicate where radiation pressure equals ideal gas pressure (ρ_{r-i}) and where ideal gas pressure equals electron degeneracy pressure (ρ_{i-d}) for a mass of pure hydrogen.

To examine further the distinction between high-mass and low-mass stars, it is useful to plot the density–temperature relation inside ZAMS stars, as shown in Figure 6.8. For each of our model stars, the ρ–T relation is plotted from the photosphere (lower left) to the star's center (upper right). The lines where radiative pressure equals ideal gas pressure (ρ_{r-i}) and where ideal gas pressure equals electron degeneracy pressure (ρ_{i-d}) are repeated from Figure 3.3. Although these lines are calculated for pure hydrogen, they nevertheless correctly tell us that radiation pressure is important only for the most massive stars. When radiation pressure dominates, stars become prone to severe instabilities; thus, we expect massive stars to be more likely to undergo extreme radial pulsations and strong outbursts of stellar wind. At the low-mass end of the stellar distribution, we see that the lowest-mass stars (with $M_\star < 0.3\,M_\odot$) have significant degeneracy pressure. When degeneracy pressure becomes dominant, a gas ball can maintain a high central pressure without the need for a high temperature. Thus, we can predict the existence of low-mass gas balls that do not undergo nuclear fusion; these objects, called "brown dwarfs," are in fact seen in the solar neighborhood. (We discuss brown dwarfs further in Section 7.6.)

Note from Figure 6.8 that as stellar mass $M_\star$ increases, the density ρ at a fixed temperature T always decreases. This has significant implications for ionization and opacity in high-mass stars, as compared to low-mass stars. In Section 4.1, we introduced the useful approximation that a species is half-ionized at a temperature $T_{1/2} \sim 0.1 I/k$, where I is the ionization energy. Thus, at a temperature T, a species with ionization energy

$$I \sim 10kT \sim 100\,\text{eV} \left(\frac{T}{10^5\,\text{K}} \right) \tag{6.71}$$

will be partially ionized and thus will be a potential source of bound–free opacity. We have seen that low-mass stars (0.3 and $1\,M_\odot$) are highly opaque at $T \approx 40\,000\,\text{K}$, where species with $I \sim 40\,\text{eV}$ are partially ionized. Since the first and second ionization energies of helium (24.6 and 54.4 eV) are about this large, helium contributes greatly to the opacity in the envelopes of low-mass stars; there is also a contribution from hydrogen, which is not yet fully ionized. However, at a fixed T, higher stellar mass $M_\star$ implies a lower density ρ. Since bound–free opacity is proportional to ρ, the envelopes of higher-mass stars are simply not dense enough to have significant bound–free opacity. For example, in the layer with $T = 40\,000\,\text{K}$, a $1\,M_\odot$ ZAMS star has a density (and thus an opacity) $\sim 10\,000$ times that of its $3\,M_\odot$ cousin.

By using our knowledge of the relevant microphysics, we have developed a general knowledge of the inner workings of a main sequence star: how energy is transported to the photosphere from deeper within the star, how the energy supply is replenished by nuclear fusion, and how fusion alters the star's composition. However, to understand the full picture of stellar evolution, there are more questions we need to address. Main sequence stars do not spring into existence out of nowhere. Thus, in Chapter 7 we examine how stars form out of the tenuous gas of the interstellar medium. In addition, main sequence stars are not immortal; their supply of fuel for fusion is finite. Thus, in Chapter 8 we examine what happens when stars run out of hydrogen in their central core.

Exercises

6.1   The Lane–Emden equation has analytic solutions for some values of n, given the standard boundary conditions $\theta = 1$ and $d\theta/d\xi = 0$ at $\xi = 0$.
   (a) Derive the analytic solution for $\theta(\xi)$ in the case $n = 0$. What is the value $\xi_\star$ for which $\theta = 0$?
   (b) Demonstrate by substitution that $\theta = (1 + \xi^2/3)^{-1/2}$ is a valid solution in the case $n = 5$. Show that although $\xi_\star$ is infinite in this case, the total mass of the gas sphere is finite.

6.2   In deriving the homology relations, we made the reasonable assumption that in Kramers' law, $\kappa = \kappa_0 \rho^\alpha T^{-\beta}$, and in the specific energy generation rate, $\epsilon = \epsilon_0 \rho T^\nu$, the values of κ_0 and ϵ_0 are fixed by microphysics and cannot be varied. Suppose, however, that Maxwell's daemon (a powerful but mischievous being) changes the values of κ_0 and ϵ_0.
   (a) For fixed values of $M_\star$ and $\bar{\mu}$, what is the dependence of $R_\star$, $L_\star$, and T_{eff} on κ_0 and ϵ_0? Assume the star remains homologous as κ_0 and ϵ_0 change.

(b) For a Sun-like star ($\alpha = 1$, $\beta = 3.5$, $\nu = 4$), does the luminosity $L_\star$ increase or decrease if the daemon increases ϵ_0?

(c) For a more massive star ($\alpha = \beta = 0$, $\nu = 18$), does $L_\star$ increase or decrease if the daemon increases ϵ_0?

6.3 The following questions should be answered approximately, using homology relations.

(a) Suppose that a star begins as 1 $M_\odot$ of pure hydrogen. If hydrogen burning occurs at $T_c = 1.5 \times 10^7$ K, what are the radius $R_\star$ and luminosity $L_\star$ of this star on the zero age main sequence? How long will it be powered by hydrogen burning?

(b) A star begins as 1 $M_\odot$ of pure helium. If helium burning occurs at $T_c = 10^8$ K, what are the radius $R_\star$ and luminosity $L_\star$ of this star when helium burning is ignited? How long will it be powered by helium burning?

(c) A star begins as 1 $M_\odot$ of pure carbon. If carbon burning occurs at $T_c = 10^9$ K, what are the radius $R_\star$ and luminosity $L_\star$ of this star when carbon burning is ignited? How long will it be powered by carbon burning?

(d) What are the relative effective temperatures $T_{\text{eff,H}}$, $T_{\text{eff,He}}$, and $T_{\text{eff,C}}$ of the hydrogen, helium, and carbon stars?

6.4 The very first stars in the universe started on the zero age main sequence with the primordial abundances $X \approx 0.75$, $Y \approx 0.25$, $Z \approx 0$ provided by Big Bang Nucleosynthesis. These first stars are called "population III" stars.[6] Suppose a metal-free population III star generates energy by the pp chain and has radiative energy transport with electron scattering opacity.

(a) Derive the mass–radius, mass–luminosity, and mass–central temperature relations for such a star, including the dependence on $\bar{\mu}$. Assuming the gas is fully ionized, how does the dependence on $\bar{\mu}$ translate into a dependence on X?

(b) If a population III star with $M_\star = 1 M_\odot$ has $T_c = 1.5 \times 10^7$ K, compute the mass at which $T_c = 10^8$ K, the ignition temperature for the triple alpha process. What do you think will happen in an initially metal-free star when T_c reaches 10^8 K?

6.5 The pressure scale height h of a star (Equation 4.67) is important in convection theory, since it determines the distance traveled by a fluid element. The scale height is large in deep stellar interiors, and small near the star's photosphere. To quantify this statement, solve the following problems:

(a) Use the ideal gas law to write the pressure scale height h as a function of T, $\bar{\mu}$, and $g = GM/r^2$.

(b) For the Sun, compute the value of h at the photosphere; then demonstrate that h diverges to infinity at the star's center.

---

[6] Stars with solar or supersolar metallicity are population I, while stars with subsolar metallicity are population II; by extension, stars with practically no metals at all are population III.

7

Star Formation: Before the Main Sequence

I'm gonna be a star, you know why?
Because I have nothing left to lose.

Lady Gaga (1986–)
"Marry the Night" [2011]

The Sun is 4.568 Gyr old; to state this more precisely, a time $t_\odot = 4.568$ Gyr has elapsed since the Sun formed via the gravitational collapse of a molecular cloud core. Thus, the Sun has existed for one-third of the total history of the universe. Some stars are older than the Sun; some are younger. In Chapter 6, our examination of main sequence models neglected the question of how stars form. From one viewpoint, this neglect was reasonable; if the timescale for star formation is much shorter than the nuclear burning time t_{nuc}, then little nuclear fuel is processed before the main sequence phase, and zero age main sequence stars can be safely treated as chemically homogeneous. However, to have a full picture of the "life cycle" of stars, and to verify that the timescale for star formation is indeed short, we need to understand how stars form out of the lower-density gas of the interstellar medium.

The formation of stars can be thought of as a two-stage process. First is a **hydrodynamic** stage, in which cool molecular gas in the interstellar medium falls inward on a timescale comparable to the freefall time. In the regions of interstellar gas where stars form, the freefall time is a few Myr or less. Eventually, the central region of the infalling gas can no longer efficiently radiate away the thermal energy that it gains as the gas is compressed. At this point, the gas destined to become a star is called a **protostar**; it consists of a central hydrostatic core surrounded by infalling gas. The hydrodynamic stage ends when gas no longer falls onto the central core (usually because the surrounding gas is blown away by the protostar's radiation pressure). During the next **hydrostatic** stage of star formation, the gaseous ball destined to become a star is called a **pre-main sequence** (PMS) star. A PMS star is in a state of nearly perfect hydrostatic equilibrium, only very gradually shrinking on the Kelvin–Helmholtz time (Equation 2.21) as it radiates away its gravitational potential energy. The hydrostatic stage ends when the center of the PMS star becomes hot enough for hydrogen burning to be ignited.

Many questions must be addressed in order to understand the complete process of star formation. First, we must ask why stars form in the first place. Even the densest clumps of cold gas in the interstellar medium have a characteristic density of only $\sim 10^6$ particles per cubic centimeter; at this density, $1\,M_\odot$ of gas will have a radius $r \sim 0.02\,\mathrm{pc} \sim 10^6\,R_\odot$. What inspires such a region to collapse by a factor of one million in radius, after being stable for long periods of time? For the early hydrodynamic phase of star formation, we must ask what regulates the timescale for the infall of gas, and what sets the characteristic mass of protostars. (Why don't protostars have the mass of a brick, or of a supermassive black hole?) For the later hydrostatic phase of star formation, we must similarly ask what sets the time that elapses before the PMS star starts its hydrogen burning phase as a full-fledged star. We also want to know how protostars and PMS stars evolve in luminosity and effective temperature; the path from cold interstellar gas to a hot self-gravitating fusion reactor is not necessarily a straightforward one.

7.1 Molecular Clouds and Instability

Within our galaxy (and other nearby galaxies) star formation is seen to occur in the densest regions of the interstellar medium: the molecular clouds. A typical molecular cloud has a temperature $T \sim 15\,\mathrm{K}$; its mean density, conventionally expressed as its number density of hydrogen nuclei, is $n_\mathrm{H} \sim 100\,\mathrm{cm}^{-3}$. Figure 7.1 shows part of a giant molecular cloud known as the Taurus Molecular Cloud. Giant molecular clouds have masses in the range $10^4\,M_\odot$–$10^7\,M_\odot$; the Taurus Molecular Cloud, with $M \approx 1.5 \times 10^4\,M_\odot$, is at the low end of this range. The diameters of giant molecular clouds range from tens to hundreds of parsecs. As you can deduce from looking at Figure 7.1, molecular clouds are inhomogeneous; their densest regions, known as **molecular cloud cores**, have temperatures as low as $T \sim 8\,\mathrm{K}$ and densities as high as $n_\mathrm{H} \sim 10^6\,\mathrm{cm}^{-3}$.

Molecular clouds get their name from the fact that most of their hydrogen is in the form of H_2. A single H_2 molecule can be photodissociated by an ultraviolet photon. In most of the interstellar medium, there is enough ultraviolet light from hot stars to ensure that hydrogen is primarily atomic rather than molecular. However, in a dense molecular cloud (such as the Taurus Molecular Cloud), ultraviolet photons from outside the cloud are all absorbed while photodissociating H_2 molecules in a relatively thin surface layer. The H_2 molecules deeper within the cloud are thus shielded from photodissociation.

In the UV-free interior of molecular clouds, the hydrogen is almost all in its molecular form. This means that 1000 H nuclei will produce 500 H_2 molecules;

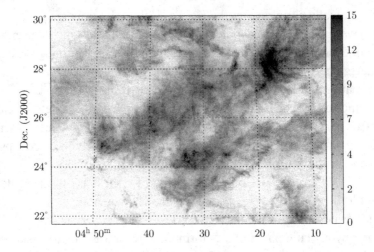

Figure 7.1 Image of ^{12}CO emission from a portion of the Taurus Molecular Cloud ($d \sim$ 140 pc); area shown corresponds to $\sim$28 × 20 pc at that distance. Shaded bar on right shows the integrated intensity scale in K km s^{-1}. [Narayanan *et al.* 2008]

at solar abundance, these molecules will be accompanied by 99 He atoms, and only $\sim$1 other gas particle, on average, drawn from molecules like CO, OH, and CH, and atoms like Ne and Ar. Ignoring the small contribution of metals, the mean molecular mass within a molecular cloud is $\bar{\mu} = 2.33$, the number density of gas particles is $n = 0.60 n_H$, and the mass density is

$$\rho = m_{AMU} \bar{\mu} n = 2.3 \times 10^{-22} \, \text{g cm}^{-3} \left(\frac{n_H}{100 \, \text{cm}^{-3}} \right). \tag{7.1}$$

This implies a freefall time for the molecular cloud (Equation 2.7) of

$$t_{ff} = \left(\frac{3\pi}{32 G \rho} \right)^{1/2} = 4.4 \, \text{Myr} \left(\frac{n_H}{100 \, \text{cm}^{-3}} \right)^{-1/2}. \tag{7.2}$$

In a dense molecular cloud core, the freefall time can be as short as $t_{ff} \sim$ 0.1 Myr. Once a region starts a freefall collapse, its increasing density makes the freefall time steadily decrease. Thus, freefall is an accelerating process, and tends to amplify existing density contrasts within a molecular cloud. Since the freefall time for a molecular cloud is much shorter than the age of our galaxy, this suggests that molecular clouds are ordinarily supported by some combination of ideal gas pressure, rotation, turbulent pressure, and magnetic pressure. However, this state of support is only marginally stable; shock waves from supernovae and other disturbances can trigger gravitational collapse to form new stars.

Molecular clouds are usually opaque at visible wavelengths, due to the presence of interstellar dust. On average, about half the metal atoms in a molecular cloud condense into solid dust grains, with radius $a \sim 0.1 \, \mu$m or less. Carbon forms graphite grains with bulk density $\rho_{\text{grain}} \approx 2.2\,\text{g cm}^{-3}$, while O, Si, Mg, and Fe form silicate grains with bulk density $\rho_{\text{grain}} \approx 3.7\,\text{g cm}^{-3}$. If a cloud has a mass density ρ, with a fraction $f_{\text{dust}} \sim 0.01$ of that mass being condensed into solid grains, then the number density of dust grains is[1]

$$n_{\text{dust}} = \frac{3\rho f_{\text{dust}}}{4\pi a^3 \rho_{\text{grain}}} \approx 1.8 \times 10^{-10}\,\text{cm}^{-3} \left(\frac{n_{\text{H}}}{100\,\text{cm}^{-3}}\right) \left(\frac{f_{\text{dust}}}{0.01}\right), \qquad (7.3)$$

assuming spherical dust grains with $a \approx 0.1\,\mu\text{m}$ and bulk density $\rho_{\text{grain}} \approx 3\,\text{g cm}^{-3}$.

Each dust grain has a cross section for photon absorption of

$$\sigma_{\text{abs}} = Q_{\text{abs}}\pi a^2, \qquad (7.4)$$

where the frequency-dependent efficiency coefficient Q_{abs} accounts for the fact that the absorption cross section is not usually identical with the geometric cross section. In particular, for wavelengths $\lambda \gg 2\pi a$, the efficiency coefficient is $Q_{\text{abs}} \ll 1$. In the visible range of the spectrum, $0.40\,\mu\text{m} < \lambda < 0.75\,\mu\text{m}$, a silicate dust grain with $a \sim 0.1\,\mu\text{m}$ will have $Q_{\text{abs}} \sim 0.2$. The mean free path for light passing through a dusty molecular cloud is then

$$\lambda_{\text{mfp}} = \frac{1}{n_{\text{dust}}\sigma_{\text{abs}}} = \frac{4\rho_{\text{grain}}a}{3Q_{\text{abs}}} \frac{1}{f_{\text{dust}}\rho}. \qquad (7.5)$$

For silicate dust grains with $a \approx 0.1\,\mu\text{m}$, this can be written as

$$\lambda_{\text{mfp}} \approx 35\,\text{pc} \left(\frac{0.2}{Q_{\text{abs}}}\right) \left(\frac{0.01}{f_{\text{dust}}}\right) \left(\frac{100\,\text{cm}^{-3}}{n_{\text{H}}}\right). \qquad (7.6)$$

For a molecular cloud core with $n_{\text{H}} \sim 10^6\,\text{cm}^{-3}$, the mean free path drops to $\lambda_{\text{mfp}} \sim 700\,\text{au}$. The densest portions of molecular clouds are therefore opaque at visible wavelengths, and appear as "dark nebulae" in projection against the background star field. Although the early steps of star formation are concealed from our eyes by a veil of dust, we can still use our knowledge of physics to make general statements about how stars form.

The structure of a main sequence star, as we saw in Chapter 2, can be treated as a hydrostatic problem, yielding time-independent equations of stellar structure. However, the formation of a protostar from a molecular cloud must be treated as a hydrodynamic problem. That is, the density $\rho(\vec{x}, t)$, pressure $P(\vec{x}, t)$, and other properties of the gas have an explicit dependence on time, while the bulk velocity of the gas, $\vec{v}(\vec{x}, t)$, can be non-zero.

In following the evolution of interstellar gas that will eventually become a star, it is useful to take the viewpoint of a **Lagrangian** observer. To do this, identify

---

[1] According to ISO 14644 guidelines, a class N cleanroom requires $n_{\text{dust}} \leq 10^N\,\text{m}^{-3}$ for dust grains with $a \geq 0.1\,\mu\text{m}$. Ordinary office space is usually class 9, the high-bay cleanroom at NASA Goddard is class 7, and a molecular cloud core with gas density $n_{\text{H}} \sim 5 \times 10^5\,\text{cm}^{-3}$ is class 0. Interstellar space, by terrestrial standards, is a clean but ill-lighted place.

an individual parcel of gas small enough to be approximated as having uniform ρ, P, and $\vec{v}$. An observer who follows the flow of the gas, and hence is located at the gas parcel's position at every moment, is a Lagrangian observer. The rate of change of a gas property Q, as measured by a Lagrangian observer, is

$$\frac{DQ}{Dt} \equiv \frac{\partial Q}{\partial t} + \vec{v} \cdot \vec{\nabla} Q. \tag{7.7}$$

If the gas parcel is not in motion ($\vec{v} = 0$), then the Lagrangian derivative DQ/Dt reduces to the usual partial derivative $\partial Q/\partial t$ with respect to time.

A gas parcel, as it flows along, must satisfy the basic equations of hydrodynamics. To start, there is the **mass continuity equation**. In Lagrangian form, this is

$$\frac{D\rho}{Dt} = -\rho \vec{\nabla} \cdot \vec{v}. \tag{7.8}$$

The mass continuity equation states that, thanks to conservation of mass, the density of the gas parcel can increase only if the bulk velocity converges at its location ($\vec{\nabla} \cdot \vec{v} < 0$), and can decrease only if the velocity diverges ($\vec{\nabla} \cdot \vec{v} > 0$).

The next basic equation of hydrodynamics is the **momentum equation**. In Lagrangian form, this is

$$\frac{D\vec{v}}{Dt} = -\frac{1}{\rho}\vec{\nabla}P + \vec{g}, \tag{7.9}$$

where P is the total pressure acting on the gas parcel, and $\vec{g}$ is the gravitational acceleration. The momentum equation states that, thanks to conservation of momentum, the bulk velocity of the gas parcel can change only as a result of a pressure gradient or a gravitational force accelerating the gas. (We are assuming an ideal gas, with no viscous forces.) In the case of a static, spherically symmetric system, Equation 7.9 reduces to the hydrostatic equation of stellar structure (Equation 2.12).

A glance at the complex structure of a molecular cloud (Figure 7.1) reveals that spherical symmetry is not necessarily a useful approximation as stars start to form. Filamentary structures hint that magnetic fields can play a role in molecular clouds, while abrupt changes in density point to the presence of shocks. Although star formation must therefore be a complicated non-linear process when viewed in detail, we can still gain some physical insight by looking at the earliest stages of collapse, when the gas destined to become a star is still only slightly denser than its surroundings. Consider a region of gas that can be approximated as having uniform density ρ_0 and pressure P_0, with velocity $v_0 = 0$. Into this uniform gas, we introduce small perturbations of the form

$$\rho(\vec{x}, t) = \rho_0 + \rho_1(\vec{x}, t), \tag{7.10}$$

$$P(\vec{x}, t) = P_0 + P_1(\vec{x}, t), \tag{7.11}$$

$$\vec{v}(\vec{x}, t) = \vec{v}_1(\vec{x}, t), \tag{7.12}$$

where $|\rho_1| \ll \rho_0$ and $|P_1| \ll P_0$. We now substitute these values for ρ, P, and $\vec{v}$ into the mass continuity equation and momentum equation (Equations 7.8 and 7.9), keeping only those terms that are linear in the perturbed quantities ρ_1, P_1, and $\vec{v}_1$. The linearized mass continuity equation then becomes

$$\frac{\partial \rho_1}{\partial t} = -\rho_0 \vec{\nabla} \cdot \vec{v}_1. \tag{7.13}$$

The linearized momentum equation becomes

$$\frac{\partial \vec{v}_1}{\partial t} = -\frac{1}{\rho_0} \vec{\nabla} P_1 + \vec{g}_1, \tag{7.14}$$

where the gravitational acceleration $\vec{g}_1$ resulting from the density perturbation ρ_1 is given by the Poisson equation,

$$\vec{\nabla} \cdot \vec{g}_1 = -4\pi G \rho_1. \tag{7.15}$$

By converting the mass continuity and momentum equations into linearized perturbation equations, we are restricting ourselves to the regime where density perturbations are still small in amplitude. What do we get in return for this sacrifice? The payoff starts when we take the time derivative of Equation 7.13 and the divergence of Equation 7.14, then combine the two equations to yield

$$\frac{\partial^2 \rho_1}{\partial t^2} = \nabla^2 P_1 - \rho_0 \vec{\nabla} \cdot \vec{g}_1. \tag{7.16}$$

Using the Poisson equation in the form of Equation 7.15, this can be written as

$$\frac{\partial^2 \rho_1}{\partial t^2} - \nabla^2 P_1 = -4\pi G \rho_0 \rho_1. \tag{7.17}$$

We expect low-amplitude fluctuations in density and pressure to be nearly isothermal in molecular clouds. This is because the cloud temperature $T \sim 15\,\mathrm{K}$ is set by a balance between heating by cosmic rays (which can penetrate the dark interior of the cloud) and cooling by emission from dust grains and molecules. Since the cooling rate increases steeply with increasing temperature, a small fluctuation in temperature produces a change in the cooling rate that rapidly restores the equilibrium temperature T_0. Assuming a constant temperature T_0 and a pressure $P = nkT_0 \propto \rho$, we can rewrite Equation 7.17 as

$$\frac{\partial^2 \rho_1}{\partial t^2} - \left(\frac{dP}{d\rho}\right) \nabla^2 \rho_1 = -4\pi G \rho_0 \rho_1. \tag{7.18}$$

Equation 7.18 is a wave equation, describing traveling fluctuations in density and pressure, otherwise known as "sound waves," moving at the **isothermal sound speed**,

$$c_s = \left(\frac{dP}{d\rho}\right)^{1/2} = \left(\frac{kT_0}{m_{\mathrm{AMU}}\bar{\mu}}\right)^{1/2} \tag{7.19}$$

$$= 0.231\,\mathrm{km\,s}^{-1} \left(\frac{2.33}{\bar{\mu}}\right)^{1/2} \left(\frac{T_0}{15\,\mathrm{K}}\right)^{1/2}, \tag{7.20}$$

scaling to properties expected in a molecular cloud with solar abundance.

The wave equation, written in terms of the sound speed c_s, is

$$\frac{\partial^2 \rho_1}{\partial t^2} - c_s^2 \nabla^2 \rho_1 = -4\pi G \rho_0 \rho_1. \tag{7.21}$$

This equation is complicated by the presence of the term on the right-hand side, representing the self-gravity of the density perturbations. To analyze the effect of gravity mathematically, consider a sine wave perturbation:

$$\rho_1 \propto \exp\left[i(\omega t - kx)\right], \tag{7.22}$$

where the wavenumber k is related to the wavelength λ by the relation $k = 2\pi/\lambda$. Substituting Equation 7.22 into the wave equation, we find the **dispersion relation** for the wave:

$$\omega^2 = k^2 c_s^2 - 4\pi G \rho_0. \tag{7.23}$$

The angular frequency ω is real, and thus the sound wave propagates stably, when $k > k_J$, where

$$k_J \equiv 2\sqrt{\pi}\frac{(G\rho_0)^{1/2}}{c_s}. \tag{7.24}$$

This wavenumber corresponds to a length scale known as the **Jeans length**,[2]

$$\lambda_J \equiv \frac{2\pi}{k_J} = \sqrt{\pi}\frac{c_s}{(G\rho_0)^{1/2}}. \tag{7.25}$$

Perturbations with a wavelength longer than the Jeans length will grow exponentially in amplitude. For molecular gas of solar abundance, the Jeans length is

$$\lambda_J = 3.4\,\mathrm{pc}\left(\frac{T_0}{15\,\mathrm{K}}\right)^{1/2}\left(\frac{n_{\mathrm{H},0}}{100\,\mathrm{cm}^{-3}}\right)^{-1/2}, \tag{7.26}$$

where $n_{\mathrm{H},0}$ is the mean number density of hydrogen nuclei. The Jeans mass M_J is usually defined as the mass of a sphere whose diameter is equal to the Jeans length:[3]

$$M_J = \frac{\pi}{6}\rho_0 \lambda_J^3 = 69\,\mathrm{M}_\odot\left(\frac{T_0}{15\,\mathrm{K}}\right)^{3/2}\left(\frac{n_{\mathrm{H},0}}{100\,\mathrm{cm}^{-3}}\right)^{-1/2}. \tag{7.27}$$

In a dense molecular cloud core, with $T \sim 8\,\mathrm{K}$ and $n_{\mathrm{H}} \sim 10^6\,\mathrm{cm}^{-3}$, the Jeans mass is as small as $M_J \sim 0.3\,\mathrm{M}_\odot$. However, throughout most of a molecular cloud, the Jeans mass is intermediate between the mass of an individual star and the mass of a young open cluster of stars. (The Pleiades, for example, has an age $t \approx 125\,\mathrm{Myr}$ and mass $M \sim 800\,\mathrm{M}_\odot$.)

To clarify the physical meaning of the Jeans length and Jeans mass, consider a gas sphere of mean density ρ_0. If the inward gravitational force dominates over

[2] The Jeans length is named after James Jeans, who studied the instability of gaseous clouds in 1902.
[3] There exists an alternate definition in which M_J is equal to the mass of a sphere whose *radius* is equal to λ_J. Beware of hidden factors of eight!

the outward force from gas pressure, the sphere will collapse in a time comparable to the freefall time, $t_{\rm ff} \approx 0.54(G\rho_0)^{-1/2}$. However, the sphere will be stabilized against collapse if it can build up a pressure gradient large enough to establish hydrostatic equilibrium. The time it takes a pressure gradient to steepen is the time it takes a pressure disturbance (also known as a sound wave) to travel from the surface of the cloud to the center, $t_s = R/c_s$, where R is the radius of the sphere. Comparing the timescales involved, the sphere will be stable against collapse when $t_s < t_{\rm ff}$, or

$$R < \frac{0.54 c_s}{(G\rho_0)^{1/2}} \sim 0.3\lambda_{\rm J}, \tag{7.28}$$

referring back to the definition of the Jeans length $\lambda_{\rm J}$ in Equation 7.25. Thus, regions with a diameter comparable to or larger than the Jeans length are unstable to collapse; this is because the speed of sound is too slow to complete the job of restoring hydrostatic equilibrium.

7.2 Isothermal Collapse and Fragmentation

A typical region of a molecular cloud, with density $n_{\rm H} \sim 100\,{\rm cm}^{-3}$ and temperature $T \sim 15\,{\rm K}$, has a Jeans mass $M_{\rm J} \sim 70\,{\rm M}_\odot$, from Equation 7.27. This is larger than the median mass $M_\star \sim 0.2\,{\rm M}_\odot$ of stars in our galaxy. In addition, it is seen that stars form in open clusters or in looser associations of stars. This indicates that a process of fragmentation must break up regions larger than a Jeans mass into objects comparable in mass to stars.

To see how this process of fragmentation works, start with a region of a molecular cloud whose mass M is slightly larger than the local Jeans mass. The total number of gas particles in this region is

$$N = \frac{M}{m_{\rm AMU}\bar{\mu}} = 5.1 \times 10^{58} \left(\frac{2.33}{\bar{\mu}}\right) \left(\frac{M}{100\,{\rm M}_\odot}\right), \tag{7.29}$$

where we have scaled to a mass somewhat greater than our typical Jeans mass of $70\,{\rm M}_\odot$. In this region, the total thermal energy is

$$E_{\rm th} = \frac{3}{2}NkT = 1.6 \times 10^{44}\,{\rm erg} \left(\frac{2.33}{\bar{\mu}}\right) \left(\frac{M}{100\,{\rm M}_\odot}\right) \left(\frac{T}{15\,{\rm K}}\right). \tag{7.30}$$

Since the region is larger than its Jeans mass, it starts to collapse. If the collapse were adiabatic, with no energy flowing out of (or into) the region, the first law of thermodynamics tells us that the gas would be heated at the rate

$$\frac{dE_{\rm th}}{dt} = -P\frac{dV}{dt} = -\frac{NkT}{V}\frac{dV}{dt}, \tag{7.31}$$

where V is the volume of the collapsing region. If we approximate the region as being roughly spherical, with radius R, the heating rate becomes

$$\frac{dE_{\text{th}}}{dt} = -\frac{3NkT}{R}\frac{dR}{dt} = -\frac{2E_{\text{th}}}{R}\frac{dR}{dt}. \tag{7.32}$$

For the region to remain isothermal at temperature T as it collapses, it must radiate energy at the rate

$$L_{\text{iso}} = -\frac{dE_{\text{th}}}{dt} = \frac{2E_{\text{th}}}{R}\frac{dR}{dt}. \tag{7.33}$$

If the region remains cool instead of undergoing adiabatic heating, it collapses at nearly the freefall rate, with $dR/dt \sim R/t_{\text{ff}}$. This leads to a luminosity for isothermal collapse of

$$L_{\text{iso}} \approx \frac{2E_{\text{th}}}{t_{\text{ff}}} \tag{7.34}$$

$$\approx 6.0 \times 10^{-4}\,L_{\odot}\left(\frac{M}{100\,M_{\odot}}\right)\left(\frac{T}{15\,\text{K}}\right)\left(\frac{n_{\text{H},0}}{100\,\text{cm}^{-3}}\right)^{1/2},$$

assuming solar abundance. As long as the actual luminosity of a collapsing region is comparable to or greater than this value, it will collapse on a freefall time.

A molecular cloud can radiate energy by making use of the dust grains that are present at number density n_{dust}. (Molecular gas can also radiate energy through line emission from CO and other molecules, but usually the thermal emission from dust is the dominant source of luminosity.) The mean free path of a photon before being absorbed by a dust grain is (Equation 7.5)

$$\lambda_{\text{mfp}} \propto n_{\text{dust}}^{-1} \propto n_{\text{H}}^{-1}. \tag{7.35}$$

Thus, as a region of mass M collapses, its optical depth τ from dust absorption is

$$\tau \sim \frac{R}{\lambda_{\text{mfp}}} \propto \frac{(M/n_{\text{H}})^{1/3}}{n_{\text{H}}^{-1}} \propto M^{1/3}n_{\text{H}}^{2/3}. \tag{7.36}$$

As the density of the collapsing region increases, it becomes more opaque. The resulting luminosity from the dust's thermal emission (the *actual* luminosity as opposed to the theoretically calculated isothermal luminosity) can be written as

$$L_{\text{act}} = 4\pi R^2 \sigma_{\text{SB}} T^4 f_{\text{op}}, \tag{7.37}$$

where $f_{\text{op}} \leq 1$ is an efficiency factor that accounts for the fact that the collapsing region is not a good blackbody until $\tau \gg 1$ at the far-infrared wavelength of emission. In terms of the mass M and density n_{H}, the luminosity of the collapsing region is

$$L_{\text{act}} = 3.27 \times 10^5\,L_{\odot}\left(\frac{M}{100\,M_{\odot}}\right)^{2/3}\left(\frac{n_{\text{H}}}{100\,\text{cm}^{-3}}\right)^{-2/3}\left(\frac{T}{15\,\text{K}}\right)^4 f_{\text{op}}, \tag{7.38}$$

assuming solar abundance.

The collapsing gas stays cool as long as its actual luminosity, from Equation 7.38, is comparable to or greater than the isothermal luminosity, from Equation 7.35. Suppose that a collapsing region of mass M starts with a density $n_H = n_{init}$. As the collapse begins, any value of the efficiency factor greater than the extremely modest value

$$f_{op} \approx 1.85 \times 10^{-9} \left(\frac{M}{100\,M_\odot} \right)^{1/3} \left(\frac{n_{init}}{100\,cm^{-3}} \right)^{7/6} \left(\frac{T}{15\,K} \right)^{-3} \tag{7.39}$$

will result in $L_{act} > L_{iso}$, with a resulting freefall collapse. However, if the collapse is nearly isothermal, with $T \approx$ constant, then the Jeans mass will be, from Equation 7.27,

$$M_J \propto T^{3/2} n_H^{-1/2} \propto n_H^{-1/2}. \tag{7.40}$$

Consider, then, what happens when the density of an isothermally collapsing region rises from n_{init} to $4n_{init}$. The Jeans mass, from Equation 7.40, then decreases by a factor of two; the collapsing region, originally just barely larger than the Jeans mass, becomes unstable to splitting into a pair of fragments, each with mass $M_{frag} \sim M/2$. After N rounds of fragmentation, we expect to have 2^N fragments, each of mass $M_{frag} \sim 2^{-N} M$ and density $n_{frag} \sim 2^{2N} n_{init}$. These fragments can continue to cool only if their efficiency factor is greater than

$$f_{op} \approx 1.85 \times 10^{-9} \left(\frac{2^{-N} M}{100\,M_\odot} \right)^{1/3} \left(\frac{2^{2N} n_{init}}{100\,cm^{-3}} \right)^{7/6} \tag{7.41}$$

$$\approx 1.85 \times 10^{-9} 2^{2N} \left(\frac{M}{100\,M_\odot} \right)^{1/3} \left(\frac{n_{init}}{100\,cm^{-3}} \right)^{7/6}, \tag{7.42}$$

assuming collapse and fragmentation occurs at the temperature $T \approx 15\,K$ typical of molecular clouds. Although it is true that the efficiency f_{op} tends to increase as fragments become denser, the efficiency needed for effective cooling reaches $f_{op} = 1$ when the number of fragmentations (from Equation 7.42) is

$$N_{frag} \approx 14.5 - 0.55 \log \left(\frac{M}{100\,M_\odot} \right) - 1.94 \log \left(\frac{n_{init}}{100\,cm^{-3}} \right). \tag{7.43}$$

Thus, after 14 rounds of fragmentation, given plausible initial conditions, the 15th round would require an unphysically high value for the efficiency ($f_{op} > 1$). Starting with an initial mass $M \approx 100\,M_\odot$, this implies a minimum fragment mass $M_{frag} \sim 2^{-14}(100\,M_\odot) \sim 0.006\,M_\odot$. These small fragments, about 8% of the mass of the smallest stars, represent the end of isothermal freefall collapse. (Although the densest regions of molecular clouds have a smaller Jeans mass and hence can have a smaller initial mass M, they also undergo fewer rounds of fragmentation. Thus, the value of M_{frag} does not depend very strongly on the initial density, and will generally be smaller than a star's mass.)

Given our assumed initial conditions, we expect a typical fragment mass $M_{\text{frag}} \sim 0.006\,M_\odot$ at the end of fragmentation. The mean density of each fragment is $\rho_{\text{frag}} \approx 2^{28}\rho_{\text{init}} \sim 6 \times 10^{-14}\,\text{g cm}^{-3}$, implying a freefall time $t_{\text{ff}} \sim 200\,\text{yr}$. This density, still much smaller than the mean density of a main sequence star, implies a radius of $R_{\text{frag}} \approx 2^{-14}R_{\text{init}} \sim 5000\,R_\odot$. The fragment still has $T_{\text{frag}} \approx T_{\text{init}} \approx 15\,\text{K}$; since it is now close to a blackbody, with $f_{\text{op}} \approx 1$, its luminosity is $L_{\text{frag}} \propto R_{\text{frag}}^2 T_{\text{frag}}^4 \sim 0.001\,L_\odot$. Since the fragment is smaller in mass and much larger in radius than a star, its gravitational potential energy is small, with (Equation 2.20)

$$W_{\text{frag}} \approx -\frac{GM_{\text{frag}}^2}{R_{\text{frag}}} \sim -3 \times 10^{40}\,\text{erg}, \tag{7.44}$$

about 10^{-8} times the gravitational potential energy of the massive, compact Sun. At this stage, the Kelvin–Helmholtz time of the fragment is thus comparable to its freefall time, with $t_{\text{KH}} = -W_{\text{frag}}/L_{\text{frag}} \sim 200\,\text{yr}$.

As the fragment shrinks on its Kelvin–Helmholtz time, its central temperature increases. At $T \sim 1200\,\text{K}$, silicate dust grains sublime, while at $T \sim 2000\,\text{K}$, the more refractory graphite grains sublime. (Vaporizing the dust doesn't make the fragment transparent again; it is now dense enough for broad molecular absorption bands to render it opaque.) The central temperature continues to rise. As $T \rightarrow 30\,000\,\text{K}$, the mean particle energy approaches the dissociation energy of an H_2 molecule, and hydrogen switches from being molecular to being atomic. As $T \rightarrow 10^5\,\text{K}$, the mean particle energy approaches the ionization energy of a hydrogen atom, and hydrogen becomes ionized. As $T \rightarrow 4 \times 10^5\,\text{K}$, the mean particle energy approaches the second ionization energy of helium, and the interior becomes almost entirely ionized. As the fragment is compressed, its freefall time decreases, while its Kelvin–Helmholtz time tends to increase. When $t_{\text{KH}} \gg t_{\text{ff}}$, it makes sense to think of the fragment as a gaseous sphere that is nearly in hydrostatic equilibrium, but which is still accreting matter from its surroundings; at this stage, we call the fragment a **protostar**.

As an aside, note that our analysis of isothermal collapse assumed that fragments keep cool by emitting thermal radiation from dust grains, with assistance from molecular emission. A molecular cloud with no metals would have no dust, and would have H_2 as its only type of molecule. Since H_2, thanks to its lack of a permanent electric dipole, is inefficient at emitting radiation, a molecular cloud without metals would not cool effectively, and thus would not undergo multiple rounds of fragmentation. It is plausible, therefore, that the first stars in the universe, made from primordial gas with $Z_p \sim 3 \times 10^{-9}$, were all highly massive, with a mass comparable to the Jeans mass of the gas from which they formed.

7.3 Protostars

Consider a protostar whose nearly hydrostatic central region, called the **protostellar core**, has a mass $M_{\text{pro}}(t)$ and radius $R_{\text{pro}}(t)$. This mass increases with time as the protostellar core accretes matter from its infalling envelope (the surrounding gas that is destined to fall onto the protostellar core). The radius R_{pro} of the protostellar core will increase or decrease as necessary to maintain hydrostatic equilibrium. The rate $\dot{M}$ at which the protostellar core accretes mass from the infalling envelope is not constant with time, nor do we expect the infall of matter to be spherically symmetric. The infalling gas carries angular momentum as well as mass; particularly in the later stages of accretion, an accretion disk will generally form around the protostellar core.

Although the accretion of matter is a complex, time-dependent, anisotropic process, we can make some general statements about the growth of the protostellar core by considering the time-averaged value of $\dot{M}$. Observations of star-forming regions imply that all stars, regardless of their final mass $M_\star$, take about the same time to grow by accretion; this time is typically $t_{\text{acc}} \sim 0.1\,\text{Myr}$. This implies an average accretion rate of

$$\dot{M} = 10\,\text{M}_\odot\,\text{Myr}^{-1} \left(\frac{t_{\text{acc}}}{0.1\,\text{Myr}}\right)^{-1} \left(\frac{M_\star}{1\,\text{M}_\odot}\right). \tag{7.45}$$

Thus, a protostar that accretes at a rate $\dot{M} = 1\,\text{M}_\odot\,\text{Myr}^{-1}$, equivalent to delicately nibbling a Jupiter mass every millennium, will become a low-mass star with $M_\star \sim 0.1\,\text{M}_\odot$. By contrast, a protostar that gobbles a Jupiter per year will grow to an ultramassive star with $M_\star \sim 100\,\text{M}_\odot$.[4]

Accretion at a rate $\dot{M}$ yields a protostellar luminosity of

$$L_{\text{acc}} = \xi \frac{GM_{\text{pro}}\dot{M}}{R_{\text{pro}}}, \tag{7.46}$$

where the dimensionless factor $\xi \leq 1$ is the efficiency with which the gravitational potential energy of the infalling matter is transferred to the protostellar core. In practice, the efficiency ξ can be considerably smaller than unity. For example, in spherically symmetric inflow, much of the energy is radiated away in accretion shocks far above the protostar's photosphere. Numerical simulations of accretion give $\xi \approx 0.1$ as the upper bound for the efficiency; accretion with ξ near this maximum value is called **hot accretion**. However, lower accretion rates can result in the infalling material shedding almost all its energy before it is incorporated into the protostar; in this **cold accretion** scenario, the efficiency can be as low as $\xi \sim 0.001$. Numerically, the luminosity expected for a protostellar core is

$$L_{\text{acc}} = 31.4\,\text{L}_\odot \left(\frac{\xi\dot{M}}{1\,\text{M}_\odot\,\text{Myr}^{-1}}\right) \left(\frac{M_{\text{pro}}}{1\,\text{M}_\odot}\right) \left(\frac{R_{\text{pro}}}{1\,\text{R}_\odot}\right)^{-1}, \tag{7.47}$$

scaling to hot accretion at a rate that will produce a $1\,\text{M}_\odot$ star over a time $t_{\text{acc}} \approx 0.1\,\text{Myr}$.

---

[4] Since the planet Jupiter has a mass $M_{\text{Jup}} = 0.954 \times 10^{-3}\,\text{M}_\odot$, "Jupiter mass" is used by astronomers as a convenient near-synonym for "milli-solar mass."

Observationally determining the actual luminosity of a protostar is hampered by the fact that it is hidden at visible and near-infrared wavelengths by the dusty infalling envelope. However, after $\sim 0.1\,\mathrm{Myr}$, radiation pressure from the luminous protostar blows away most of the surrounding gas and dust. The newly unveiled object is a gas sphere (or spheroid, if it rotates rapidly), in a state very close to hydrostatic equilibrium. It differs from a main sequence star, however, in the fact that it is not yet powered by hydrogen burning in its core. Thus, the gas sphere is now called a **pre-main sequence star**, or PMS star. In contrast to a protostar (powered by accretion) or a main sequence star (powered by fusion), a PMS star is powered by the Kelvin–Helmholtz mechanism; its gravitational potential energy is converted to thermal energy, which is then radiated away into space.

7.4 Pre-Main Sequence Stars

Just as we could build models of main sequence stars without knowing the details of the star formation process, we can build models of PMS stars without knowing the details of the earlier protostellar phase. Consider a pre-main sequence star with a mass $M_\star$ that is nearly constant; in other words, the PMS star is no longer accreting matter at a significant rate, and it does not have a strong wind.[5] The timescale for a PMS star to gradually shrink is its Kelvin–Helmholtz time (Equation 2.21),

$$ t_{\mathrm{KH}} \sim \frac{GM_\star^2}{R_\star L_\star} \sim 30\,\mathrm{Myr} \left(\frac{M_\star}{1\,\mathrm{M_\odot}} \right)^2 \left(\frac{R_\star}{1\,\mathrm{R_\odot}} \right)^{-1} \left(\frac{L_\star}{1\,\mathrm{L_\odot}} \right)^{-1}. \tag{7.48} $$

Since the Kelvin–Helmholtz time is now much longer than the freefall time, the PMS star undergoes **quasi-static** contraction; at any given instant, the static equations of stellar structure give an excellent approximation to the interior structure of a pre-main sequence star.

In Equation 7.48, we simply scale the luminosity $L_\star$ to units of the solar luminosity, since we don't know, at this point, what an appropriate luminosity would be for a PMS star. We do know that at the photospheric radius $R_\star$, the luminosity must obey the relation

$$ L_\star = 4\pi R_\star^2 \sigma_{\mathrm{SB}} T_{\mathrm{eff}}^4, \tag{7.49} $$

where T_{eff} is the effective temperature of the photosphere. However, we still need to determine how the luminosity and effective temperature evolve with time as the PMS star shrinks. For instance, if the luminosity $L_\star$ remained constant with time, then the effective temperature would have to increase as $T_{\mathrm{eff}} \propto R_\star^{-1/2}$, and the PMS star would move from right to left on a Hertzsprung–Russell diagram.

---

[5] We use the same subscript ($\star$) for a PMS star as for a main sequence star; this is because most properties of a hydrostatic gas ball are indifferent to its energy source. Thus, many results from this section apply to fusion-powered stars as well as to PMS stars powered by gravitational potential energy.

By contrast, if the effective temperature remained constant with time, then the luminosity would have to decrease as $L_\star \propto R_\star^2$, and the PMS star would plummet straight downward on the H–R diagram.

What is the actual trajectory we expect for an object of constant mass $M_\star$ on the Hertzsprung–Russell diagram? (For simplicity, we will use the theorists' H–R diagram, which plots $\log L_\star$ versus $\log T_{\text{eff}}$.) From the definition of effective temperature (Equation 7.49), we can write

$$\log L_\star = C_1 + 2 \log R_\star + 4 \log T_{\text{eff}}, \tag{7.50}$$

where $C_1 = \log(4\pi\sigma_{\text{SB}})$. However, the time evolution of $R_\star$ and $L_\star$ for a PMS star of fixed mass $M_\star$ is non-trivial to compute. In the 1950s, Louis Henyey and his collaborators gained some insight into the evolution of PMS stars by assuming that they formed a homologous family, all in hydrostatic equilibrium and all utilizing radiative energy transport in their interiors. As shown in Equation 6.50, these assumptions imply a luminosity–mass–radius relation

$$L_\star \propto M_\star^{3-\alpha+\beta} R_\star^{3\alpha-\beta} \tag{7.51}$$

when the opacity is given by a generalized Kramers' law,

$$\kappa = \kappa_0 \rho^\alpha T^{-\beta}. \tag{7.52}$$

Note that since Equation 7.51 makes no assumptions about the energy source, it applies to stars and PMS stars equally, as long as energy is transported radiatively.

The opacity in the interior of a PMS star is dominated by bound–free and free–free absorption, with $\alpha = 1$ and $\beta = 3.5$ in Kramers' law. In this case, Equation 7.51 reduces to

$$L_\star \propto M_\star^{5.5} R_\star^{-0.5}, \tag{7.53}$$

resulting in a PMS star that becomes more luminous as it contracts. Equation 7.53 can be rewritten as

$$\log R_\star = C_2 - 2 \log L_\star + 11 \log M_\star. \tag{7.54}$$

Substituting Equation 7.54 into Equation 7.50 yields the relation among $L_\star$, T_{eff}, and $M_\star$ for a gas ball with radiative energy transport in its interior:

$$\log L_\star = C_3 + 0.8 \log T_{\text{eff}} + 4.4 \log M_\star \qquad \text{[radiative]}. \tag{7.55}$$

This corresponds to a PMS star evolving up and to the left on the H–R diagram, with $L_\star \propto R_\star^{-0.5}$ increasing with time, and $T_{\text{eff}} \propto L_\star^{1.25} \propto R_\star^{-0.625}$ also increasing. A path of this kind on the Hertzsprung–Russell diagram is called a **Henyey track**.

Although Equation 7.55 gives an analytic solution for the observable evolution of PMS stars, it depends on the assumption of a homologous family of objects, all with radiative energy transport in their interiors. It was soon realized, after the pioneering work of Henyey, that this assumption was not universally correct.

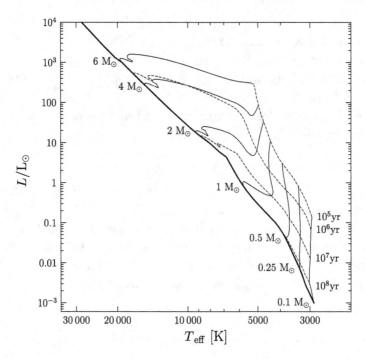

Figure 7.2 Thick solid line = zero-age main sequence. Thin solid lines = trajectories of PMS stars approaching the main sequence. Dashed lines = isochrones connecting PMS stars at the ages 0.1, 1, 10, and 100 Myr.

To illustrate the limitations, consider the results shown in Figure 7.2. This plot shows the pre-main sequence evolution of stars on the H–R diagram, using modern numerical simulations that include all sources of opacity and all means of energy transport. Notice that the highest-mass PMS stars ($M_\star \geq 4\,M_\odot$) have evolutionary tracks that are fairly well described by a Henyey track (Equation 7.55). However, the lowest-mass PMS stars ($M_\star \leq 0.5\,M_\odot$) have evolutionary tracks that are nearly vertical, with an effective temperature that is nearly constant as the PMS star shrinks and becomes less luminous. What is the fundamental physical difference between high-mass and low-mass PMS stars that reveals itself as a difference in their evolution on a Hertzsprung–Russell diagram? (And why do intermediate mass stars switch from a nearly vertical track at the early stages of their evolution to a Henyey track at later stages?)

The fundamental physical difference of low-mass PMS stars is that they transport energy by *convection* in their interiors, rather than by radiation. It is only in their atmospheres that the burden of energy transport is handed over from rising blobs of hot gas to photons. To predict the observable properties of a low-mass PMS star, we therefore need to understand the structure of both its convective interior and its radiative atmosphere. A star (or other hydrostatic gas ball) with convective energy transport is well modeled as a polytrope with $n = 1.5$, as discussed in Section 6.2. This means that a convective PMS star has the polytropic relation

$$P = P_c \left(\frac{\rho}{\rho_c}\right)^{5/3},$$

(7.56)

where the central pressure is (Equation 6.27)

$$P_c = 0.771 \frac{GM_\star^2}{R_\star^4} = 8.69 \times 10^{15} \, \text{dyn cm}^{-2} \left(\frac{M_\star}{1 \, M_\odot}\right)^2 \left(\frac{R_\star}{1 \, R_\odot}\right)^{-4}$$

(7.57)

and the central density is

$$\rho_c = 1.430 \frac{M_\star}{R_\star^3} = 8.44 \, \text{g cm}^{-3} \left(\frac{M_\star}{1 \, M_\odot}\right) \left(\frac{R_\star}{1 \, R_\odot}\right)^{-3}.$$

(7.58)

For simplicity, assume that energy transport is handed over from convection to radiation at the star's photosphere, where the pressure is (combining Equations 7.56–7.58)

$$P_{\text{pho}} = 0.425 GM_\star^{1/3} R_\star \rho_{\text{pho}}^{5/3}.$$

(7.59)

However, since the gas at the photosphere is well described by the ideal gas law, we have an additional relation that tells us the gas pressure at the photosphere:

$$P_{\text{pho}} = \frac{k}{m_{\text{AMU}}} \frac{\rho_{\text{pho}} T_{\text{eff}}}{\bar{\mu}},$$

(7.60)

where T_{eff} is the effective temperature of the photosphere.

Equation 7.59 tells us the pressure P_{pho} by looking at the convective interior just below the photosphere. Equation 7.60 also tells us P_{pho}, but by looking at the ideal gas atmosphere just above the photosphere. We can derive one more equation for P_{pho}, this time by making use of the fact that the atmosphere is not perfectly transparent. The photospheric radius $R_\star$ is defined as the radius where the optical depth looking upward is $\tau = 2/3$. Thus, we can write

$$\tau = \int_{R_\star}^{\infty} \kappa \rho \, dr = \frac{2}{3}.$$

(7.61)

This can also be written as

$$\tau = \bar{\kappa} \int_{R_\star}^{\infty} \rho \, dr = \frac{2}{3},$$

(7.62)

where $\bar{\kappa}$ is a density-weighted mean opacity. The atmosphere is in hydrostatic equilibrium, which requires

$$\frac{dP}{dr} = -\rho \frac{GM}{r^2} = -\rho g,$$

(7.63)

where g is the gravitational acceleration in the PMS star's atmosphere. However, the atmosphere has a scale height that is small compared to $R_\star$, and a mass that is tiny compared to $M_\star$. Thus, we can assume that $g = GM_\star/R_\star^2$ is constant through the atmosphere, and that the pressure at the photospheric radius is

$$P_{\text{pho}} = \frac{GM_\star}{R_\star^2} \int_{R_\star}^{\infty} \rho \, dr. \tag{7.64}$$

This means that we can write the pressure at the photosphere (Equation 7.64) in terms of the atmospheric opacity (Equation 7.62),

$$P_{\text{pho}} = \frac{2G \, M_\star}{3\bar{\kappa} \, R_\star^2}, \tag{7.65}$$

since both optical depth and pressure depend on the mass column density of gas overhead.

To compute the atmospheric opacity $\bar{\kappa}$ with extreme accuracy, we would need to know the detailed density and temperature structure of the atmosphere above the photospheric radius $R_\star$. Fortunately, we do not need extreme accuracy for an order-of-magnitude analysis. Since the density in the atmosphere drops exponentially with height, the density-weighted mean opacity $\bar{\kappa}$ is nearly equal to the opacity at $R_\star$,

$$\bar{\kappa} \approx \kappa_0 \rho_{\text{pho}}^\alpha T_{\text{eff}}^{-\beta}, \tag{7.66}$$

assuming a generalized Kramers' law opacity. In combination with Equation 7.65, this gives us another relation for the pressure at the photospheric radius:

$$P_{\text{pho}} \approx \frac{2G}{3\kappa_0} M_\star R_\star^{-2} \rho_{\text{pho}}^{-\alpha} T_{\text{eff}}^\beta. \tag{7.67}$$

We now have three separate equations for the photospheric pressure of a low-mass PMS star. The convective interior has a polytropic relation (Equation 7.59) that lets us write

$$\log P_{\text{pho}} = K_1 + \frac{1}{3} \log M_\star + \log R_\star + \frac{5}{3} \log \rho_{\text{pho}}. \tag{7.68}$$

The atmosphere has an ideal gas pressure (Equation 7.60) that lets us write

$$\log P_{\text{pho}} = K_2 + \log \rho_{\text{pho}} + \log T_{\text{eff}}. \tag{7.69}$$

Since the atmosphere is in hydrostatic equilibrium and has Kramers' law opacity, we also know (from Equation 7.67) that

$$\log P_{\text{pho}} = K_3 + \log M_\star - 2 \log R_\star - \alpha \log \rho_{\text{pho}} + \beta \log T_{\text{eff}}. \tag{7.70}$$

Finally, the definition of effective temperature tells us (Equation 7.50) that

$$\log L_\star = C_1 + 2 \log R_\star + 4 \log T_{\text{eff}}. \tag{7.71}$$

Using Equations 7.68–7.71, we can eliminate three variables. Targeting $R_\star$, ρ_{pho}, and P_{pho} for elimination, we end with a relation among $L_\star$, $M_\star$, and T_{eff}. After some mildly tedious algebra,[6] the result is found to be

$$\log L_\star = C + A_T \log T_{eff} + B_M \log M_\star \quad \text{[convective]}, \quad (7.72)$$

with

$$A_T = 2\frac{3 + 9\alpha - 2\beta}{3\alpha - 1} \quad (7.73)$$

and

$$B_M = -2\frac{3 + \alpha}{3\alpha - 1}. \quad (7.74)$$

Thus, the evolution of a convective PMS star on the Hertzsprung–Russell diagram depends crucially on the opacity of the thin atmospheric layer through which its light escapes.

In the temperature interval $3000\,\text{K} < T_{eff} < 8000\,\text{K}$, the opacity of stellar atmospheres is dominated by the H^- ion. For this opacity source the Kramers' law parameters are, from Equation 4.36, $\alpha = 0.5$ and $\beta = -9$. This large negative value of β represents a steep *decrease* in opacity as the temperature drops and fewer free electrons are available to form H^- ions. If we assume the atmosphere of the PMS star has H^- opacity, we then find $A_T = 102$ and $B_M = -14$ in Equation 7.72. The extraordinarily strong dependence of $L_\star$ on T_{eff} means that convective PMS stars with photospheric H^- opacity follow nearly vertical tracks downward in an H–R diagram. The fact that $B_M/A_T \approx -0.14$ also implies that the effective temperature of a convective PMS star is not strongly dependent on mass, with $T_{eff} \propto M_\star^{0.14}$ for PMS stars with roughly similar luminosity. The normalized relation among luminosity, mass, and temperature is

$$T_{eff} \approx 4140\,\text{K} \left(\frac{L}{1\,L_\odot}\right)^{0.01} \left(\frac{M}{1\,M_\odot}\right)^{0.14} \quad (7.75)$$

for a PMS star with a convective interior and H^- opacity in its photosphere.

When a low-mass PMS star is moving almost straight downward in a Hertzsprung–Russell diagram, it is said to be following a **Hayashi track**. Around the year 1960, Chushiro Hayashi realized that low-mass pre-main sequence stars had purely convective interiors, and thus did not follow the Henyey tracks characteristic of higher-mass PMS stars. Detailed numerical calculations, with results shown in Figure 7.2, reveal that PMS stars with $M \leq 0.5\,M_\odot$ do indeed follow Hayashi tracks. However, PMS stars with $0.5\,M_\odot < M_\star < 4\,M_\odot$ trace out a more complicated path on the H–R diagram. Initially, they follow Hayashi tracks, maintaining a nearly constant effective temperature while their radius shrinks. However, as they undergo quasi-static contraction, their central temperature rises

---

[6] The adverb modifying "tedious" may vary according to your personal tolerance for tedium.

as $T_c \propto R_\star^{-1}$, while their central density rises as $\rho_c \propto R_\star^{-3}$. This means that their central opacity, well described by Kramers' law at this stage, decreases as $\kappa \propto \rho_c T_c^{-3.5} \propto R_\star^{0.5}$. Eventually, the interior opacity for PMS stars in this mass range drops to the point where radiation becomes more effective than convection at transporting energy. When this occurs, the PMS star develops a radiative core and switches from a Hayashi track to a Henyey track.

Higher-mass PMS stars that leave their Hayashi track always move to the left (higher T_{eff}) on an H–R diagram; they are never seen to move to the right (lower T_{eff}). This is because the region rightward of a Hayashi track is a "forbidden zone"; gas spheres of mass $M_\star$ cannot be in stable hydrostatic equilibrium if they have T_{eff} lower than the Hayashi track value (Equation 7.75). To see why, consider a PMS star on its Hayashi track; it has a fully convective interior, and thus has $\nabla_{\text{act}} = \nabla_{\text{ad}}$ to an accuracy of $\sim 10^{-8}$ in its interior, as detailed in Section 4.3. If we take this PMS star and increase its T_{eff}, moving it leftward on the H–R diagram, radiative transport will take over in its interior, implying $\nabla_{\text{act}} = \nabla_{\text{rad}} < \nabla_{\text{ad}}$. But what happens when we take the PMS star and *decrease* T_{eff}, moving it rightward on the H–R diagram? In this case, instead of having $\nabla_{\text{act}} = \nabla_{\text{ad}} + 10^{-8}$, implying stable convective transport, or having $\nabla_{\text{act}} < \nabla_{\text{ad}}$, implying stable radiative transport, it has $\nabla_{\text{act}} > \nabla_{\text{ad}}$ at a level much higher than 1 part in 10^8. In this case, the temperature gradient ∇_{act} is referred to as being **superadiabatic**. From Equations 4.68 and 4.90, we find that a superadiabatic temperature gradient drives rapid convection, with blobs of hot gas rising at a speed of

$$v \sim (\nabla_{\text{act}} - \nabla_{\text{ad}})^{1/2} c_s. \tag{7.76}$$

Thus, once ∇_{act} exceeds ∇_{ad} by more than a minuscule amount, high-speed convection efficiently carries thermal energy upward toward the photosphere, decreasing the temperature gradient and bringing the PMS star back to the slower, stable convection found on the Hayashi track.

7.5 Birthline and Deuterium Burning

In the previous two sections, we presented a simple view of pre-stellar evolution: first a protostar is powered by hydrodynamic accretion onto a hydrostatic core, then a quasi-static PMS star is powered by the Kelvin–Helmholtz mechanism. This idealized, gravitationally powered scenario provides useful insights, but doesn't contain all the physics. Look back, for instance, at the model pre-main sequence stars whose tracks on the H–R diagram are illustrated in Figure 7.2. The model PMS stars are postulated to begin with a large radius and high luminosity, in the upper right of the H–R diagram. However, this initial state is somewhat arbitrary, in that it isn't found by following the hydrodynamic processes through which the earlier protostar formed and evolved.

Where does a PMS star actually make its debut on the Hertzsprung–Russell diagram, after the dusty gas that shrouded the protostellar core has been blown away? Let's start our investigation by going back to the protostar phase, in which a central hydrostatic core of mass M_{pro} and radius R_{pro} has an accretion rate $\dot{M}$. Assuming hot accretion with $\xi = 0.1$, the resulting luminosity is (Equation 7.47)

$$L_{\mathrm{acc}} = 31.4\,L_{\odot} \left(\frac{\dot{M}}{10\,M_{\odot}\,\mathrm{Myr}^{-1}} \right) \left(\frac{M_{\mathrm{pro}}}{1\,M_{\odot}} \right) \left(\frac{R_{\mathrm{pro}}}{1\,R_{\odot}} \right)^{-1}. \tag{7.77}$$

By using the relation $L \propto R^2 T_{\mathrm{eff}}^4$, we can rewrite Equation 7.77 in the form of a mass–radius relation,

$$R_{\mathrm{pro}} = 3.15\,R_{\odot} \left(\frac{\dot{M}}{10\,M_{\odot}\,\mathrm{Myr}^{-1}} \right)^{1/3} \left(\frac{M_{\mathrm{pro}}}{1\,M_{\odot}} \right)^{1/3} \left(\frac{T_{\mathrm{eff}}}{5772\,\mathrm{K}} \right)^{-4/3}, \tag{7.78}$$

or a mass–luminosity relation,

$$L_{\mathrm{acc}} = 9.95\,L_{\odot} \left(\frac{\dot{M}}{10\,M_{\odot}\,\mathrm{Myr}^{-1}} \right)^{2/3} \left(\frac{M_{\mathrm{pro}}}{1\,M_{\odot}} \right)^{2/3} \left(\frac{T_{\mathrm{eff}}}{5772\,\mathrm{K}} \right)^{4/3}. \tag{7.79}$$

Although the protostar's core is hidden from us by dust (and we therefore cannot measure T_{eff} directly), we expect that its interior is cool and opaque enough to be convective. Thus, the hydrostatic core of a protostar should lie on the Hayashi track for an object of mass M_{pro}. Using Equation 7.75 to find the appropriate T_{eff} as a function of M_{pro}, we can rewrite the mass–radius relation for the accreting, convective protostellar core as

$$R_{\mathrm{pro}} \approx 4.8\,R_{\odot} \left(\frac{\dot{M}}{10\,M_{\odot}\,\mathrm{Myr}^{-1}} \right)^{0.32} \left(\frac{M_{\mathrm{pro}}}{1\,M_{\odot}} \right)^{0.14}. \tag{7.80}$$

This relation implies that the central temperature of the protostellar core, $T_c \propto M_{\mathrm{pro}}/R_{\mathrm{pro}}$, should increase as the core becomes more massive, with $T_c \propto M_{\mathrm{pro}}^{0.86}$. Making similar use of the Hayashi relation for T_{eff}, the mass–luminosity relation for an accreting, convective protostellar core is

$$L_{\mathrm{acc}} \approx 6.5\,L_{\odot} \left(\frac{\dot{M}}{10\,M_{\odot}\,\mathrm{Myr}^{-1}} \right)^{0.68} \left(\frac{M_{\mathrm{pro}}}{1\,M_{\odot}} \right)^{0.86}. \tag{7.81}$$

This relation implies that the mass-to-light ratio of the protostellar core, $\Upsilon_{\mathrm{acc}} \equiv M_{\mathrm{pro}}/L_{\mathrm{acc}}$, should gradually increase as the core becomes more massive, with $\Upsilon_{\mathrm{acc}} \propto M_{\mathrm{pro}}^{0.14}$.

On a Hertzsprung–Russell diagram, the location where PMS stars first appear when they stop accreting matter is called the **birthline**. If the universe were arranged for the convenience of astronomers, all protostars would have the same accretion time t_{acc} and efficiency $\xi = 0.1$; at the end of the accretion phase, when $M_{\mathrm{pro}} = M_{\star}$, the accretion would then be cut off by instantaneously whisking away the surrounding dusty gas. In this case, the radius of the newly unveiled PMS star would be, from Equation 7.80,

$$R_{BL} \approx 4.8 \, R_\odot \left(\frac{t_{acc}}{0.1 \, \text{Myr}} \right)^{-0.32} \left(\frac{M_\star}{1 \, M_\odot} \right)^{0.46}, \tag{7.82}$$

using Equation 7.45 for the accretion rate $\dot{M}$ as a function of t_{acc} and the mass $M_\star$ of the PMS star. Similarly, the luminosity of the unveiled PMS star would be, from Equation 7.81,

$$L_{BL} \approx 6.5 \, L_\odot \left(\frac{t_{acc}}{0.1 \, \text{Myr}} \right)^{-0.68} \left(\frac{M_\star}{1 \, M_\odot} \right)^{1.54}. \tag{7.83}$$

In this idealized case, newly visible PMS stars would indeed lie along a mathematical line on the H–R diagram, with $L_{BL} \propto M_\star^{1.54}$, from Equation 7.83, and $T_{eff} \propto M_\star^{0.15} \propto L_{BL}^{0.10}$, from Equation 7.75. On this ideal birthline, a 0.3 $M_\odot$ PMS star would begin with $L_{BL} \approx 1 \, L_\odot$ and $T_{eff} \approx 3500 \, \text{K}$; it would then follow a Hayashi track to its main sequence values of $L_{MS} \approx 0.008 \, L_\odot$ and $T_{eff} \approx 3500 \, \text{K}$.

The universe (alas) is not arranged for the convenience of astronomers. Protostars have different values of the accretion timescale t_{acc} and efficiency ξ. Thus, the "birthline" is really a "birthzone" of finite breadth, with longer accretion times and lower efficiencies corresponding to lower birthline luminosity at a given mass $M_\star$. Another complicating factor is that protostars have an additional source of energy: nuclear fusion. Although hydrogen burning, by definition, doesn't ignite until the end of the PMS star phase, there are other fusion sources that are more easily tapped. In particular, consider deuterium burning via the proton–deuteron reaction (Equation 5.50)

$$p + D \rightarrow {}^3\text{He} + \gamma. \tag{7.84}$$

The deuterium-burning rate becomes significant at $T \approx 10^6 \, \text{K}$. At this relatively low temperature, the reaction is quite sensitive to temperature changes; its τ value of $\tau_{12} = 17.23 T_7^{-1/3}$ leads to a temperature dependence $\nu = 11.7$ at $T_7 = 0.1$. The energy release per proton–deuteron reaction, $Q = 2.493 \, \text{MeV}$, corresponds to $q_{deut} = 2.65 \times 10^{18}$ erg per gram of deuterium. Stars generally form with little deuterium; the protosolar mass fraction of deuterium, for instance, was $X_2 = 2.8 \times 10^{-5}$. However, this still corresponds to a specific nuclear energy $u_{deut} = q_{deut} X_2 = 7.4 \times 10^{13} \, \text{erg g}^{-1} \sim 8 \times 10^{-8} c^2$ at solar abundance.

When the central temperature of an accreting protostar reaches $T_c \approx 10^6 \, \text{K}$, deuterium burning is ignited. Because of the sensitivity of deuterium burning to temperature, it acts as a thermostat that keeps the central temperature at $T_c \sim 10^6 \, \text{K}$ for as long as the deuterium lasts. (A slight uptick in T_c causes a larger jump in $\epsilon \propto T_c^{11.7}$. The resulting increase in energy density and pressure causes the protostar to expand, and the resulting adiabatic cooling drives T_c back down again.) Since $T_c \propto M_{pro}/R_{pro}$, a constant central temperature implies that a protostar powered by deuterium burning must have $R_{pro} \propto M_{pro}$. By contrast, as shown in Equation 7.80, a protostar powered by hot accretion swells at the slower rate $R_{pro} \propto M_{pro}^{0.14}$.

The injection of energy provided by deuterium burning drives vigorous convective mixing; thus nearly all the deuterium in the protostellar core is available for burning. After a brief period of non-equilibrium burning, during which the protostar depletes the deuterium present at the time of ignition, it enters a steady-state epoch in which the deuterium in the accreting gas is rapidly mixed to the center of the protostar and burned. If material with a deuterium mass fraction X_2 is accreted at a rate $\dot{M}$, the steady-state luminosity from deuterium burning is

$$L_{DB} = q_{deut} X_2 \dot{M}, \tag{7.85}$$

independent of the mass of the protostellar core. If we assume a uniform accretion time t_{acc}, then the deuterium-burning luminosity can be written as

$$L_{DB} = 13.1\, L_\odot \left(\frac{X_2}{3 \times 10^{-5}} \right) \left(\frac{t_{acc}}{0.1\,\text{Myr}} \right)^{-1} \left(\frac{M_\star}{1\,M_\odot} \right), \tag{7.86}$$

where $M_\star$ is the mass of the PMS star that the accreting protostellar core will eventually become.

When the PMS star is unveiled, its luminosity from deuterium burning (Equation 7.86) is greater than its luminosity from hot accretion (Equation 7.83) if its mass is

$$M_\star < 3.6\, M_\odot \left(\frac{X_2}{3 \times 10^{-5}} \right)^{1.85} \left(\frac{t_{acc}}{0.1\,\text{Myr}} \right)^{-0.60}, \tag{7.87}$$

where X_2 is the deuterium mass fraction of the accreted matter. Thus, low-mass PMS stars will not first appear on the **hot accretion** birthline, with $L_{BL} \propto M_\star^{1.54} \propto T_{eff}^{10}$. Instead, they appear on a **deuterium-burning** birthline, with $L_{DB} \propto M_\star \propto T_{eff}^{6.8}$.

Assuming that a $0.3\,M_\odot$ PMS star was powered solely by hot accretion, we computed that its birthline luminosity was $L \approx 1.0\,L_\odot$. At the same accretion rate, however, the deuterium-burning luminosity is $L_{DB} \approx 3.7\,L_\odot$. Since the effective temperature of a convective hydrostatic sphere is dictated by the opacity of its photosphere, we can say that a $0.3\,M_\odot$ convective gas ball must have $T_{eff} \approx 3500\,\text{K}$, regardless of its energy source. This implies that the birthline radius of the $0.3\,M_\odot$ PMS star would be $R_{BL} \approx 2.7\,R_\odot$ if hot accretion were the power source, but $R_{DB} \approx 5.2\,R_\odot$ if deuterium burning occurs. Thus, the presence of deuterium burning means that a PMS star of mass $M \sim 0.3\,M_\odot$ has a longer ride down its Hayashi track before settling down as a main sequence star with radius $R \sim 0.3\,R_\odot$.

Another consequence of deuterium burning is relevant for protostars that grow to $M_{pro} > 2\,M_\odot$. A convective protostellar core that burns deuterium has $R_{pro} \propto M_{pro}$. During the era of central deuterium burning, the central density of a protostar must therefore drop as $\rho_c \propto M_{pro}R_{pro}^{-3} \propto M_{pro}^{-2}$. This implies that the central opacity, well described by Kramers' law at $T_c \sim 10^6\,$K, must also decrease, with $\kappa \propto \rho_c T_c^{-3.5} \propto M_{pro}^{-2}$. Eventually, the interior opacity drops to the point where radiation becomes more effective than convection at transporting energy. Detailed calculations show that convection first shuts down in a layer between the protostellar core's center and its photosphere; this happens when the total accreted mass reaches $M_{pro} \sim 2\,M_\odot$. This non-convecting layer is a **radiative barrier** that prevents accreted deuterium from being convected to the protostar's center. Below the radiative barrier, the central region depletes its deuterium and starts contracting on its Kelvin–Helmholtz time. The energy released heats the radiative barrier to the point where **shell deuterium burning** ignites in the layer above. Meanwhile, the deuterium-free central region becomes hotter and less opaque; the now fully radiative core contracts until it reaches a temperature $T_c \sim 10^7\,$K and hydrogen burning begins. In this way, the most massive stars ignite hydrogen burning while they are still accreting matter at a significant rate; thus, they never have a distinct pre-main sequence phase after accretion ends and before hydrogen burning begins.

7.6 Initial Mass Function

In a region with active star formation, the newly formed stars are found to have a range of masses, reflecting the range of accretion rates during the protostellar phase. The **initial mass function** $N(M_\star)$ for newly formed stars is defined such that $N(M_\star)dM_\star$ is the number of stars in a population with masses in the range $M_\star \rightarrow M_\star + dM_\star$. In 1955, Edwin Salpeter found that the initial mass function was well fitted by a power law,

$$N(M_\star)dM_\star \propto M_\star^{-2.35}dM_\star, \tag{7.88}$$

for stars in the mass range $0.4\,M_\odot < M_\star < 10\,M_\odot$. This power-law form for the initial mass function, known as the Salpeter mass function, is often a useful approximation. There are shortcomings, however, to the Salpeter mass function. For instance, what happens at $M_\star < 0.4\,M_\odot$? If we simply extrapolate the power law to a minimum mass M_{min}, we find that the total mass of stars diverges as $M_{min}^{-0.35}$ in the limit $M_{min} \rightarrow 0$. Obviously, there must be a low-mass cutoff to the initial mass function.

The **Chabrier mass function** is a useful functional form for the initial mass function. In the Chabrier mass function, a power-law distribution at high mass gives way to a lognormal distribution at low mass. The Chabrier mass function can be written as

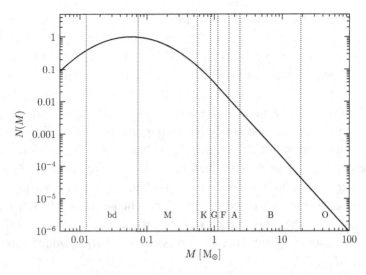

Figure 7.3 Chabrier mass function (arbitrarily normalized) with $M_b = 1\,M_\odot$, $M_c = 0.2\,M_\odot$, $\sigma = 1.1$, and $\beta = 2.3$. Vertical lines give the approximate division between main sequence spectral types, taken from Table 1.1. The mass range for brown dwarfs is labeled "bd."

$$N(M_\star) \propto M_\star^{-1} \exp\left(-\frac{[\ln(M_\star/M_c)]^2}{2\sigma^2}\right) \qquad [M_\star \leq M_b], \qquad (7.89)$$

$$\propto M_\star^{-\beta} \qquad\qquad\qquad\qquad [M_\star > M_b]. \qquad (7.90)$$

The parameters of the Chabrier mass function are observed to vary from one star-forming region to another. However, typical values are $M_b = 1\,M_\odot$ for the transition mass, $M_c = 0.2\,M_\odot$ for the characteristic mass of the lognormal portion, $\sigma = 1.1$ for the width of the lognormal portion, and $\beta = 2.3$ for the slope at the high-mass end. A Chabrier mass function with these parameters is shown in Figure 7.3. Note that the Chabrier mass function has a maximum at $M_\star = \exp(-\sigma^2)M_c \approx 0.06\,M_\odot$. However, its full width at half maximum stretches from $M_\star \approx 0.016\,M_\odot$ to $M_\star \approx 0.22\,M_\odot$; the breadth of the Chabrier function indicates that the multi-step process of star formation does not produce a strongly preferred mass. It should also be noted that the power-law tail of the Chabrier mass function positively skews the distribution; although its mode is $M_\star \approx 0.06\,M_\odot$, the median mass is $M_\star \approx 0.2\,M_\odot$ and the mean mass is $M_\star \approx 0.5\,M_\odot$.

By the time a low-mass PMS star starts burning hydrogen and becomes a zero age main sequence star, its mass $M_\star$ is already fixed. Thus, the shape of the Chabrier mass function has nothing to do with the process of nuclear fusion. As it turns out, the lowest-mass objects in the Chabrier mass function do not burn hydrogen at all. Since the central temperature of a gas ball goes as $T_c \propto MR^{-1}$, you might think that once its mass M is fixed, you can raise T_c to an arbitrarily high value by continuing to squeeze down the radius R. However, while the central density of the object grows as $\rho_c \propto MR^{-3}$ as it shrinks, the critical density at which degeneracy pressure takes over (Equation 3.78) grows only as $\rho_{i\text{-d}} \propto T_c^{3/2} \propto M^{3/2}R^{-3/2}$. Thus, at some radius $R_{\text{deg}} \propto M^{-1/3}$, electron degeneracy pressure dominates at the gas ball's center, and it can no longer shrink. The central temperature when degeneracy becomes important is $T_{\text{deg}} \propto MR_{\text{deg}}^{-1} \propto M^{4/3}$. Detailed calculations indicate that gas balls with a mass $M < 0.074\,\mathrm{M_\odot}$ are held up by electron degeneracy pressure before reaching the temperature $T_c \approx 10^7\,\mathrm{K}$ required for hydrogen burning to start.[7] A gas ball too low in mass to become a star but too massive to be labeled a "planet" is called a **brown dwarf**. The nearest brown dwarfs to the Sun are the two members of the Luhman 16 system, at a distance $d = 2.0\,\mathrm{pc}$; Luhman 16A has a mass $M \approx 34\,M_{\text{Jup}}$, while Luhman 16B has $M \approx 29\,M_{\text{Jup}}$.

The name "brown dwarf," introduced by Jill Tarter in 1975, was not intended as a literal color descriptor. The youngest brown dwarfs, in fact, have spectral type M and are red in color. Older, cooler brown dwarfs require an extension of the Harvard OBAFGKM spectral types. Brown dwarfs with effective temperature in the range $1300\,\mathrm{K} < T_{\text{eff}} < 2400\,\mathrm{K}$ have spectral type L, marked by strong absorption lines of metal hydrides such as FeH and of alkali metals such as Na I and K I.[8] Cooler brown dwarfs, with $600\,\mathrm{K} < T_{\text{eff}} < 1300\,\mathrm{K}$, have spectral type T, marked by strong absorption from methane. The coldest (and hardest to detect) brown dwarfs, with $T_{\text{eff}} < 600\,\mathrm{K}$, are assigned spectral type Y. A brown dwarf with $M = 30\,M_{\text{Jup}}$, comparable in size to the members of the Luhman 16 system, will take $\sim 40\,\mathrm{Myr}$ after the end of accretion to cool to $T_{\text{eff}} = 2400\,\mathrm{K}$ and become an L dwarf; it will take $\sim 400\,\mathrm{Myr}$ to become a T dwarf, and $\sim 4\,\mathrm{Gyr}$ to become a Y dwarf. Luhman 16A and 16B are both near the boundary between spectral types L and T, with the more massive component being slightly hotter; this implies an age for the system of $t \sim 400\,\mathrm{Myr}$.

The smallest gas spheres, with mass $M < 13\,M_{\text{Jup}} = 0.0124\,\mathrm{M_\odot}$, never attain the central temperature $T_c \sim 10^6\,\mathrm{K}$ required for deuterium burning. Thus, one proposed definition for the boundary between brown dwarfs and gas giant planets is that brown dwarfs undergo a deuterium-burning phase, while planets do not. If we look at the Chabrier mass function over the entire mass range for brown dwarfs and stars ($0.0124\,\mathrm{M_\odot} < M < 100\,\mathrm{M_\odot}$), we find that although brown dwarfs with $M < 0.074\,\mathrm{M_\odot}$ make up $\sim 17\%$ of the individual objects, they provide only $\sim 1.5\%$ of the total mass.

---

[7] The lower mass limit for stars is somewhat metallicity dependent; the quoted mass $M = 0.074\,\mathrm{M_\odot} = 78 M_{\text{Jup}}$ assumes solar metallicity.

[8] Because of strong absorption by sodium and potassium at green and yellow wavelengths, some brown dwarfs would actually appear magenta to the human eye.

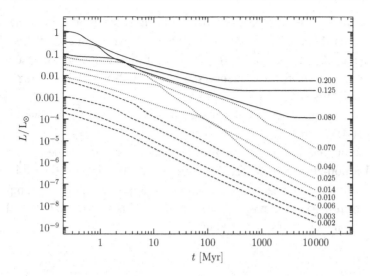

Figure 7.4 Luminosity evolution of low-mass stars (solid lines), brown dwarfs (dotted lines), and gas giant planets (dashed lines). Curves are labeled with the mass of the object in solar masses. [Data from Burrows *et al.* 1993, 1997]

When accretion ends, the distinction between the highest-mass brown dwarfs and the lowest-mass stars is not immediately apparent. Figure 7.4 shows the evolution in luminosity of low-mass stars, brown dwarfs, and (non-deuterium-burning) planets after they have stopped accreting. Low-mass stars (indicated by solid lines in Figure 7.4) have a deuterium-burning phase that lingers for a few million years after accretion stops; during the deuterium-burning phase, the luminosity of the low-mass PMS star is much higher than its eventual zero age main sequence luminosity. For brown dwarfs, the deuterium-burning phase is longer lasting, but at a lower luminosity. For brown dwarfs just above the deuterium-burning limit at $M = 0.0124\,M_\odot$, the luminosity never has a pronounced deuterium-burning plateau, when the luminosity is nearly constant and deuterium burning provides the overwhelming majority of the brown dwarf's luminosity. For instance, the $M = 0.014\,M_\odot$ brown dwarf in Figure 7.4 derives more than half its luminosity from deuterium burning during the interval $t = 10$–$180\,$Myr. However, the share of its luminosity provided by deuterium burning peaks at 85% when $t \approx 40\,$Myr; even at this peak, a significant minority of the power comes from the release of gravitational potential energy.

Exercises

7.1 Show that for the slope of the Chabrier initial mass function to be continuous at the transition mass M_b, it is required that $\sigma^2(\beta - 1) = \ln(M_b/M_c)$.

7.2   We have assumed that hydrostatic equilibrium is a good approximation during both the pre-main sequence stage and the main sequence stage of a star's
existence. Let's test that approximation for the following two cases:

(a) A 1 $M_\odot$ pre-main sequence star contracts from 7.0 $R_\odot$ to 0.87 $R_\odot$ over
the course of 30 Myr.

(b) The main sequence Sun expands from 0.87 $R_\odot$ to 1.0 $R_\odot$ over the course
of 4.57 Gyr.

These numbers imply a non-zero (but small) surface velocity, and in general a non-zero surface acceleration. For concreteness, assume a constant
acceleration a for each case. Compute the acceleration and compare it to
the mean surface gravity $g = GM_\star/R_\star^2$. (This factor quantifies the mean
deviation from hydrostatic balance.)

8

Evolved Stars: After the Main Sequence

I see my life go drifting like a river
From change to change; I have been many things

William Butler Yeats (1865–1939)
"Fergus and the Druid" [1925 version]

A star remains on the main sequence as long as it is powered by fusion of hydrogen into helium in its central core. If a fraction $f_{nuc} \geq 0.1$ of a star's mass is available for hydrogen burning, Equation 1.33 tells us that stars with a mass less than $\sim 0.9 \, M_\odot$ have a main sequence lifetime t_{MS} longer than the current age of the universe, $t_0 \approx 13.7 \, Gyr$. However, higher-mass stars have smaller values of t_{MS}, with the most massive stars having main sequence lifetimes as short as a few million years. Thus, we live in a sufficiently old universe that post-main sequence stars exist and are observable. (Since we have already used the abbreviation "PMS" to refer to *pre*-main sequence stars, in this chapter our brief nickname for *post*-main sequence stars will be "evolved" stars.)

The study of post-main sequence evolution is aided by observations of star clusters, which contain stars of nearly uniform age and metallicity. In a cluster of stars with age t_{cl}, there is a characteristic **turnoff mass**, M_{TO}, representing the mass of a star whose main sequence lifetime t_{MS} is equal to the age of the cluster. For older clusters, where the turnoff mass is $M_{TO} < 4 \, M_\odot$, Equation 1.33 yields

$$M_{TO} \approx 2.0 \, M_\odot \left(\frac{t_{cl}}{1 \, Gyr} \right)^{-0.29} \left(\frac{f_{nuc}}{0.1} \right)^{0.29}. \tag{8.1}$$

For younger clusters, Equation 1.34, which reflects the shallower mass–luminosity relation for more massive stars, yields a turnoff mass of

$$M_{TO} \approx 5.3 \, M_\odot \left(\frac{t_{cl}}{100 \, Myr} \right)^{-0.5} \left(\frac{f_{nuc}}{0.2} \right)^{0.5}. \tag{8.2}$$

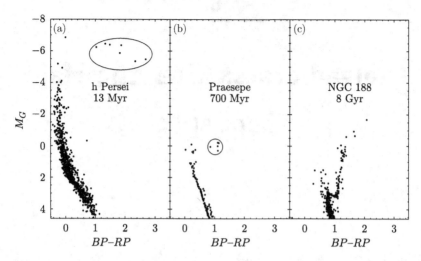

Figure 8.1 Hertzsprung–Russell diagrams for open star clusters with different ages. (a) h Persei ($t_{cl} \approx 13\,\mathrm{Myr}$). (b) Praesepe ($t_{cl} \approx 700\,\mathrm{Myr}$). (c) NGC 188 ($t_{cl} \approx 8\,\mathrm{Gyr}$). [Data from Cantat-Gaudin *et al.* 2018]

Note that for a younger cluster with a higher turnoff mass, we scale to a higher value of f_{nuc}. As we discuss in Section 8.1, higher-mass stars have convective cores that permit them to burn more of their hydrogen.

In a given cluster, stars with $M_\star < M_{TO}$ are on the main sequence; in fact, if the cluster is young enough, the lowest-mass stars will still be in their pre-main sequence phase. Stars with $M_\star > M_{TO}$ are no longer on the main sequence. In the solar neighborhood, as shown in the H–R diagram of Figure 1.10, the obvious groups of post-main sequence, or evolved, stars are red giants and red clump stars. (White dwarfs, since they are no longer powered by fusion, are more usefully classified as stellar remnants rather than stars.) There are only ∼10% as many evolved stars as main sequence stars in the solar neighborhood; this suggests that the post-main sequence lifetime of a star is short: typically only ∼10% of its main sequence lifetime t_{MS}. For an individual cluster, as opposed to the grab bag of stars in the solar neighborhood, we thus expect that the evolved stars in the cluster's H–R diagram show the behavior of stars with a mass only slightly greater than the turnoff mass M_{TO}.

Figure 8.1 includes H–R diagrams for three star clusters with very different ages.[1] The youngest of the three systems is the open cluster h Persei, at a distance $d \approx 2.3\,\mathrm{kpc}$. Its age is only $t_{cl} \approx 13\,\mathrm{Myr}$, leading to a turnoff mass of $M_{TO} \sim 15\,M_\odot$. In the H–R diagram for h Per (Figure 8.1(a)), the post-main sequence stars are in the circled region; they are slightly brighter than the main sequence turnoff but are significantly cooler. These evolved stars are now yellow supergiants and red supergiants.

---

[1] In Figure 8.1, we omit stars dimmer than $M_G = 4.6$, corresponding to main sequence stars with $M_\star \approx 1\,M_\odot$. The long-lived stars of the lower main sequence are not (at the moment) of interest to us.

The next system in Figure 8.1 is the nearby open cluster Praesepe, at $d = 186\,\mathrm{pc}$. Its age is $t_{\mathrm{cl}} \approx 700\,\mathrm{Myr}$, leading to $M_{\mathrm{TO}} \sim 2.2\,\mathrm{M_\odot}$. The H–R diagram for Praesepe (Figure 8.1(b)) shows a group of post-main sequence stars, circled in the diagram, that lie near the red clump at $M_G \approx 0.5$ and $BP - RP \approx 1.2$. Although these evolved red clump stars are much less luminous than the evolved red supergiants in h Per, they share the characteristic of being significantly cooler than the main sequence turnoff for the cluster. The scarcity of stars in the region between the main sequence and the evolved stars is called the **Hertzsprung gap**. The emptiness of the Hertzsprung gap indicates that the transition between the main sequence and the star's next stage of existence (either a red supergiant or a red giant) must occur relatively rapidly.

The oldest system in Figure 8.1 is the unusually old open cluster NGC 188, at $d \approx 2.0\,\mathrm{kpc}$. Its age is $t_{\mathrm{cl}} \approx 8\,\mathrm{Gyr}$, leading to a turnoff mass $M_{\mathrm{TO}} \sim 1.1\,\mathrm{M_\odot}$. In NGC 188 (Figure 8.1(c)) we see both a distinct main sequence and a red giant branch of evolved stars. However, the space between the main sequence and the red giant branch is bridged by a population of stars, with no perceptible gap. The "bridge" stars in this region are subgiants of luminosity class IV.

Looking at H–R diagrams for clusters of different ages, we can trace out the post-main sequence evolution of stars. Relatively low-mass stars swell into red giants, then eventually become white dwarfs after they exhaust their fuel for fusion. Early in their existence, when they are white-hot, white dwarfs are easily visible and can be placed on an H–R diagram like that of Figure 1.10. More massive stars become supergiants, then end their post-main sequence career more dramatically, by exploding and producing either an ultra-compact remnant or no remnant at all. Understanding the physical factors that dictate these very different fates is a key goal of stellar evolution theory. In addition, the study of how evolved stars produce elements heavier than helium, then disperse them into interstellar space, is key to understanding chemical evolution in our galaxy. Let's start our study of post-main sequence life by seeing what happens when a star exhausts its central hydrogen; afterward, we will trace its evolution through all the later stages of post-main sequence existence.

8.1 Building a Helium Core

The evolving hydrogen profile $X(M, t)$ of a main sequence star depends on how energy is transported within the star's central regions. If convection is the means of transport, then the turbulent motion of convection ensures that the gas is well-mixed; this implies that the mass fractions X, Y, and Z are uniform within a convection zone. If radiation is the means of energy transport, there are other potential sources of mixing, but they are generally slow for Sun-like stars with radiative cores. Consider, for instance, the effect of **diffusion** on the hydrogen profile $X(M)$. At a given temperature, the thermal speed of protons is smaller than that of electrons by a factor $(m_e/m_p)^{1/2} = 0.0233$, while their mean free path λ between electrostatic collisions is the same. Thus, the thermal diffusivity for the sluggish protons is

$$D_{\text{th},p} = \frac{1}{3}v_{\text{th}}\lambda = 0.0233 D_{\text{th},e} \tag{8.3}$$

$$\approx 7.5 \,\text{cm}^2 \,\text{s}^{-1} \left(\frac{\rho}{160 \,\text{g cm}^{-3}}\right)^{-1} \left(\frac{T}{1.6 \times 10^7 \,\text{K}}\right)^{5/2},$$

where we have scaled to properties appropriate for the Sun's center. As a result, the typical distance over which protons can diffuse during the Sun's main sequence lifetime t_{MS} is

$$r_{\text{diff}} = (D_{\text{th},p}t_{\text{MS}})^{1/2} \tag{8.4}$$

$$\approx 0.02 \,\text{R}_\odot \left(\frac{\rho}{160 \,\text{g cm}^{-3}}\right)^{-1/2} \left(\frac{T}{1.6 \times 10^7 \,\text{K}}\right)^{5/4} \left(\frac{t_{\text{MS}}}{10 \,\text{Gyr}}\right)^{1/2}.$$

Since this distance is smaller than the radius of the Sun's fusion core, $R_{\text{core}} \approx 0.2 \,\text{R}_\odot$, we don't expect diffusion to greatly affect the hydrogen profile in the radiative zone of a Sun-like star. However, since r_{diff} is not infinitesimally small, diffusion does have to be taken into account when doing detailed modeling of the Sun.

As shown in Figure 8.2, zero age main sequence stars with $M_\star < 0.3 \,\text{M}_\odot$ are fully convective, all the way up to their photospheres. This means that hydrogen from throughout the star is convected to the star's center, raising the fraction of the mass available for fusion to $f_{\text{nuc}} \sim 0.7$, and extending the already prodigious main sequence lifetimes of these low-mass dwarfs. The convection also takes the freshly made helium from the star's center and spreads it throughout the star, making stars with $M_\star < 0.3 \,\text{M}_\odot$ nearly uniform in composition.

At the other end of the stellar mass range, Figure 8.2 shows that ZAMS stars with $M_\star > 1.6 \,\text{M}_\odot$ produce the majority of their luminosity in a central convective core. Thus, the central convective region of these higher-mass stars remains nearly uniform in chemical composition, with a gradient in composition found only at the boundary between the inner convective core and the outer radiative zone. The more massive a star, the larger its central convective core relative to its total mass; this increases the fraction f_{nuc} of the star's mass available for fusion.

The convective and radiative zones shown in Figure 8.2 constitute a single snapshot at the very start of the stars' main sequence life. By following a modeled star through time, we can see the evolution of both the location of convection zones and the star's composition profile. For example, the evolving hydrogen profile $X(M,t)$ of a $1 \,\text{M}_\odot$ star is shown in Figure 8.3(a). Initially, this Sun-like star has a uniform hydrogen mass fraction $X = 0.7$. As hydrogen burning progresses in the star's core, however, a smooth gradient is set up from $X = 0.7$ at

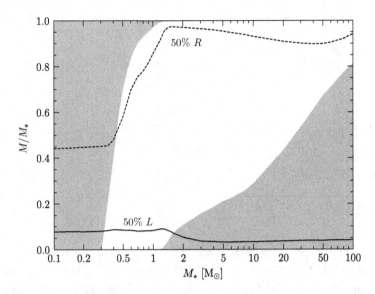

Figure 8.2 Convection regions (gray) in zero age main sequence stars with $X = 0.70$ and $Z = 0.02$. Solid line: mass shell inside which 50% of the star's luminosity is enclosed. Dashed line: mass enclosed at $r = 0.5R_\star$.

$M > 0.5\,\mathrm{M}_\odot$ to a minimum value of X at the star's center. Until $t \approx 8\,\mathrm{Gyr}$, the specific energy generation rate $\epsilon(X, \rho, T)$ increases monotonically as you approach the star's center, just as in the ZAMS Sun (Figure 6.7(e)); this is because the increase in temperature and density as $M \to 0$ more than compensates for the decrease in X. At $t \approx 8\,\mathrm{Gyr}$, however, the central value of X becomes sufficiently tiny that the maximum value of ϵ moves away from the star's center. Over the next 2 Gyr, the $1\,\mathrm{M}_\odot$ star gradually transitions from having a central hydrogen-burning core to having a hydrogen-burning shell outside a nearly hydrogen-free core. By $t = 10\,\mathrm{Gyr}$, the star has $X < 0.001$ within a central region of mass $M \approx 0.02\,\mathrm{M}_\odot$.

The hydrogen profile evolves differently for the $4\,\mathrm{M}_\odot$ star depicted in Figure 8.3(b). When this relatively massive star is on the zero age main sequence, it has a central convective core with mass $M \approx 0.2M_\star$, as shown in Figure 8.2. Thus, the value of X initially drops uniformly over the central $M \approx 0.2M_\star$. However, the opacity in the hot central region is largely due to electron scattering, with $\kappa_e = 0.20(1+X)\,\mathrm{cm^2\,g^{-1}}$. Thus, as the CNO bi-cycle keeps cycling along and the central value of X drops, the opacity in the central region decreases. As radiation consequently becomes more effective as a means of energy transport, the size of the central convective core decreases. By the time X drops to zero at the center of the $4\,\mathrm{M}_\odot$ star, at $t \approx 150\,\mathrm{Myr}$, the size of the convective core, with uniform X, has decreased to $M \approx 0.1M_\star$.

A star that has just begun to fuse hydrogen into helium at its center is a zero age main sequence (ZAMS) star; a star that has just stopped fusing hydrogen into

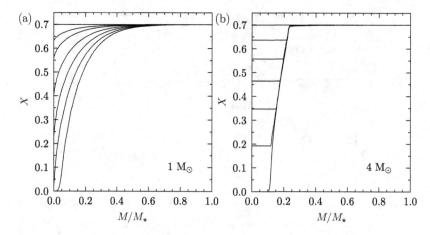

Figure 8.3 (a) Hydrogen mass fraction in a $1\,M_\odot$ main sequence star. Lines correspond to ages $t \approx 0$, 1, 2, 4, 6, 8, and 10 Gyr. (b) Hydrogen mass fraction in a $4\,M_\odot$ main sequence star. Lines correspond to ages $t \approx 0$, 25, 50, 75, 100, 125, and 150 Myr.

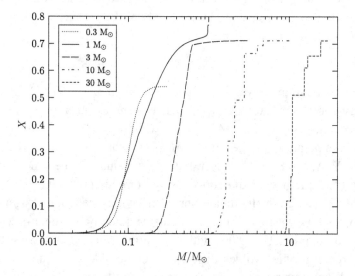

Figure 8.4 Hydrogen mass fraction X in non-rotating stars on the terminal age main sequence. Convective overshoot is omitted from these stellar models.

helium at its center is a **terminal age main sequence** (TAMS) star. The hydrogen profiles of TAMS stars are shown in Figure 8.4. All these stars began with a uniform hydrogen mass fraction $X = 0.71$. On the terminal age main sequence, the value of X is increased in the outer layers of the $1\,M_\odot$ star because of the diffusive settling of heavier elements. (More massive stars lack diffusive settling because their main sequence lifetimes are too short; the $0.3\,M_\odot$ star lacks diffusive settling because its convection effectively smooths out composition gradients.)The interiors of TAMS stars all show decreases in X from hydrogen burning. The fully hydrogen-depleted cores of massive TAMS stars can be quite large. On the terminal age main sequence, a 3 $M_\odot$ star has a convective, hydrogen-depleted core with a mass $\sim6\%$ of its total mass, while a 10 $M_\odot$ TAMS star has a core with $\sim10\%$ of its total mass, and a 30 $M_\odot$ TAMS star has a core with $\sim30\%$ of its total mass. The large convective core of the 30 $M_\odot$ star is particularly effective at extending its main sequence lifetime. The 1 $M_\odot$ and 3 $M_\odot$ stellar models have lifetimes, from ZAMS to TAMS, of $t_{MS} = 10.2\,\mathrm{Gyr}$ and 0.32 Gyr. Using the approximation of Equation 1.33, this implies they burn a fraction $f_{nuc} \sim 0.1$ of their initial mass on the main sequence. The 10 $M_\odot$ and 30 $M_\odot$ stellar models have $t_{MS} = 20.8\,\mathrm{Myr}$ and 5.5 Myr; from Equation 1.34, this implies that the 10 $M_\odot$ star has $f_{nuc} \approx 0.15$ on the main sequence, but the 30 $M_\odot$ star has $f_{nuc} \approx 0.35$.

In Figure 8.4, note that the hydrogen profiles of the two most massive TAMS stars show "ledges" of constant X, in contrast to the smooth gradient in X seen for the lower-mass ZAMS stars. This reflects the fact that turbulence is a chaotic process, prone to instabilities. In the higher-mass stars, the convective core doesn't shrink at a steady rate, but is prone to occasional glitches. Since chaotic processes like turbulence are challenging to model numerically, the computed composition profile and main sequence lifetime of a stellar model can be quite sensitive to its treatment of convection. In some stellar mass ranges, for instance, different scenarios for the treatment of convective boundaries can change the main sequence lifetime of a star by as much as a factor of two. This is a caveat to keep in mind when we make inferences about the behavior of evolved massive stars based on numerical models.

8.2 Schönberg–Chandrasekhar Limit

When a star ends its main sequence lifetime, it has a helium core (with some metals mixed in) surrounded by a hydrogen-rich envelope. The temperature of the core is still high enough for hydrogen burning to take place; however, there is no hydrogen to burn within the core. Instead, hydrogen burning is transferred to a thin layer above the helium core. This "shell hydrogen burning" takes place via the CNO bi-cycle.

When shell hydrogen burning begins, the helium core lying below has a luminosity L_{core} that is insignificantly small. At this point, the helium core isn't hot enough to ignite the triple alpha process, so it has no nuclear fusion source available. In addition, the helium core can't effectively tap its gravitational potential energy at this point because the hydrogen-burning layer above it acts as a warm heating pad that keeps the core's temperature and pressure high. A small core luminosity implies a small energy flux and thus a small temperature gradient; if the core is radiative, then $dT/dr \propto L/r^2$ (from Equation 6.6) and if the core is convective, then dT/dr must be even smaller than in the radiative case. The core can therefore be well approximated as being isothermal, with uniform temperature T_{core}. An isothermal gas whose pressure is given by an ideal gas law has $P \propto \rho T \propto \rho$, which is equivalent to a polytropic index $n = \infty$. As Schönberg[2] and Chandrasekhar demonstrated in 1942, an isothermal core, with its soft equation of state, cannot have an arbitrarily large mass and still remain in thermal equilibrium. For a given stellar mass $M_\star$, there is a maximum permissible core mass M_{max} for a central isothermal core; above this mass, the pressure at the core's surface is unable to support the weight of the hydrogen-rich envelope above it while still maintaining thermal equilibrium. The Schönberg–Chandrasekhar (S–C) limit for the mass of an isothermal core can be found using the **virial theorem**.

For a spherical star, the virial theorem can be derived starting with the equation of hydrostatic equilibrium:

$$\frac{dP}{dM} = -\frac{GM}{4\pi r^4}. \tag{8.5}$$

If we multiply by volume $V = (4\pi/3)r^3$ and integrate from the center to the surface of the helium core, we find

$$\int_{P_c}^{P_s} V dP = -\frac{1}{3}\int_0^{M_{core}} \frac{GMdM}{r}, \tag{8.6}$$

where P_c is the pressure at the center of the helium core, and P_s is the pressure at the surface of the helium core. The right-hand side of Equation 8.6 is $1/3$ times the gravitational potential energy of the helium core (Equation 2.18). The left-hand side can be integrated by parts, meaning that Equation 8.6 can be rewritten as

$$[PV]_0^{V_{core}} - \int_0^{V_{core}} PdV = \frac{1}{3}W_{core}. \tag{8.7}$$

Since $PV = 0$ at the core's center, we find that

$$P_s V_{core} - \int_0^{V_{core}} PdV = -\frac{\alpha}{3}\frac{GM_{core}^2}{R_{core}}, \tag{8.8}$$

where α is a numerical factor of order unity. A stiff, uniform-density core, for instance, would have $\alpha = 0.6$; a soft isothermal core, however, will be more centrally concentrated and have $\alpha \sim 1$. If the core is isothermal, with $T = T_{core}$, we can write

$$P = \frac{\rho}{m_{AMU}\bar{\mu}_{core}}kT_{core}. \tag{8.9}$$

---

[2] The Brazilian polymath Mário Schenberg generally used the spelling "Schönberg" for his scientific papers and "Schenberg" for his art criticism.

Inserting this back into Equation 8.8, we find

$$P_s V_{\text{core}} - \frac{kT_{\text{core}}}{m_{\text{AMU}}\bar{\mu}_{\text{core}}} \int_0^{V_{\text{core}}} \rho \, dV = -\frac{\alpha}{3}\frac{GM_{\text{core}}^2}{R_{\text{core}}}, \tag{8.10}$$

which becomes

$$P_s V_{\text{core}} = \frac{kT_{\text{core}}}{m_{\text{AMU}}\bar{\mu}_{\text{core}}} M_{\text{core}} - \frac{\alpha}{3}\frac{GM_{\text{core}}^2}{R_{\text{core}}}. \tag{8.11}$$

If the surface pressure P_s is zero, this becomes the usual form of the virial equation for a self-gravitating gas, which equates twice the thermal energy to the absolute value of the gravitational potential energy.

Dividing Equation 8.11 by the core volume V_{core}, we find the pressure P_s required at the core's surface to keep it in equilibrium:

$$P_s = \frac{3}{4\pi}\frac{kT_{\text{core}}}{m_{\text{AMU}}\bar{\mu}_{\text{core}}}\frac{M_{\text{core}}}{R_{\text{core}}^3} - \frac{\alpha G}{4\pi}\frac{M_{\text{core}}^2}{R_{\text{core}}^4}. \tag{8.12}$$

Differentiating Equation 8.12 with respect to the core radius R_{core}, we find that for a fixed core mass M_{core}, the equilibrium pressure at the core surface, P_s, has a maximum when the core radius has a critical value of

$$R_{\text{cr}} = \frac{4\alpha}{9}GM_{\text{core}}\frac{m_{\text{AMU}}\bar{\mu}_{\text{core}}}{kT_{\text{core}}} \tag{8.13}$$

$$\approx 0.13\,R_\odot\,\alpha\left(\frac{M_{\text{core}}}{0.1\,M_\odot}\right)\left(\frac{T_{\text{core}}}{10^7\,\text{K}}\right)^{-1}, \tag{8.14}$$

assuming a core of ionized helium, with $\bar{\mu}_{\text{core}} \approx 1.3$. By substituting this critical core radius back into Equation 8.12, we find that the maximum possible pressure at the surface of an isothermal core in thermal equilibrium is

$$P_{s,\text{max}} \approx \frac{0.68}{\alpha^3 G^3}\frac{1}{M_{\text{core}}^2}\left(\frac{kT_{\text{core}}}{m_{\text{AMU}}\bar{\mu}_{\text{core}}}\right)^4 \tag{8.15}$$

$$\approx 9.7 \times 10^{15}\,\text{dyn cm}^{-2}\,\alpha^{-3}\left(\frac{M_{\text{core}}}{0.1\,M_\odot}\right)^{-2}\left(\frac{T_{\text{core}}}{10^7\,\text{K}}\right)^4. \tag{8.16}$$

Is this maximum pressure enough to support the weight of the hydrogen-rich envelope pressing down on the inert helium core? Let's do some calculations.

The hydrogen envelope is thick and massive, while the helium core is compact and lower in mass than the envelope. To lowest order, the structure of the hydrogen-rich envelope is unaffected by the presence of the helium core. The pressure P_{env} at the base of the heavy hydrogen-rich envelope will be nearly the same as the central pressure of a star with mass $M_\star$ and radius $R_\star$ that has no central helium core. The pressure at the base of the envelope will thus be

$$P_{\text{env}} \sim \frac{GM_\star^2}{R_\star^4}, \tag{8.17}$$

where the $\sim$ in the above equation conceals a numerical factor that is as small as 0.77 for a stiffish $n = 1.5$ polytropic envelope, but as large as 11 for a softer $n = 3$ polytrope. Similarly, the temperature T_{env} at the base of the heavy hydrogen-rich envelope would be nearly the same as the central temperature of a star with mass $M_\star$ and radius $R_\star$. Thus,

$$T_{env} \sim \frac{m_{AMU} \bar{\mu}_{env}}{k} \frac{GM_\star}{R_\star}, \tag{8.18}$$

where the $\sim$ in the above equation conceals a numerical factor that is as small as 0.54 for an $n = 1.5$ polytrope, but as large as 0.86 for a softer $n = 3$ polytrope.

In thermal equilibrium, the temperature T_{core} of the isothermal core must equal the temperature T_{env} at the base of the envelope. Thus, the maximum equilibrium pressure at the surface of the isothermal core, from Equation 8.15, can be rewritten, using Equation 8.18, as

$$P_{s,max} \approx \frac{0.68}{\alpha^3 G^3} \frac{1}{M_{core}^2} \left(\frac{k T_{env}}{m_{AMU} \bar{\mu}_{core}} \right)^4 \tag{8.19}$$

$$\sim \frac{1}{\alpha^3 G^3} \frac{1}{M_{core}^2} \left(\frac{\bar{\mu}_{env}}{\bar{\mu}_{core}} \right)^4 \left(\frac{GM_\star}{R_\star} \right)^4. \tag{8.20}$$

For the isothermal helium core to remain in equilibrium, this maximum pressure must be at least as large as the actual pressure P_{env} at the base of the envelope, as given in Equation 8.17. Thus, the S–C criterion for instability is

$$\frac{GM_\star^2}{R_\star^4} > \frac{1}{\alpha^3 G^3} \frac{1}{M_{core}^2} \left(\frac{\bar{\mu}_{env}}{\bar{\mu}_{core}} \right)^4 \left(\frac{GM_\star}{R_\star} \right)^4, \tag{8.21}$$

or

$$\frac{M_{core}}{M_\star} > \frac{1}{\alpha^{3/2}} \left(\frac{\bar{\mu}_{env}}{\bar{\mu}_{core}} \right)^2 \qquad \text{[unstable].} \tag{8.22}$$

A more careful calculation of the S–C criterion, assuming a soft $n = 3$ envelope and retaining all factors of order unity, results in the instability criterion

$$\frac{M_{core}}{M_\star} > 0.38 \left(\frac{\bar{\mu}_{env}}{\bar{\mu}_{core}} \right)^2 \qquad \text{[unstable].} \tag{8.23}$$

The isothermal inert core is made mostly of ionized helium ($Y \approx 0.98$, $Z \approx 0.02$), and thus has $\bar{\mu}_{core} \approx 1.3$. The overlying envelope is made mostly of ionized hydrogen ($X \approx 0.71$, $Y \approx 0.27$, $Z \approx 0.02$), and thus has $\bar{\mu}_{env} \approx 0.6$. The S–C limit for an isothermal helium core is then

$$\frac{M_{core}}{M_\star} \approx 0.38 \left(\frac{0.6}{1.3} \right)^2 \approx 0.08. \tag{8.24}$$

An isothermal helium core comprising more than $\sim$8% of a star's total mass will be unstable and begin to contract.

Numerical models show that stars with $M_\star > 2\,M_\odot$ have convective cores with $M_{\text{core}} > 0.08M_\star$ at the time they deplete their central hydrogen. Thus, a high-mass star never has an isothermal helium core at all. Instead, a temperature gradient is set up that increases the pressure gradient and keeps the non-isothermal core in hydrostatic equilibrium. However, the temperature gradient dT/dr necessarily creates an energy flux L/r^2 to maintain thermal equilibrium, and the helium core thereafter shrinks on its Kelvin–Helmholtz timescale. Although this timescale is long compared to the core's freefall time, it is shorter than the preceding main sequence lifetime of the high-mass star. Thus, we expect a star with $M_\star > 2\,M_\odot$ to evolve relatively rapidly once it leaves the main sequence. On an H–R diagram, the post-main sequence evolution of a massive star is manifested as a rapid transit from left to right across the Hertzsprung gap. That is, as the helium core shrinks and becomes hotter, the outer envelope of the star responds by expanding and becoming cooler. In short, the star becomes a red supergiant (if it is very massive) or a red giant (if its mass is slightly less extreme).

Stars with $M_\star < 2\,M_\odot$ have a different evolutionary path once they leave the main sequence. These lower-mass stars have helium cores with $M_{\text{core}} < 0.08M_\star$ when they exhaust their central hydrogen. Thus, they have a relatively undramatic phase of their existence when they have an isothermal helium core in equilibrium, surrounded by a thick hydrogen-burning shell. During this phase, relatively low-mass stars are slowly evolving subgiants; for instance, see the thickly populated subgiant bridge in the old cluster NGC 188 (Figure 8.1). It is only when the mass of the helium core has grown to $M_{\text{core}} \approx 0.08M_\star$ that the core develops a significant temperature gradient and starts to contract on its Kelvin–Helmholtz time. As with higher-mass stars, the shrinking and heating of the core is accompanied by the expansion and cooling of the outer envelope. In short, the star becomes a red giant.

The question of why stars become red giants (or red supergiants) is more complex than it might seem at first glance. Many attempts to explain the low surface temperature of red giants exist in the scientific literature. We could simply point to the results of numerical models and say, "Look! We put in all the relevant physics and got stars that are large and cool during their shell hydrogen-burning phase." However, a more physical and intuitive explanation is quite desirable.

First, we note that stars become red giants when they have two energy sources: the Kelvin–Helmholtz mechanism in an inert core, and hydrogen burning in a surrounding shell. The hydrogen burning occurs via the temperature-sensitive CNO bi-cycle, at a temperature $T_{\text{CNO}} \sim 3 \times 10^7\,\text{K}$. As the core contracts, its temperature increases. This in turn increases the temperature of the adjacent hydrogen-burning shell. However, a small uptick in the shell temperature T translates to a steep increase in its energy generation rate $\epsilon_{\text{CNO}} \propto T^{18}$. The resulting increase in energy density and pressure makes the shell expand. This expansion drives down the shell's density, and also decreases the temperature T by adiabatic cooling. The shell temperature is restored to its original value T_{CNO}. (This is an example of nuclear fusion acting as a thermostat.) However, the shell's density after expansion is now less than its original value. Thus, although thermal equilibrium is restored in the shell, it is now lower in pressure. To maintain hydrostatic equilibrium, this pressure must be equal to $P_{env} \sim GM_\star^2/R_\star^4$ (Equation 8.17). The star thus restructures itself to match the new lower pressure by increasing its radius $R_\star$.

Thus, there is a generic mechanism that forces shell-burning stars to expand, as long as their central core produces energy by the Kelvin–Helmholtz mechanism. Initially, the expanding star's effective temperature decreases as its radius increases; during this stage, the star moves rightward on an H–R diagram. However, when the swollen star reaches its Hayashi track, its value of T_{eff} remains nearly constant, and it becomes more luminous as it expands; during this stage, the star moves up the red giant branch on an H–R diagram.

The contraction of the inert helium core doesn't last forever. As the core shrinks, it becomes hotter. If the core reaches a temperature of $T_c \approx 10^8$ K, it will ignite helium burning via the triple alpha process, thus adding another source of fusion energy. However, if the core becomes dense enough, it will be supported by electron degeneracy pressure before reaching the helium ignition temperature. As we computed in Section 7.6, a shrinking self-gravitating sphere of mass M will be supported by degeneracy pressure when it reaches a radius $R_{deg} \propto M^{-1/3}$ and a central temperature $T_{deg} \propto MR_{deg}^{-1} \propto M^{4/3}$. Brown dwarfs with mass $M = 0.074 \, M_\odot$ or less are held up by degeneracy pressure before they reach the temperature $T \approx 10^7$ K needed to ignite hydrogen burning. Scaling with mass, we find that a core with $M_{core} < 0.4 \, M_\odot$ will be held up by degeneracy pressure before reaching helium ignition at $T \approx 10^8$ K. Similarly, a core with $M_{core} < 1.6 \, M_\odot$ will be held up before reaching carbon ignition at $T \approx 6 \times 10^8$ K. (Detailed numerical modeling gives a somewhat lower minimum core mass $M_{core} \approx 1 \, M_\odot$ for carbon ignition. It remains generally true, however, that each stage of stellar fusion has a higher ignition temperature, and thus requires a higher-mass core for ignition to occur.)

8.3 Shell Hydrogen Burning and Red Giants

Keeping in mind all the possible energy sources (the Kelvin–Helmholtz mechanism as well as different fusion reactions), and keeping in mind all the possible pressure sources (degeneracy as well as ideal gas pressure and radiation pressure), we can numerically model the post-main sequence evolution of stars. This permits a more detailed and quantitative study of evolved stars. Figure 8.5, for example, shows the evolution of our benchmark stars (1, 3, 10, and 30 $M_\odot$) on a theorists' H–R diagram. To keep the diagram from resembling the scrawls of a sugar-crazed

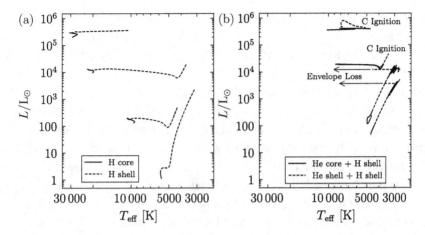

Figure 8.5 Evolution of stars with mass $M_\star = 1, 3, 10,$ and $30\,M_\odot$. (a) Core hydrogen burning (solid line) and shell hydrogen burning (dashed line) evolution. (b) Core helium + shell hydrogen burning (solid line) and shell helium + shell hydrogen burning (dashed line) evolution.

toddler, we have divided it into two panels: Figure 8.5(a) shows the earlier stages of evolution while Figure 8.5(b) shows the later stages.

Figure 8.5(a) shows the evolution of stars with core hydrogen burning, otherwise known as main sequence stars, as solid lines. For all stellar masses, the luminosity $L_\star$ increases gradually on the main sequence, as the central mean molecular mass increases and the central temperature consequently increases to maintain hydrostatic equilibrium. When core hydrogen is exhausted on the terminal age main sequence, the star begins its shell hydrogen-burning stage, shown with dashed lines. Initially, as we have seen, the shell hydrogen-burning star moves rightward on the H–R diagram, maintaining a nearly constant luminosity as its radius swells and its effective temperature drops. For the lower-mass stars, the rightward motion stops only when the evolving star reaches the Hayashi track for its mass. The still-swelling star then evolves upward (to higher luminosity) along the Hayashi track, thus avoiding the "forbidden zone" at low temperature.

In typical astronomy fashion, the labels used for post-main sequence stars are heterogeneous. Massive evolved stars, with $M_\star = 10\,M_\odot$ or more, are called blue supergiants if they are still near the main sequence, yellow supergiants if they are between the main sequence and the Hayashi track, and red supergiants if they are at or near the Hayashi track. At lower masses, evolved stars between the main sequence and the Hayashi track are called subgiants. When the lower-mass stars reach the Hayashi track (called the "red giant branch" in this context), they are called red giants, or RGB stars.

Figure 8.5(b) shows the evolution of post-main sequence stars once helium burning is ignited in their core at $T_c \approx 10^8$ K. The solid lines show the evolution of stars with core helium burning and shell hydrogen burning. The dashed lines show the evolution once helium is exhausted in the core and the star is powered by fusion in two nested shells; the inner shell is where helium burning occurs, while the outer shell is where hydrogen burning occurs. (There is an additional source of energy from the Kelvin–Helmholtz mechanism as the inert carbon/oxygen core contracts.)

It will not surprise you to learn that the labels used for these more evolved stars are also heterogeneous. Massive evolved stars that have entered the helium-burning phases of their life do not form a distinct observational class; since they remain at a roughly constant luminosity, they are still labeled as red or yellow supergiants, depending on their observed effective temperature. Lower-mass stars, once they start burning helium in their cores, run down the Hayashi track until they become red clump stars.[3] When helium is exhausted in their cores, low-mass stars run back up their Hayashi track. During this phase, the Hayashi track is called the "asymptotic giant branch," and the double shell-burning stars running up the Hayashi track are called asymptotic giant branch stars, or AGB stars. The complex set of labels applied to evolved stars is an unsurprising consequence of having observed classes of stars defined before the underlying physical processes that explain them were understood.

Now that we have laid out the labels that astronomers have applied to different classes of evolved stars, let's look at some of the more intriguing physical processes that occur within red giants. (In the next sections we will return to the more evolved core helium-burning stars and AGB stars.) As a red giant's envelope expands to reduce the pressure in the hydrogen-burning shell, the temperature of the star's photosphere drops. Just below the photosphere, the star develops a convective zone. As the star climbs up the red giant branch, its increasing luminosity causes the convective zone to spread deeper into the star's interior, as radiation becomes increasingly unable to transport the energy flux. When the outer convection zone of an evolved star extends deep enough, it has the consequence of dredging up nuclear burning products from the interior and mixing them with the gas of the envelope. One signature of material that has undergone nuclear processing is that it lacks lithium. The isotope ^{7}Li is produced in Big Bang Nucleosynthesis with an abundance $n_{Li}/n_H = 5.6 \times 10^{-10}$, customarily written as $A(Li) \equiv \log(n_{Li}/n_H) + 12 = 2.75$. Inside stars, however, lithium burning occurs by the reaction

$$^7Li + p \rightarrow \, ^8Be \rightarrow 2\,^4He, \tag{8.25}$$

with an ignition temperature $T \sim 2.5 \times 10^6$ K. Since this is lower than the ignition temperatures for the pp chain and CNO bi-cycle, any region of a star that has undergone hydrogen burning will be severely depleted in lithium. If convection goes deep enough to dredge up material that has undergone lithium

---

[3] The stellar models used in Figure 8.5 assume solar metallicity; low-mass, low-metallicity stars that burn helium in their cores are called "horizontal branch" stars, for reasons described below.

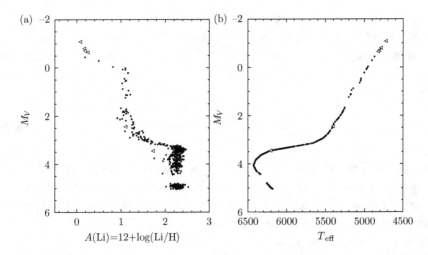

Figure 8.6 (a) Photospheric Li abundance for stars in the globular cluster NGC 6397. Triangles mark upper limits on the inferred Li abundance. (b) Position of the observed stars on an H–R diagram. [Data from Lind *et al.* 2009]

burning, the observational signature will be a dilution of the photospheric ^{7}Li abundance.

The convective mixing that brings fusion-processed material to the surface of red giants is called the **first dredge-up**; this is the first time in a star's history that it has an outer convection zone that reaches as deep as the central region where fusion has occurred. To see the effects of the first dredge-up, consider the old, metal-poor globular cluster NGC 6397. At $d \approx 2.4$ kpc, this is one of the closest globular clusters. The estimated age of NGC 6397 is $t_{cl} \approx 13.4$ Gyr, leading to a turnoff mass of $M_{TO} \sim 0.9\,M_\odot$.

Figure 8.6(b) shows an H–R diagram for NGC 6397. Stars dimmer than $M_V \approx 5$ were excluded from this sample, so we see a bit of the main sequence, the sub-giant bridge, and the red giant branch. Figure 8.6(a) shows the value of A(Li) deduced from the lithium lines in the stellar spectra. The main sequence stars have a lithium abundance of A(Li) ≈ 2.3. This value, implying a number density $\sim$35% of the BBN value, represents the protostellar lithium abundance of the stars in NGC 6397. However, in the middle of the subgiant sequence, at $M_V \sim 3.3$, there is a dramatic drop in lithium abundance, to A(Li) ~ 1.2. This represents the mixing of lithium-free material from deeper in the star into the outer layers, and is a signature of the first dredge-up.[4]

The first dredge-up allows processed gas to reach an evolved star's surface. For this processed gas to be incorporated into the next generation of stars, we need a

---

[4] In Figure 8.6(a), the second drop in Li abundance at $M_V \sim 0$ happens when the hydrogen-burning shell reaches the chemical discontinuity caused by the first dredge-up. An instability is set up that carries Li from the envelope down to hotter layers where it is destroyed.

mechanism to return matter to the interstellar medium. One such mechanism is a **stellar wind**. During a star's main sequence lifetime, stellar wind losses are insignificant except for the most massive, luminous O stars. For instance, the Sun's measured mass loss rate $\dot{M}_{wind} \approx 2.0 \times 10^{-8} \, M_\odot \, \text{Myr}^{-1}$ implies a fractional mass loss of

$$\frac{\Delta M}{M_\odot} \sim \frac{\dot{M}_{wind} t_{MS}}{M_\odot} \sim 2 \times 10^{-4}. \qquad (8.26)$$

However, as shown in Figure 1.11, hot main sequence stars of spectral type O can have high mass loss rates. A star with initial mass $M_\star = 30 \, M_\odot$ has a main sequence lifetime $t_{MS} \approx 5.5 \, \text{Myr}$. If it has a mass loss rate $\dot{M} \sim 1 \, M_\odot \, \text{Myr}^{-1}$ on the main sequence (typical for such a massive star), it will lose $\sim 20\%$ of its mass while still on the main sequence.

The culprit for the high mass loss rate from O stars is radiation pressure. As Figure 6.8 shows, the most massive stars have radiation pressure comparable to ideal gas pressure in their outer regions. A star supported by radiation pressure has a unique Eddington luminosity, $L_{Edd} \propto M_\star/\kappa$. If the total luminosity $L_\star$ of the massive star rises above L_{Edd}, then the low-density atmosphere of the star will become unbound and be blown away in a radiatively driven wind. Note that the criterion $L_\star > L_{Edd}$ can be reached either by an increase in the star's luminosity $L_\star$ or by an increase in the opacity κ of the star's atmosphere, which causes a decrease in L_{Edd}. High-metallicity O stars have enhanced opacity resulting from bound–free and bound–bound absorption; this strengthens the coupling between the radiation and the gas, and drives particularly strong winds.

Lower-mass stars, with no significant contribution from radiation pressure on the main sequence, do not develop strong stellar winds until they reach the red giant branch. The mass loss rate from red giants is not always large enough to give rise to a strong P Cygni line. Thus, the value of $\dot{M}_{wind}$ for red giants can be difficult to measure directly. However, it can be deduced indirectly. Consider an old globular cluster such as NGC 6397 (Figure 8.6), which has a turnoff mass $M_{TO} \sim 0.9 \, M_\odot$. In the same cluster, there also exists a population of horizontal branch stars, which have evolved as far as the core helium-burning phase. (These horizontal branch stars are hotter than the main sequence turnoff at $T_{eff} \approx 6400 \, \text{K}$, and thus are not shown in Figure 8.6.) The mass of the horizontal branch stars in NGC 6397 is $M_{HB} \sim 0.7 \, M_\odot$. This seems paradoxical at first glance. Stars that are now burning helium in their cores must have turned off the main sequence at some time in the past; since more massive stars have shorter main sequence lifetimes, the stars that have now evolved to the core helium-burning phase must have had $M_\star > M_{TO} \sim 0.9 \, M_\odot$ when they were on the main sequence. The apparent paradox is resolved if the stars that are now on the horizontal branch lost at least $\Delta M \sim 0.2 \, M_\odot$ to a stellar wind during their ascent of the red giant branch. Given a red giant lifetime of $t_{RGB} \sim 1 \, \text{Gyr}$ for stars with initial mass $M_\star \sim 1 \, M_\odot$, this implies a mean mass loss rate of $\dot{M}_{\mathrm{RGB}} \sim 2 \times 10^{-4}\,\mathrm{M_\odot\,Myr^{-1}}$ for red giants, four orders of magnitude greater than the Sun's main sequence mass loss rate.

It is possible for an evolved star to have a "superwind" with a mass loss rate $\dot{M} \sim 100\,\mathrm{M_\odot\,Myr^{-1}}$ or more. The star VY Canis Majoris, discussed in Section 1.3, is a prominent example of an evolved star with a superwind. However, these truly extravagant mass loss rates are a late development in a star's evolution, occurring on the asymptotic giant branch.

8.4 Helium Ignition and Core Helium Burning

Stars with a mass greater than $M_\star \approx 2.2\,\mathrm{M_\odot}$ have core masses large enough that they reach the temperature for helium ignition, $T_c \approx 10^8\,\mathrm{K}$, before degeneracy pressure plays an important role. These high-mass stars begin helium burning in a relatively peaceful manner. At the other end of the stellar mass range, stars with $M_\star < 0.4\,\mathrm{M_\odot}$ have cores that are fully supported by degeneracy pressure before the central temperature reaches $10^8\,\mathrm{K}$. These low-mass stars will never be powered by helium burning. Stars in the intermediate mass range, $0.4\,\mathrm{M_\odot} < M_\star < 2.2\,\mathrm{M_\odot}$, do eventually burn helium in their cores. However, helium ignition comes only after a relatively long subgiant lifetime, and the ignition, when it does arrive, happens with a **helium flash**.

To follow the path to helium ignition in this intermediate mass range, consider a TAMS star with $M_\star = 1\,\mathrm{M_\odot}$. Its central inert helium core has $M_{\mathrm{core}} \approx 0.02\,\mathrm{M_\odot}$. This is well below the S–C limit, so the star evolves into a subgiant with a broad hydrogen-burning shell surrounding a stable, nearly isothermal helium core. During this phase, the star's luminosity is $L_\star \sim 3\,\mathrm{L_\odot}$, supplied almost entirely by shell hydrogen burning. Thus, H will be converted to He and added to the helium core at a rate

$$\dot{M} = \frac{L_\star}{\eta_{\mathrm{nuc}}c^2} = 0.028\,\mathrm{M_\odot\,Gyr^{-1}}\left(\frac{L_\star}{3\,\mathrm{L_\odot}}\right). \tag{8.27}$$

At this rate, it takes $\sim$2 Gyr before the star's core reaches its S–C limit of $0.08\,\mathrm{M_\odot}$. When the core exceeds the S–C limit, it shrinks on its Kelvin–Helmholtz time, while the star's envelope starts to swell. As the star runs up the red giant branch, the core becomes steadily denser and hotter (as well as more massive, as the hydrogen-burning shell piles on more helium). The race between degeneracy, driven by increasing central density, and helium ignition, driven by increasing central temperature, is won in this case by degeneracy. For a 1 $\mathrm{M_\odot}$ star, and for all stars in the range 0.4–2.2 $\mathrm{M_\odot}$, the core becomes supported by electron degeneracy pressure before helium is ignited in the core. When helium burning starts in a degenerate core, it does so with a helium flash.

Why a "flash?" Consider what happens when helium burning begins while the core is supported by ideal gas pressure, $P \propto \rho T$. The gamma rays produced by the triple alpha process heat the gas. The increase in T then produces an increase in P and the core expands. However, the expansion of the core results in adiabatic cooling of the gas. The resulting drop in T results in a steep decrease in $\epsilon_{3\alpha}$. (This is another example of nuclear fusion acting as a thermostat.) Now, however, consider what happens when helium burning begins while the core is supported by non-relativistic electron degeneracy pressure, $P_{deg} \propto \rho^{5/3}$. In this case, the gamma rays produced by the triple alpha process still heat the gas – the atomic nuclei acquire faster thermal speeds – but the pressure of the gas is unaffected. Thus, the core fails to expand, and there is no adiabatic cooling. The higher T thus leads to a higher $\epsilon_{3\alpha}$, which leads to a higher T, which leads to a higher $\epsilon_{3\alpha}$, and so forth. It's a positive feedback loop that ends only when the temperature rises high enough for ideal gas pressure to dominate over electron degeneracy pressure. The helium flash itself, during which the core undergoes a thermal runaway, lasts only a short time. Numerical simulations indicate that the helium flash luminosity is $L > 10^9 \, L_\odot$ for roughly a week, reaching $L \sim 10^{11} \, L_\odot$ for a few seconds near the peak of the runaway. However, there are no immediate outward manifestations of this injection of $\sim 10^{48}$ erg of energy. Most of the energy goes to expand the core, while the remainder is absorbed by the star's massive, opaque envelope. After the degeneracy of the helium-burning core is lifted, the star settles into a stable state on its Kelvin–Helmholtz time.

An initially surprising feature of the helium flash is that the region of highest temperature is not necessarily at the core's center. The lowered central temperature is caused by a phenomenon that will become more dramatic in the later stages in massive star evolution: **neutrino cooling**. So far, we have regarded neutrinos as a minor by-product of nuclear fusion reactions. However, under some conditions, neutrinos can be produced in large numbers even in the absence of fusion reactions. Since these neutrinos interact only weakly, they rapidly leave the star, taking energy with them. One process by which neutrinos can remove thermal energy from a star is called the **Urca process**. Suppose that element E, with Z protons and $A - Z$ neutrons, undergoes an electron capture, becoming a nucleus of element D:

$$^{A}_{Z}\text{E} + e^- \rightarrow {}^{A}_{Z-1}\text{D} + \nu_e. \tag{8.28}$$

If this isotope of element D is unstable to beta decay, it undergoes the reaction

$$^{A}_{Z-1}\text{D} \rightarrow {}^{A}_{Z}\text{E} + e^- + \bar{\nu}_e. \tag{8.29}$$

There is no net nuclear processing from this pair of reactions; however, the neutrino and anti-neutrino carry away energy. The two reactions of the Urca process thus act as a source of neutrino cooling.[5] The Urca process depends sensitively

---

[5] The Urca process was named by George Gamow and Mário Schönberg after a casino in Rio de Janeiro. As Gamow later explained, "the Urca Process results in a rapid disappearance of thermal energy from the interior of a star, similar to the rapid disappearance of money from the pockets of the gamblers on the Casino da Urca."

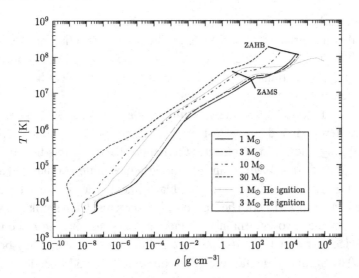

Figure 8.7 Temperature versus density within zero age horizontal branch (ZAHB) stars; that is, stars that have just started core helium burning. For comparison, the central temperature and density of zero age main sequence (ZAMS) stars are also shown. Gray lines: properties of 1 M$_\odot$ and 3 M$_\odot$ red giants at the instant of helium ignition.

on density, as do other similar processes that result in neutrino emission. Within a degenerate helium core, the density gradient is sufficiently steep that the higher neutrino cooling rate at the center causes a temperature inversion.

Once helium burning is ignited in the core, the star's internal structure is rearranged. If degeneracy was present before, it is now lifted. In the core, helium is being converted to carbon and oxygen by the triple alpha process. Since it has a fusion energy source, the core is no longer even approximately isothermal. Although the temperature at the core's center is $T_c \sim 10^8$ K (the temperature required for helium burning), the temperature at the core's surface is significantly lower. Since the hydrogen-burning shell is no longer required to be extremely hot, its energy generation rate drops. Thus, the luminosity of a 1 M$_\odot$ star decreases from $L_\star \sim 2500$ L$_\odot$ right before the helium flash (Figure 8.5(a)) to ~ 50 L$_\odot$ when it settles down to stable helium burning in its core (Figure 8.5(b)). The adjustment in the interior structure, and in the resulting luminosity, is milder for higher-mass stars; since they are not degenerate at helium ignition, this leads to a less severe structural shake-up.

Figure 8.7 shows the interior structure of model stars that have settled down to stable core helium burning. For each star, the central temperature is higher than in its main sequence precursor. The triple alpha reaction is very sensitive to temperature, with $\epsilon_{3\alpha} \propto T^{40}$; this has two practical consequences. First, a small range in T produces a very large range in L, so that all core helium-burning stars have similar central temperatures. Second, energy generation is strongly concentrated in the hottest region where helium fuel is available. Packing so much energy generation into a small volume requires a large energy flux to maintain thermal equilibrium. Even given the relatively low electron-scattering opacity κ_e in the hot core, this flux is too great to be transported by radiation. Thus, we expect helium-burning regions to be convective rather than radiative.

Note also in Figure 8.7 that the structure of the $1\,M_\odot$ and $3\,M_\odot$ stars are significantly different as red giants just before helium ignition (compare the solid and dashed gray lines). However, after the helium flash disrupts the structure of the $1\,M_\odot$ star, it settles down to an interior structure quite similar to that of the $3\,M_\odot$ star (compare the solid and dashed black lines). The similar interior structure of the $1\,M_\odot$ and $3\,M_\odot$ core helium-burning stars results in similar observable properties. On the zero age main sequence, a $3\,M_\odot$ star has an effective temperature twice that of a $1\,M_\odot$ star; as core helium-burning stars, they both have $T_{\text{eff}} \approx 5000$ K. Similarly, on the zero age main sequence, a $3\,M_\odot$ star is more than 100 times as luminous as a $1\,M_\odot$ star; in the core helium-burning stage, however, a $3\,M_\odot$ star has barely twice the luminosity of a $1\,M_\odot$ star.

Within a core helium-burning star, the envelope is not as strongly coupled to the (now convective) core as it was during the ascent of the red giant branch, when envelope expansion was closely linked to core shrinkage. In analogy with the (core hydrogen-burning) main sequence, there should be a relationship between the helium core mass M_{core} and the helium-burning luminosity L_{core}. We can therefore already draw an interesting conclusion: stars with $M_\star > 2.2\,M_\odot$ have a core mass M_{core} that is an increasing function of $M_\star$, and thus should have L_{core} also increasing with $M_\star$. However, a star with $M_\star < 2.2\,M_\odot$ must wait to ignite helium until its degenerate helium core grows to a mass $M_{\text{core}} \sim 0.5\,M_\odot$, a critical mass which to first approximation is independent of $M_\star$. Thus, all low-mass stars should have similar helium-burning luminosities.

If a star were powered *solely* by helium burning in its core, it would be analogous in many ways to a main sequence star, which is powered solely by hydrogen burning in its core. This would embolden us to use homology relations from Section 6.3 to compare the two types of star. For instance, a solar-metallicity ZAMS star has $\bar{\mu}_{\text{core}} \approx 0.6$ in its core, while a star beginning its core helium-burning phase has $\bar{\mu}_{\text{core}} \approx 1.3$. The homology relations in Equations 6.51 and 6.52 tell us that luminosity has a steep dependence on mean molecular mass (for instance, $L_\star \propto \bar{\mu}^4$ when electron scattering opacity dominates). Thus, we expect that a star powered only by core helium burning will be far more luminous than a ZAMS star of the same mass. However, when a star has a helium-burning core, its central temperature is more than four times the value it had as a hydrogen-burning ZAMS star (see Figure 8.7). Thus, the homology relation $T_c \propto \bar{\mu}M_\star/R_\star$ tells us that a star powered solely by core helium burning must also be more compact than its main sequence progenitor. The combination of large $L_\star$ and small

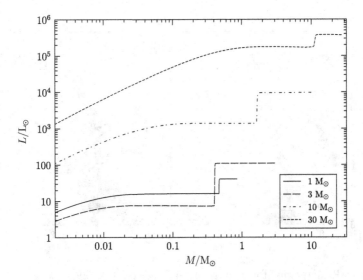

Figure 8.8 Enclosed luminosity versus mass for stars that have started core helium burning. Initial stellar masses are 1, 3, 10, and 30 M$_\odot$.

$R_\star$ implies an exceptionally high effective temperature for stars powered by core helium burning.

In reality, a normal star that burns helium in its core is also burning hydrogen in a shell. We do see some unusual stars, called **Wolf–Rayet stars**, that have helium-burning cores but whose hydrogen-rich envelopes have been lost, through either a strong stellar wind or tidal stripping by a binary companion. Wolf–Rayet stars are in fact very hot and compact, with effective temperatures as high as $T_{\rm eff} \sim 200\,000$ K. However, most core helium-burning stars are much cooler and larger than Wolf–Rayet stars. This is because of the presence of their hydrogen-burning shell, which ensures that the structure of a core helium-burning star is no longer even roughly homologous to that of a main sequence star.

The luminosity of normal core helium-burning stars is shown as a function of enclosed mass in Figure 8.8. The luminosity of the thin hydrogen-burning shell is visible as the abrupt jump in $L(M)$ outside the core. Both the 1 M$_\odot$ and 30 M$_\odot$ star have a shell hydrogen-burning luminosity comparable to the core helium-burning luminosity. For the 1 M$_\odot$ star, $L_{\rm shell} \sim L_{\rm core} \sim 25$ L$_\odot$, while for the 30 M$_\odot$ star, the luminosity is much higher, with $L_{\rm shell} \sim L_{\rm core} \sim 2 \times 10^5$ L$_\odot$. By contrast, the intermediate-mass stars, with $M_\star = 3$ M$_\odot$ and 10 M$_\odot$, have a shell luminosity an order of magnitude greater than the core luminosity. Given that $L_{\rm shell} \geq L_{\rm core}$, we expect the core helium-burning stars to be similar in structure (and similar in location on an H–R diagram) to red giant stars. In both cases, the luminosity of the hydrogen-burning shell dominates over the luminosity of the central core. (Whether the small core luminosity is supplied by the Kelvin–Helmholtz

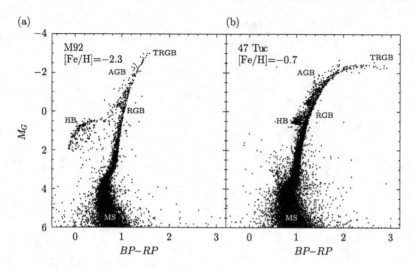

Figure 8.9 H–R diagrams for two globular clusters. (a) M92 ($t_{cl} \approx 12.5\,\text{Gyr}$). (b) 47 Tucanae ($t_{cl} \approx 11.4\,\text{Gyr}$). The main sequence (MS), red giant branch (RGB), tip of the red giant branch (TRGB), horizontal branch / red clump (HB), and asymptotic giant branch (AGB) are labeled. [Data from Vasiliev and Baumgardt 2021]

mechanism or by helium burning is not relevant to the observable properties of the star.)

Looking back at Figure 8.5(b), we find that the model stars do stay on the cool side of the H–R diagram during their core helium-burning phase. In particular, the $1\,M_\odot$ star has a luminosity of $L_\star \approx 50\,L_\odot$ and effective temperature $T_{\text{eff}} \approx 4700\,\text{K}$ during this phase. This places the $1\,M_\odot$ stellar model in the red clump that we see, for instance, in the solar neighborhood H–R diagram (Figure 1.10). It is no coincidence, we now realize, that the red clump falls on top of the red giant branch; the similar luminosity and color for these two stages of stellar evolution result from the fact that they have similar internal structure, with a hydrogen-burning shell producing half or more of the total luminosity.

It is possible, however, for a core helium-burning star to have an effective temperature much hotter than the value $T_{\text{eff}} \sim 5000\,\text{K}$ seen for stars in the red clump. For example, consider the two globular clusters whose H–R diagrams are given in Figure 8.9. The cluster M92 has an age of $t_{cl} \approx 12.5\,\text{Gyr}$, implying a turnoff mass $M_{\text{TO}} \sim 1\,M_\odot$. The cluster 47 Tucanae is slightly younger, with $t_{cl} \approx 11.4\,\text{Gyr}$; this gives it a turnoff mass just 3% greater than that of M92. Despite their similar turnoff masses, the H–R diagrams of the two clusters are noticeably different from each other. For instance, 47 Tuc (Figure 8.9(b)) has a distinct red clump at $M_G \sim 0.5$ and $BP - RP \sim 1$. By contrast, M92 (Figure 8.9(a)) lacks a red clump. Instead, it has a population of stars that are comparable in luminosity to the red clump of 47 Tuc, but which have a range of bluer colors, stretching as far as $BP - RP \sim -0.2$. These relatively blue stars occupy the **horizontal**

branch of the H–R diagram. In M92, the horizontal branch is seen to droop at its blue end. This is because although all stars on the horizontal branch are comparable in luminosity, the hottest stars produce much of their light at ultraviolet wavelengths. Thus, the adjective "horizontal," like many other terms used in astronomical nomenclature, shouldn't be taken literally.

The terms "red clump" and "horizontal branch" are empirical names, derived from the observed properties of stars. However, the deep interior of red clump stars and horizontal branch stars are essentially the same; they are all powered by core helium burning and shell hydrogen burning. What then causes the observed difference between the bluer stars in M92 and the redder stars in 47 Tuc? The major difference between the two clusters is that M92 is far poorer in metals, with an iron abundance [Fe/H] $= -2.3$, while 47 Tuc has [Fe/H] $= -0.7$. The higher-metallicity stars of 47 Tuc have high enough opacity to form a thick convective outer envelope, similar to that of a red giant star with the same luminosity. However, the lower-metallicity stars of M92 have lower opacity and thus do not form thick outer convection zones.

Horizontal branch stars (and red clump stars, as well) that started with $M_\star <$ 2.2 M$_\odot$ now all have roughly the same helium core mass, $M_{\mathrm{core}} \sim 0.5$ M$_\odot$. Thus, a lower total stellar mass $M_\star$ on the horizontal branch implies a lower envelope mass. In horizontal branch stars, the low envelope mass leads to a low luminosity for its hydrogen-burning shell. Furthermore, the low metal content starves the CNO bi-cycle of seed nuclei, further lowering the impact of shell hydrogen burning. Thus, horizontal branch stars move closer to the locus of pure core helium-burning stars, on the hot side of the H–R diagram.

Red clump and horizontal branch stars remain quiescently burning helium in their cores for approximately 100 Myr. Once helium is exhausted in a star's core, helium will be ignited in a shell surrounding the carbon/oxygen core. Thus, there will be two shell sources: an inner helium-burning shell and an outer hydrogen-burning shell. The minimum temperature for ignition of carbon burning, $T \sim$ 6×10^8 K, is much higher than the 10^8 K ignition temperature for helium burning. Just as when hydrogen was first exhausted in the core, we can again think of the core as an inert ash pile wrapped by a nuclear burning shell. This triggers a repeat of the processes that create red giants; the inert core becomes smaller and denser, while the hydrogen-rich envelope becomes larger and cooler. This time, the star is referred to as going up the **asymptotic giant branch** of the H–R diagram, labeled "AGB" in Figure 8.9.

During the asymptotic giant branch phase, we might expect an evolutionary sequence analogous to that seen in red giants; sufficiently massive stars will have carbon/oxygen cores that are already large enough to ultimately ignite carbon, while lower-mass stars will need to grow their cores until they can become hot enough to ignite carbon burning. It is true that 10 M$_\odot$ and 30 M$_\odot$ stellar models have large helium cores that become large carbon/oxygen cores, and that the C/O cores are initially large enough to ignite carbon burning. However, models show that stars with an initial mass less than $\sim 8\,M_\odot$ never ignite carbon, and end as white dwarfs. The failure of lower-mass stars to ignite carbon is explained by the fact that helium burning ($\eta_{He} \approx 0.000\,83$) has a much lower efficiency than hydrogen burning ($\eta_H = 0.007\,12 \sim 9\eta_{He}$). Consider, for instance, a star with initial mass $M_\star = 1\,M_\odot$. When it starts helium burning in its core, it has a core mass $M_{core} \approx 0.5\,M_\odot$, and its core helium-burning luminosity is about the same as its shell hydrogen-burning luminosity (see Figure 8.8). During the time it takes to burn the $0.5\,M_\odot$ of helium in its core, the shell burns only $\sim 0.06\,M_\odot$ of hydrogen; during the time it takes the core to burn the additional $0.06\,M_\odot$ of helium, the shell burns only $\sim 0.007\,M_\odot$ of hydrogen. This process converges on a carbon/oxygen core mass just $\sim 13\%$ larger than the mass of the original helium core. The only ways to form a very large carbon/oxygen core are to start with a very large helium core or to have a very high shell hydrogen-burning luminosity.

8.5 Asymptotic Giant Branch

A star in its asymptotic giant branch phase (an AGB star, for short) has two shells that produce nuclear fusion energy. The inner shell fuses helium into carbon and oxygen via the triple alpha process at $T \sim 10^8$ K; the outer shell fuses hydrogen into helium via the CNO bi-cycle at $T \sim 3 \times 10^7$ K. In general, these two shells eat outward through the AGB star at different rates; this can lead to instabilities. For instance, the outer hydrogen-burning shell eats outward through the star at the rate

$$\dot{M}_H = \frac{L_H}{X\eta_H c^2}, \tag{8.30}$$

where L_H is the luminosity of the hydrogen-burning shell, $X \approx 0.7$ is the hydrogen mass fraction just outside the shell, and $\eta_H = 0.007\,12$ is the efficiency of energy production for hydrogen burning. The inner helium-burning shell eats at the rate

$$\dot{M}_{He} = \frac{L_{He}}{Y\eta_{He} c^2}, \tag{8.31}$$

where L_{He} is the luminosity of the helium-burning shell, $Y \approx 0.98$ is the helium mass fraction just outside the shell, and $\eta_{He} \approx 0.000\,83$ is the efficiency of energy production for helium burning. Thus,

$$\frac{\dot{M}_{He}}{\dot{M}_H} = \frac{X}{Y}\frac{\eta_H}{\eta_{He}}\frac{L_{He}}{L_H} \approx 6\frac{L_{He}}{L_H}. \tag{8.32}$$

If the luminosity of the inner helium-burning shell is more than $\sim 16\%$ of the luminosity of the outer hydrogen-burning shell, the two shells will converge, with a decreasing mass of helium between them.

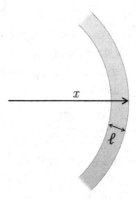

Figure 8.10 Nuclear fusion occurs in a shell of thickness ℓ, whose outer surface is at a distance x from the star's center.

For most of the time, an AGB star has $L_H > 6L_{He}$, and an increasingly massive layer of inert helium is built up between the outer hydrogen-burning shell and the inner helium-burning shell. The inert helium layer, with no fusion energy source, contracts gravitationally and becomes hotter. However, since the triple alpha rate is strongly dependent on temperature ($\epsilon_{3\alpha} \propto T^{40}$), an increase in temperature at the base of the inert helium layer leads to greatly increased L_{He} within the helium-burning shell. The helium-burning shell gobbles outward through the inert helium layer until it catches up with the hydrogen-burning shell. The proximity of the hotter helium-burning shell to the cooler hydrogen-burning shell causes the hydrogen-burning shell to get hotter (driving up L_H) and the helium-burning shell to get cooler (driving down L_{He}). Now $L_H > 6L_{He}$, and the dance between the two shells begins again.

A major cause for instability within AGB stars is that a star whose nuclear energy comes from one or more thin shells (this includes red giants and AGB stars) is subject to thermal instability. To see why thin shells are prone to this **shell flash** instability, while plump cores are not, consider the situation shown in Figure 8.10. Fusion occurs within a shell whose thickness is $\ell \ll x$, where x is the distance from the surface of the shell to the star's center. Since the shell is thin, we can write its mass as

$$M_{shell} \approx 4\pi x^2 \ell \rho, \qquad (8.33)$$

where ρ is the mass density of the shell. Suppose we slightly increase the thickness of the shell by moving the outer surface from x to $x+dx$, where $dx \ll \ell \ll x$. If we increase the volume of the shell in this way while keeping its mass constant, the density will naturally go down:

$$\frac{d\rho}{\rho} = -\frac{dx}{\ell} = -\frac{dx}{x}\frac{x}{\ell}. \qquad (8.34)$$

Both before and after the shell is thickened, it is in hydrostatic equilibrium. The mass interior to x is $M(x)$; the mass of the stellar envelope outside x is $M_{env} \equiv M_\star - M(x)$, where $M_\star$ is the total mass of the star. If the density of the outer envelope falls off rapidly with radius, we can approximate the pressure at x as the product of the gravitational acceleration,

$$g(x) = \frac{GM(x)}{x^2}, \tag{8.35}$$

and the mass column density of the outer envelope,

$$\Sigma(x) \approx \frac{M_{env}}{4\pi x^2}. \tag{8.36}$$

(This is equivalent to the approximation used in Equation 7.64 when we computed the pressure at a star's photosphere.) The pressure at x must therefore be

$$P(x) \sim g(x)\Sigma(x) \sim \frac{GM(x)M_{env}}{4\pi x^4}. \tag{8.37}$$

If we then move the upper surface of the shell upward by a small amount dx, the pressure required to maintain hydrostatic equilibrium goes down slightly:

$$\frac{dP}{P} = -4\frac{dx}{x}. \tag{8.38}$$

Now, by combining Equations 8.34 and 8.38, we find a relation between the density change and pressure change as the thin shell expands slightly:

$$\frac{dP}{P} = \frac{4\ell}{x}\frac{d\rho}{\rho}. \tag{8.39}$$

Thus, for a thin shell, with $\ell < 0.25x$, the fractional change in pressure as the shell expands will be smaller than the fractional change in density. If the pressure in the shell is ordinary ideal gas pressure, with $P \propto \rho T$, then the fractional change in temperature as the shell expands will be

$$\frac{dT}{T} = \frac{dP}{P} - \frac{d\rho}{\rho} = \left(\frac{4\ell}{x} - 1\right)\frac{d\rho}{\rho}. \tag{8.40}$$

For a shell with $\ell < 0.25x$, an expansion of the shell leads to an *increase* in temperature. The CNO bi-cycle and the triple alpha process are both highly sensitive to temperature. Thus, an increase in temperature, caused by expanding the shell, will lead to an increase in ϵ, which leads to a further increase in temperature. A positive feedback loop will cause the shell to keep increasing in temperature until the thickness of the shell expands to $\ell > 0.25x$, at which point it will be restored to equilibrium. Because the triple alpha process is so exquisitely sensitive to temperature, AGB stars, with their helium-burning shells, are particularly subject to shell flashes.

The comparison between helium flashes and shell flashes is an interesting one. Helium flashes occur because electron degeneracy pressure is completely insensitive to changes in temperature. In this case, the temperature goes upward while the pressure and density remain constant. The helium flash runaway ends when the ideal gas pressure exceeds the electron degeneracy pressure. Shell flashes, by contrast, occur because the pressure in a thin shell is almost completely insensitive to changes in the shell density. In this case, the temperature goes upward and the density goes downward while the ideal gas pressure ($P \propto \rho T$) remains nearly constant. The shell flash runaway ends when the thin shell becomes sufficiently thick.

8.6 Making s-Process Elements

Helium-burning shells in asymptotic giant branch stars are particularly interesting because they help to populate the periodic table with elements heavier than iron and nickel. Elements beyond the iron peak on the curve of binding energy (Figure 5.1) are less strongly bound than iron peak elements. At the iron peak, the isotopes ^{56}Fe, ^{58}Fe, and ^{62}Ni have a binding energy per nucleon of $B/A \approx$ 8.79 MeV; by contrast, ^{208}Pb, the heaviest stable isotope, has $B/A \approx 7.87$ MeV. This implies that many pathways to building ^{208}Pb are endothermic. Consider, for instance, a collection of 26 atoms of ^{56}Fe that you would like to convert into 7 atoms of ^{208}Pb. Both the initial and final states have 1456 nucleons, so no additional nucleons need to be imported. However, the initial iron atoms have a total binding energy of $B = 1456(8.79\,\text{MeV}) = 12.80$ GeV, while the final lead atoms have a total binding energy of $B = 1456(7.87\,\text{MeV}) = 11.46$ GeV. Thus, transmutation of this amount of iron into lead requires adding 1.34 GeV of energy to loosen nuclear bonds. In addition, since ^{208}Pb is more neutron-rich than ^{56}Fe, a total of 102 protons must be converted into neutrons in order to make the 7 lead atoms. If this is done by beta-plus decay ($p \rightarrow n + e^+ + \nu_e$), a minimum energy of 1.8 MeV per converted proton must be added. Thus, conversion of 26 ^{56}Fe atoms into 7 ^{208}Pb atoms by this pathway is endothermic:

$$1.52\,\text{GeV} + 26\,^{56}\text{Fe} \rightarrow 7\,^{208}\text{Pb} + 102e^+ + 102\nu_e. \qquad (8.41)$$

Converting a lump of iron into lead this way requires 220 MeV per lead atom formed; 88% of the energy goes to reduce the binding energy per nucleon, and the remainder goes to convert some of the nucleons from protons to neutrons.

It is possible to convert iron to lead by an exothermic reaction, but it requires the presence of free neutrons (which are in short supply in most of the universe). Suppose, for instance, you start with a collection of 41 atoms of ^{56}Fe that you want to convert into 13 atoms of ^{208}Pb. Both the initial and final states have 1066 protons, so no additional protons need to be imported. However, since ^{208}Pb is more neutron-rich than ^{56}Fe, a total of 408 free neutrons must be imported to convert this collection of iron atoms into lead atoms. The initial 41 iron atoms have a total binding energy of $B = 41 \cdot 56(8.79\,\text{MeV}) = 20.18\,\text{GeV}$, while the final 13 lead atoms have a total binding energy of $B = 13 \cdot 208(7.87\,\text{MeV}) = 21.28\,\text{GeV}$. Thus, the conversion of 41 ^{56}Fe atoms and 408 free neutrons into 13 ^{208}Pb atoms is exothermic:

$$41\,^{56}\text{Fe} + 408\,n \rightarrow 13\,^{208}\text{Pb} + 1.10\,\text{GeV}. \tag{8.42}$$

Converting a lump of iron into lead this way releases 85 MeV per lead atom formed. We conclude that building elements beyond the iron peak can be energetically favorable, but only in regions where there is a sufficiently high number density of free neutrons. A helium-burning shell in an AGB star is such a location.

Within an AGB star, elements beyond the iron peak are made by the **slow neutron-capture process**, known as the **s-process** for short. (The contrasting rapid neutron-capture process, or **r-process**, is discussed in Section 9.4.) The s-process involves a relatively slow process of capturing free neutrons by an atomic nucleus. When the temperature of a helium-burning shell rises above $T \sim 9 \times 10^7\,\text{K}$, free neutrons are produced by the reaction

$$^{13}\text{C} + {}^4\text{He} \rightarrow {}^{16}\text{O} + n. \tag{8.43}$$

The initial ^{13}C is present because it is a product of the CN cycle (Equation 5.80). When the temperature rises above $T \sim 3 \times 10^8\,\text{K}$, additional free neutrons are produced by the chain of reactions

$$^{14}\text{N} + {}^4\text{He} \rightarrow {}^{18}\text{F} + \gamma, \tag{8.44}$$

$$^{18}\text{F} \rightarrow {}^{18}\text{O} + e^+ + \nu_e \quad [\tau = 2.6\,\text{hr}], \tag{8.45}$$

$$^{18}\text{O} + {}^4\text{He} \rightarrow {}^{22}\text{Ne} + \gamma, \tag{8.46}$$

$$^{22}\text{Ne} + {}^4\text{He} \rightarrow {}^{25}\text{Mg} + n. \tag{8.47}$$

The initial ^{14}N is present because it is a product of the CN cycle (Equation 5.81) and the NO cycle (Equation 5.90). The number density of free neutrons produced by these reactions in a helium-burning shell is typically in the range $n_{\text{fn}} = 10^6$–$10^{10}\,\text{cm}^{-3}$. The density of atomic nuclei in the shell is sufficiently high that the free neutrons are absorbed by nuclei before they have time to decay. For a heavy element beyond the iron peak, the cross section σ_{nc} for capturing a neutron is (roughly speaking) inversely proportional to the neutron's speed. Over the relevant range of speeds for free neutrons produced in AGB stars, a typical heavy element has a rate coefficient

$$\langle \sigma_{\text{nc}} v_{\text{fn}} \rangle \sim 3 \times 10^{-17}\,\text{cm}^3\,\text{s}^{-1} \tag{8.48}$$

for neutron captures. Thus, the typical time between neutron captures for a heavy nucleus in a helium-burning shell is

$$t_{nc} \sim \frac{1}{n_{fn} \langle \sigma_{nc} v_{fn} \rangle} \sim 3 \times 10^8 \text{ s} \left(\frac{n_{fn}}{10^8 \text{ cm}^{-3}} \right)^{-1}. \tag{8.49}$$

This ranges from $t_{nc} \sim 40$ d for $n_{fn} \sim 10^{10}$ cm^{-3} to $t_{nc} \sim 1000$ yr for $n_{fn} \sim 10^6$ cm^{-3}. Saying that the s-process is *slow* is a statement that the time t_{nc} between neutron captures is longer than the typical lifetime τ before a neutron-rich nucleus undergoes beta decay.

Consider, for example, an AGB star that is fairly rich in metals that have been produced in a previous generation of stars. A nucleus of ^{56}Fe present in the helium-burning shell captures a free neutron:

$$^{56}\text{Fe} + n \rightarrow \ ^{57}\text{Fe} + \gamma. \tag{8.50}$$

Since ^{57}Fe is a stable isotope, it remains until it captures another free neutron:

$$^{57}\text{Fe} + n \rightarrow \ ^{58}\text{Fe} + \gamma. \tag{8.51}$$

Since ^{58}Fe is also stable, it remains until it captures yet another free neutron:

$$^{58}\text{Fe} + n \rightarrow \ ^{59}\text{Fe} + \gamma. \tag{8.52}$$

The isotope ^{59}Fe, however, is unstable to beta decay with a lifetime $\tau = 64$ d. If the neutron capture time t_{nc} is longer than 64 d, then the ^{59}Fe will most likely decay prior to capturing another neutron:

$$^{59}\text{Fe} \rightarrow \ ^{59}\text{Co} + e^- + \bar{v}_e. \tag{8.53}$$

The cobalt nucleus then captures a free neutron to become ^{60}Co, which is unstable to beta decay with a lifetime $\tau = 7.6$ yr. If the free neutron density is sufficiently low that t_{nc} is longer than 7.6 yr, the ^{60}Co will most likely decay:

$$^{60}\text{Co} \rightarrow \ ^{60}\text{Ni} + e^- + \bar{v}_e. \tag{8.54}$$

The nickel nucleus then captures free neutrons until it becomes ^{63}Ni, which is unstable to beta decay with a lifetime $\tau = 150$ yr:

$$^{63}\text{Ni} \rightarrow \ ^{63}\text{Cu} + e^- + \bar{v}_e. \tag{8.55}$$

If the ^{63}Ni captures a free neutron during its relatively long lifetime, it will continue as far as ^{65}Ni, which is unstable with a much shorter lifetime of $\tau = 3.6$ hr:

$$^{65}\text{Ni} \rightarrow \ ^{65}\text{Cu} + e^- + \bar{v}_e. \tag{8.56}$$

The s-process continues in this way, building heavier isotopes of a given element until it reaches an isotope with a lifetime shorter than the prevailing neutron capture time t_{nc}. The unstable isotope then decays, typically by emitting an electron and transmuting itself into an atom of the next higher atomic number.

The s-process path reaches as far as ^{208}Pb and ^{209}Bi. (Although ^{209}Bi is not absolutely stable, its lifetime $\tau \sim 3 \times 10^{19}$ yr means it is effectively stable in our youthful universe.) If ^{209}Bi captures a neutron, the resulting ^{210}Bi is unstable to beta decay with a relatively short lifetime of $\tau = 7$ d. Thus, in locations where the s-process occurs, ^{210}Bi will decay to ^{210}Po. This isotope of polonium is unstable with a lifetime of $\tau = 200$ d; however, it decays by the emission of an alpha particle to create a stable isotope of lead:

$$^{210}\text{Po} \rightarrow {}^{206}\text{Pb} + {}^{4}\text{He}. \tag{8.57}$$

The ^{206}Pb captures neutrons until it reaches the unstable isotope ^{209}Pb, which undergoes beta decay with a lifetime of $\tau = 4.7$ hr:

$$^{209}\text{Pb} \rightarrow {}^{209}\text{Bi} + e^{-} + \bar{\nu}_{e}. \tag{8.58}$$

Thus, attempts to build elements heavier than bismuth by the s-process end by looping back to ^{209}Bi. We know that isotopes with $A > 209$ exist in the universe; Figure 1.4, for instance, reminds us that the solar system contains the isotopes ^{232}Th and ^{238}U. These heavy, long-lived isotopes must be made by a process other than the s-process.

The production of elements by the s-process is observationally verified by the presence of technetium (Tc) absorption lines in the spectra of some AGB stars. Technetium, with atomic number $Z = 43$, is the lightest element with no stable isotopes. All isotopes of technetium have a lifetime of $\tau < 7$ Myr. Thus, the technetium present in an AGB star (an evolved star with an age of $t \gg 7$ Myr) must have been manufactured within the star itself. Any technetium present in the molecular gas from which the star formed has long since decayed.

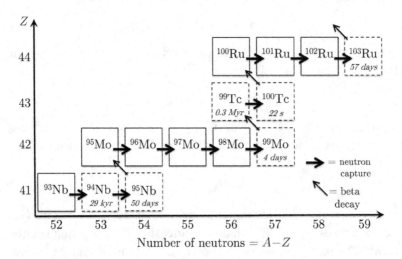

Figure 8.11 Portion of the s-process path, including niobium (Nb), molybdenum (Mo), technetium (Tc), and ruthenium (Ru). Isotopes unstable to beta decay are shown in dashed boxes.

Figure 8.11 shows the portion of the s-process path in which technetium is created. Molybdenum, with $Z = 42$, has many stable isotopes. Neutron capture, starting with ^{95}Mo, builds stable nuclei as far as ^{98}Mo. However, ^{99}Mo is unstable to beta decay, with a lifetime of $\tau = 4$ d. The ^{99}Tc that results from the decay of ^{99}Mo is also unstable to beta decay, but with a relatively long lifetime, $\tau_{99} = 0.30$ Myr. This is sufficiently long that the ^{99}Tc is much less likely to undergo beta decay than to capture a neutron and convert itself to ^{100}Tc. The beta decay of ^{100}Tc has a very much shorter lifetime, $\tau_{100} = 22$ s $\sim 2 \times 10^{-12} \tau_{99}$. We thus expect that the technetium present in the helium-burning shell of an AGB star is almost entirely in the form of ^{99}Tc. The technetium, along with other s-process elements, is then brought up to the photosphere of the AGB star by convection in the star's outer envelope.

8.7 Superwinds and Planetary Nebulae

Asymptotic giant branch stars, with two thin shells of energy generation, are more variable than red giant branch stars, and have stronger stellar winds. Measured mass loss rates for AGB stars generally lie in the range $0.01\,M_\odot\,\mathrm{Myr}^{-1} < \dot{M} < 10\,M_\odot\,\mathrm{Myr}^{-1}$. Measured wind speeds are low, compared to the fast winds emitted by hotter stars. Typical wind speeds of an AGB star lie in the range $10\,\mathrm{km\,s}^{-1} < v_{\mathrm{AGB}} < 50\,\mathrm{km\,s}^{-1}$. Since a star with initial mass $M_{\mathrm{init}} \sim 1\,M_\odot$ spends $t_{\mathrm{AGB}} \sim 1$ Myr on the asymptotic giant branch, the observed mass loss rates can strip away much of the star's mass and spread it over a distance $r \sim v_{\mathrm{AGB}} t_{\mathrm{AGB}} \sim 30$ pc. The exact mechanism driving stellar winds from AGB stars is not perfectly understood. A leading candidate is stellar pulsations, with a boost from radiation pressure acting on the dust grains that condense in the cool outer atmospheres of AGB stars.

One famous AGB star is the variable star Mira (omicron Ceti). Technically, we should refer to this star as "Mira A," since it is part of a wide binary system with another star, Mira B. At visible wavelengths, the apparent magnitude of Mira varies from $m_V \approx 9.5$ to $m_V \approx 3.5$ with a period $p = 332$ d. Since Mira is a cool star, with $T_{\mathrm{eff}} \approx 3000$ K, most of its luminosity emerges at near-infrared wavelengths. The average bolometric luminosity of Mira is $L_\star \approx 7000\,L_\odot$, and its mass is estimated to be $M_\star \approx 1.1\,M_\odot$. Mira has a slow stellar wind, with $v \sim 10\,\mathrm{km\,s}^{-1}$; however, the density of the wind is high enough to produce a mass loss rate of $\dot{M} \approx 0.3\,M_\odot\,\mathrm{Myr}^{-1}$. The star Mira is the prototype of an entire class of variable stars, known as **Mira variables**. The Mira variables are pulsating stars with red colors and with variability periods $\mathcal{P} > 100$ d. Mira variables are AGB stars that are approaching the end of their existence. The longest-period Mira variables, those with $\mathcal{P} > 2$ yr, have extremely strong winds, with $\dot{M} \sim 100\,M_\odot\,\mathrm{Myr}^{-1}$. These "superwinds," as they are called, must correspond to a short-lived epoch in

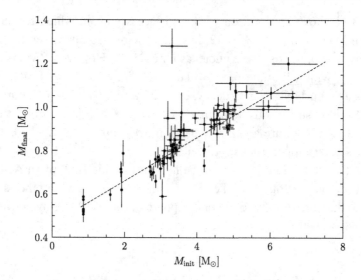

Figure 8.12 Relation between a star's initial mass and its final mass as a white dwarf. Data are shown for 78 white dwarfs from 13 star clusters with ages from 115 Myr to 10.2 Gyr. Sirius B is also shown (open circle), with an assumed age of $t = 245$ Myr for the Sirius system. [Data from Cummings *et al.* 2018]

a star's history. If Mira suddenly developed a superwind, increasing its mass loss rate by a factor of 300, it would blow itself away in just 10 000 years.

As a result of developing superwinds, AGB stars end by ejecting their outer envelope of gas (including any fusion shells) and leaving behind a naked, inert core of carbon and oxygen; this stellar remnant is a **white dwarf**. By looking at star clusters of different ages, where stars of different initial masses have reached the tip of the asymptotic giant branch, an empirical relation can be found between the initial mass M_{init} of a zero age main sequence star and the final mass M_{final} of the white dwarf left behind when the star has ejected its entire envelope (thus ending its AGB phase). In the mass range $0.8\,M_\odot < M_{init} < 7\,M_\odot$, the relation between initial and final mass can be fitted (with significant scatter) by a straight line, as shown in Figure 8.12. Numerically, the relation is

$$M_{final} = 0.10 M_{init} + 0.46\,M_\odot. \qquad (8.59)$$

Thus, a star that has $M_{init} = 1.2\,M_\odot$ on the main sequence will trim itself down to $M_{final} \sim 0.6\,M_\odot$, losing half its mass in a stellar wind (partly as a red giant, partly as an AGB star). A star that has $M_{init} = 7\,M_\odot$ on the main sequence will end with $M_{final} \sim 1.2\,M_\odot$, losing more than 80% of its initial mass (mostly on the AGB).

The winds of red giants and AGB stars have interesting chemical compositions, because of the dredge-up phases that occurred in their past. The first dredge-up happens when a star begins its ascent of the red giant branch. As seen in Figure 8.6, the gas that is dredged up to the surface of the red giant shows the lithium depletion that is characteristic of fusion-processed material. The material brought to the surface also shows the low C/N ratio that is characteristic of the CN cycle (see Figure 5.9). The second dredge-up occurs only for stars with initial masses in the range $4\,M_\odot < M_{init} < 8\,M_\odot$. In this mass range, a dredge-up happens as the star begins ascending the asymptotic giant branch. This second dredge-up brings helium, the end product of the hydrogen-burning shell, up to the surface, increasing the helium mass fraction from $Y \approx 0.27$ to $Y \approx 0.31$. The third dredge-up happens when a star with initial mass $M_{init} < 8\,M_\odot$ approaches the tip of the asymptotic giant branch. In this last dredge-up, convection can reach down as far as the helium-burning shell, bringing up carbon-rich material. Observationally, a star that has undergone the third dredge-up is a **carbon star**; that is, it contains more carbon than oxygen in its atmosphere, in contrast to the Sun, which contains more oxygen than carbon.[6] The third dredge-up that brings up carbon is also the process that brings the unstable element technetium from the helium-burning shell of an AGB star up to its atmosphere.

The last outburst of an AGB star's superwind leaves a naked carbon/oxygen core, surrounded by the slowly expanding remains of the star's envelope. The core is not hot enough to start carbon burning, so it contracts and heats up as it converts gravitational potential energy to thermal energy. The surface temperature of the carbon/oxygen core reaches $T_{eff} \sim 10^5\,K$ or higher. As such, it acts as a source of ionizing ultraviolet radiation, and converts the surrounding gas into an ionized emission nebula. An emission nebula of this type is known as a **planetary nebula**. William Herschel gave them the name "planetary" nebula because, as he explained in a 1782 letter, "these bodies appear to have a disk that is rather like a planet, that is to say, of equal brightness all over, round or somewhat oval, and about as well-defined as the disk of the planets."

The Saturn Nebula (NGC 7009), shown in Figure 8.13, was the first planetary nebula discovered by Herschel. It is a fairly typical planetary nebula (if there is such a thing as a "typical" planetary nebula).[7] The central naked core has a luminosity $L \sim 5000\,L_\odot$ and an effective temperature $T_{eff} \sim 80\,000\,K$. The measured expansion speed for the nebula indicates it reached the state seen in Figure 8.13 after $\sim$3000 yr of expansion. Planetary nebulae are short-lived; typically they are visible for $t_{pn} \sim 20\,000\,yr$. This short duration is primarily due to the expansion of the gas of which the nebula is made. The gas in the Saturn Nebula, for instance, is expanding at a speed $v_{pn} \approx 40\,km\,s^{-1} \approx 0.04\,pc\,kyr^{-1}$. By the end of 20 000 years, the gas of the planetary nebula is spread over a volume $\sim$0.8 pc in radius. The resulting low density of atoms and ions means that the surface brightness of the planetary nebula drops too low to be detected. The central naked core

---

[6] Carbon stars can be identified visually by their intense red colors; for example, the carbon star R Leporis, with $B - V \approx 4.7$, has the nickname "Hind's crimson star." The vivid red color is due to broad absorption bands from C_2 in the blue/green range of the spectrum.

[7] The ansae, or "handles," at either side of the Saturn Nebula are what reminded early observers of the ringed planet Saturn. However, the ansae of the Saturn Nebula are linear jets rather than a two-dimensional ring structure seen nearly edge on.

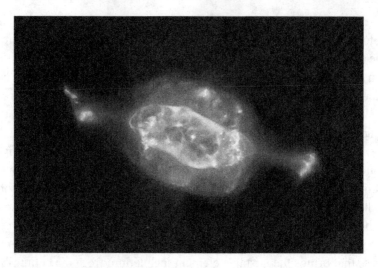

Figure 8.13 The Saturn Nebula, a planetary nebula at $d \sim 1.4\,$kpc. The field of view is $78 \times 51\,$arcsec, corresponding to $\sim 0.53 \times 0.35\,$pc at the distance of the nebula. [NASA/STScI]

continues to contract until it is supported by electron degeneracy pressure. This is the point at which it becomes a white dwarf.

Once shell flashes begin in an AGB star, the outer envelope is swiftly ejected. This has significant implications for the possibility of carbon burning within stars. A carbon/oxygen core requires a minimum mass $M_{core} \sim 1\,M_\odot$ for carbon ignition. As noted in Section 8.4, the inefficiency of helium burning means that, even under ideal conditions, a star can grow a carbon/oxygen core that is only 13% more massive than its previous helium core. However, the rapid stripping of the envelope by shell flashes means that AGB stars cannot reach this ideal maximum for the carbon/oxygen core mass. In practice, carbon ignition is possible only for stars that leave the main sequence with a helium core that already has $M_{core} > 1\,M_\odot$. From Figure 8.4, we see that terminal age main sequence stars have cores this massive when their total mass is $M_\star > 10\,M_\odot$. We therefore expect helium burning to be the end of the line for stars less massive than $M_\star \sim 10\,M_\odot$. Carbon burning, plus the more advanced burning stages outlined in Section 5.5, will be possible only in the most massive stars. The final burning stages for stars with $M_\star > 10\,M_\odot$ are dramatic, but they occur so rapidly that they barely register on the H–R diagram. We therefore present their story in the next chapter, as a prelude to the drama of supernova explosions and the exotic physics of neutron stars.

Exercises

8.1　How precisely do you need to measure the luminosity $L_\star$ of a turnoff star in order to determine a cluster age $t_{cl} = 10\,$Gyr with 10% precision? How precisely do you need to measure the effective temperature T_{eff} of a turnoff star to determine the same cluster age with the same precision?

8.2 Suppose that helium-burning cores within stars form a homologous family with specific energy generation rate $\epsilon \propto \rho^2 T^\nu$, ideal gas pressure, and radiative energy transport with electron scattering opacity. If we write the mass–radius relation for the helium-burning core as $R_{core} \propto M_{core}^a$, what is a as a function of ν? For what value of ν is R_{core} independent of M_{core}? Given $\nu \approx 40$ for the triple alpha process, do you expect high-mass cores to have a larger or smaller radius than low-mass cores?

8.3 It is possible to estimate the relative amount of energy radiated by a star in each phase of its existence. Suppose, for instance, that a star burns a fraction $f_{nuc} = 0.1$ of its hydrogen into helium on the main sequence, with efficiency $\eta_H = 0.00712$. The star then remains a red giant until it has burned a total of $0.5\,M_\odot$ of hydrogen into helium. (This assumes that the initial mass of the star was greater than $0.5\,M_\odot$.) The star then remains a core helium-burning star until its $0.5\,M_\odot$ helium core has been burned to a mix of carbon and oxygen, with efficiency $\eta_{He} = 0.00083$; you may assume that the hydrogen-burning shell has twice the luminosity of the helium-burning core.

(a) For a $1\,M_\odot$ star, what are the relative amounts of energy produced on the main sequence, red giant branch, and core helium-burning phase respectively?

(b) Repeat the calculations of part (a) for a $2\,M_\odot$ star and a $4\,M_\odot$ star.

(c) What core mass would you expect for each of these stars at the end of the core helium-burning phase?

8.4 For a massive star, assume that its luminosity from hydrogen burning (first in its core, then in a shell) remains constant over its entire fusion lifetime. You may also assume that core helium burning has about the same luminosity as shell hydrogen burning.

Massive stars appear as blue supergiants when undergoing core hydrogen burning, and as red supergiants when undergoing core helium burning. Given our assumptions above, what is the ratio of a massive star's lifetime as a red supergiant to its lifetime as a blue supergiant? In a population with ongoing star formation, do you expect the total luminosity from blue supergiants to be greater than or less than the total luminosity from red supergiants?

8.5 A long time ago in a galaxy far, far away, an instantaneous burst of star formation produced $10^{11}\,M_\odot$ of stars, with a minimum mass $M_{min} = 0.3\,M_\odot$ and a maximum mass $M_{max} = 100\,M_\odot$. Between these limits, the initial mass function was a Salpeter mass function, $N(M_\star) = KM_\star^{-2.35}$, where K is a normalization constant. At the present day, the stars with $M_\star = 1\,M_\odot$ have just run out of hydrogen in their cores.

(a) Compute the initial number of stars created, and the number still on the main sequence at the present day.

(b) Assume that all stars with initial mass $1\,M_\odot < M_\star < 8\,M_\odot$ are now white dwarfs. Compute the number of white dwarfs in the population today.

(c) Assume that all stars with initial mass $8\,M_\odot < M_\star < 20\,M_\odot$ are now neutron stars. Compute the number of neutron stars in the population today.

(d) Assume that all stars with initial mass $20\,M_\odot < M_\star < 100\,M_\odot$ are now black holes. Compute the number of black holes in the population today.

Ex-Stars

No memory of having starred
Atones for later disregard,
Or keeps the end from being hard.

Robert Frost (1874–1963)
"Provide, Provide" [1936]

As Russell and Vogt pointed out in the 1920s, the properties of a main sequence star depend crucially on its mass. After the main sequence, the star's mass is also vitally important in determining its physical properties. Will helium burning begin or not? If it begins, will it begin with a flash? Will carbon burning begin or not? The answers to these questions, as we have seen, depend primarily on the star's mass. The ultimate fate of a star is also mass-dependent. A stellar remnant can be a white dwarf supported by electron degeneracy pressure, a neutron star supported by neutron degeneracy pressure, or a black hole.

To start our study of stellar remnants, let's pause for a brief overview of how the life story of a self-gravitating gas ball depends on its initial mass M_{init} on the zero age main sequence. (Note: all the mass ranges given below are approximate, and assume an isolated, slowly rotating object of solar metallicity.)

$M_{init} < 0.074\,M_\odot$: These are brown dwarfs, never powered by fusion (except for brief deuterium burning). Brown dwarfs start as spectral type M, but drop in effective temperature as they cool. They end as balls of hydrogen and helium (and the metals they started with) supported by electron degeneracy pressure.

$0.074\,M_\odot < M_{init} < 0.4\,M_\odot$: These stars are M dwarfs on the main sequence. Stars in this mass range never become hot enough to ignite helium burning. They will eventually end as helium white dwarfs. Assuming a Chabrier initial mass function, given in Equation 7.90, about 66% of newly formed stars fall in this mass range.

$0.4\,M_\odot < M_{init} < 2.2\,M_\odot$: These are M, K, G, F, or A stars when they are on the main sequence. They go beyond hydrogen burning and start helium burning with a helium flash. These stars, after they run up the asymptotic giant branch, end as carbon/oxygen white dwarfs. About 30% of newly formed stars fall in this mass range.

$2.2\,M_\odot < M_{\text{init}} < 8\,M_\odot$: These are A or B stars when they are on the main sequence. They go beyond hydrogen burning but begin helium burning quietly, without a helium flash. These stars, after they run up the asymptotic giant branch, also end as carbon/oxygen white dwarfs. About 3% of newly formed stars fall in this mass range.

$8\,M_\odot < M_{\text{init}} < 10\,M_\odot$: These are B stars when they are on the main sequence. They go beyond hydrogen burning and helium burning to start carbon burning with a carbon flash. However, they do not become hot enough to ignite neon burning. Stars in this mass range end as oxygen/neon/magnesium white dwarfs. About 0.2% of newly formed stars fall in this mass range.

$10\,M_\odot < M_{\text{init}} < 15\,M_\odot$: These are B or O stars when they are on the main sequence. They run through the entire range of fusion reactions, from hydrogen burning through silicon burning. Their iron cores, however, are too massive to be supported by electron degeneracy pressure, and they end as neutron stars. About 0.2% of newly formed stars fall in this mass range.

$M_{\text{init}} > 15\,M_\odot$: These are O stars when on the main sequence. They run through the entire range of fusion reactions, from hydrogen burning through silicon burning. However, their final state is the result of chaotic events in their last stages of evolution, and is not a straightforward function of initial mass. Theoretical studies indicate that some of these initially massive stars end by collapsing directly to black holes. (Such an event is sometimes called a "failed supernova"; you might also call it, more positively, "successful black hole formation.") Other stars in this initial mass range temporarily form neutron stars, which later collapse to black holes when infalling matter pushes the neutron star over the upper mass limit for stability. Finally, some form neutron stars which accrete less matter and remain stable. Stars with $M_{\text{init}} > 15\,M_\odot$ represent about 0.3% of newly formed stars.

The maximum possible mass of a white dwarf, called the Chandrasekhar limit, is $M_{\text{Ch}} \approx 1.46\,M_\odot$. The maximum possible mass of a neutron star, called the Tolman–Oppenheimer–Volkoff (TOV) limit, is $M_{\text{TOV}} \sim 2.3\,M_\odot$. Thus, whether a star becomes a white dwarf, a neutron star, or a black hole depends on what its final mass is relative to the Chandrasekhar limit or the TOV limit. A star that starts on the zero age main sequence with $M_{\text{init}} < M_{\text{Ch}}$ will end as a white dwarf. (This assumes that it's not in a close binary and having mass dumped on it by its companion. For simplicity, we assume in this chapter that stars are isolated.) If a star begins its main sequence career with $M_{\text{init}} > M_{\text{Ch}}$, its fate will depend on the amount of mass that it loses both during and after its sojourn on the main sequence.

9.1 White Dwarfs

Compared to its fluffy, multi-layered AGB predecessor, a white dwarf is refreshingly simple. Since it is supported by non-relativistic electron degeneracy pressure, it is a polytrope of index $n = 1.5$:

$$P_{\text{deg,nr}} = K_{\text{nr}} \rho^{1+1/n} = K_{\text{nr}} \rho^{5/3}, \tag{9.1}$$

where the polytropic constant in this case is (compare to Equation 3.52)

$$K_{\text{nr}} = 0.603 \left(\frac{2}{\mu_e} \right)^{5/3} \frac{\hbar^2}{m_e m_{\text{AMU}}^{5/3}}. \tag{9.2}$$

In the above equation, we use the reduced Planck constant $\hbar$ instead of $h = 2\pi\hbar$, largely because it makes the numerical factor in front conveniently close to unity. We scale to $\mu_e = 2$, since that is the appropriate nucleon-to-electron ratio for the elements ^{4}He, ^{12}C, ^{16}O, ^{20}Ne, and ^{24}Mg of which white dwarfs are made. A white dwarf is a fairly stiff $n = 1.5$ polytrope in hydrostatic equilibrium. Thus, from Equation 6.27, its central pressure is

$$P_c = 0.771 \frac{G M_{\text{WD}}^2}{R_{\text{WD}}^4}. \tag{9.3}$$

However, the polytropic relation for a non-relativistic white dwarf (Equations 9.1 and 9.2) tells us that

$$P_c = 0.603 \left(\frac{2}{\mu_e} \right)^{5/3} \frac{\hbar^2}{m_e m_{\text{AMU}}^{5/3}} \rho_c^{5/3}. \tag{9.4}$$

Making use of the fact that the central density ρ_c is 5.991 times the mean density $\bar{\rho}$ for an $n = 1.5$ polytropic sphere, we can rewrite Equation 9.4 as

$$P_c = 1.095 \left(\frac{2}{\mu_e} \right)^{5/3} \frac{\hbar^2}{m_e m_{\text{AMU}}^{5/3}} \frac{M_{\text{WD}}^{5/3}}{R_{\text{WD}}^5}. \tag{9.5}$$

Combining Equation 9.3 with Equation 9.5 gives us the mass–radius relation for a non-relativistic white dwarf:

$$R_{\text{WD}} = 1.420 \left(\frac{2}{\mu_e} \right)^{5/3} \frac{\hbar^2}{G m_e m_{\text{AMU}}^2} \left(\frac{M_{\text{WD}}}{m_{\text{AMU}}} \right)^{-1/3}. \tag{9.6}$$

Scaled to a typical white dwarf mass, and assuming $\mu_e = 2$, the white dwarf's radius is

$$R_{\text{WD}} = 0.0151 \, R_\odot \left(\frac{M_{\text{WD}}}{0.6 \, M_\odot} \right)^{-1/3}. \tag{9.7}$$

Since larger white dwarf masses correspond to smaller radii, we expect the mean density of white dwarfs to increase fairly steeply with mass:

$$\bar{\rho}_{\text{WD}} = \frac{3}{4\pi} \frac{M_{\text{WD}}}{R_{\text{WD}}^3} = 0.0834 \left(\frac{\mu_e}{2} \right)^5 \frac{G^3 m_e^3 m_{\text{AMU}}^5}{\hbar^6} M_{\text{WD}}^2. \tag{9.8}$$

Scaled to a typical white dwarf mass, and assuming $\mu_e = 2$, the white dwarf's mean density is

$$\overline{\rho}_{WD} = 2.46 \times 10^5 \, \text{g cm}^{-3} \left(\frac{M_{WD}}{0.6 \, M_\odot} \right)^2 . \tag{9.9}$$

So far, we've assumed that the white dwarf is non-relativistic; that is, we assume that the Fermi momentum of the degenerate electrons is $p_F \ll m_e c$. However, as we saw in Section 3.4, degenerate electrons become relativistic at a mass density of (Equation 3.56)

$$\rho_{rel} = \frac{8\pi}{3} \frac{m_{AMU}}{\lambda_C^3} \mu_e = 0.067\,55 \left(\frac{m_e c}{\hbar} \right)^3 m_{AMU} \left(\frac{\mu_e}{2} \right) . \tag{9.10}$$

Inserting this critical density into the mass–density relation for a white dwarf (Equation 9.8), we find that degenerate electrons in the white dwarf become relativistic at a mass

$$M_{rel} \approx 0.90 \left(\frac{2}{\mu_e} \right)^2 M_L , \tag{9.11}$$

where the **Landau mass**[1] M_L is defined as

$$M_L \equiv \left(\frac{\hbar^3 c^3}{G^3 m_{AMU}^4} \right)^{1/2} = 3.74 \times 10^{33} \, \text{g} = 1.88 \, M_\odot . \tag{9.12}$$

The parameters that enter into the Landau mass remind us that the structure of a white dwarf is an interplay among gravity (G, m_{AMU}), quantum mechanics ($\hbar$), and special relativity (c). Notice, though, that the Landau mass is independent of the mass (m_e) of the particles supplying the degeneracy pressure.

If we adopt the Landau mass M_L as a natural mass scale, then the Schwarzschild radius R_L of the Landau mass can act as our length scale:

$$R_L \equiv \frac{2GM_L}{c^2} = 5.55 \, \text{km} = 7.98 \times 10^{-6} \, R_\odot . \tag{9.13}$$

In terms of M_L and R_L, the mass–radius relation for a non-relativistic white dwarf (Equation 9.6) can be rewritten as

$$R_{WD} = 0.710 R_L \left(\frac{2}{\mu_e} \right)^{5/3} \frac{m_{AMU}}{m_e} \left(\frac{M_{WD}}{M_L} \right)^{-1/3} . \tag{9.14}$$

Thus, even when the mass of a white dwarf approaches the Landau mass (at which point our assumption of non-relativistic electrons breaks down), its radius will be larger than its Schwarzschild radius by a factor $\sim 0.7 m_{AMU}/m_e \sim 1300$.

---

[1] The Landau mass is named after the physicist Lev Landau, who studied, among many other things, the effects of degeneracy on stellar structure.

As $M_{\rm WD} \rightarrow M_{\rm L}$, the degenerate electrons in a white dwarf's interior start to become relativistic. In the relativistic limit, a white dwarf is a polytrope of index $n = 3$:

$$P_{\rm deg,rel} = K_{\rm rel}\rho^{1+1/n} = K_{\rm rel}\rho^{4/3}, \tag{9.15}$$

where the polytropic constant in this case is (compare to Equation 3.59)

$$K_{\rm rel} = 0.307 \left(\frac{2}{\mu_e}\right)^{4/3} \frac{\hbar c}{m_{\rm AMU}^{4/3}}. \tag{9.16}$$

Relativistic white dwarfs, with $n = 3$, are softer than the non-relativistic variety, which have $n = 1.5$.

The increasing softness of a white dwarf, as its degenerate electrons become more nearly relativistic, has disastrous consequences. For the soft relativistic white dwarf to remain in hydrostatic equilibrium, it must have a central pressure (Equation 6.28)

$$P_c = 11.05 \frac{GM_{\rm WD}^2}{R_{\rm WD}^4}. \tag{9.17}$$

However, the polytropic relation for a relativistic white dwarf (Equations 9.15 and 9.16) tells us that

$$P_c = 0.307 \left(\frac{2}{\mu_e}\right)^{4/3} \frac{\hbar c}{m_{\rm AMU}^{4/3}} \rho_c^{4/3}. \tag{9.18}$$

Making use of the fact that the central density ρ_c is 54.18 times the mean density $\bar{\rho}$ for a soft $n = 3$ polytropic sphere, we can rewrite Equation 9.18 as

$$P_c = 9.319 \left(\frac{2}{\mu_e}\right)^{4/3} \frac{\hbar c}{m_{\rm AMU}^{4/3}} \frac{M_{\rm WD}^{4/3}}{R_{\rm WD}^4}. \tag{9.19}$$

Comparing Equation 9.17 with Equation 9.19, we find that a relativistic white dwarf is in hydrostatic equilibrium when

$$G\frac{M_{\rm WD}^2}{R_{\rm WD}^4} = 0.843 \left(\frac{2}{\mu_e}\right)^{4/3} \frac{\hbar c}{m_{\rm AMU}^{4/3}} \frac{M_{\rm WD}^{4/3}}{R_{\rm WD}^4}. \tag{9.20}$$

The factors of $R_{\rm WD}^4$ cancel out. This means that a relativistic white dwarf can be in hydrostatic equilibrium at a unique mass, known as the **Chandrasekhar limit**,[2]

$$M_{\rm Ch} = (0.843)^{3/2} \left(\frac{2}{\mu_e}\right)^2 \left(\frac{\hbar^3 c^3}{G^3 m_{\rm AMU}^4}\right)^{1/2} \tag{9.21}$$

$$= 0.774 \left(\frac{2}{\mu_e}\right)^2 M_{\rm L} = 1.46 \, M_\odot \left(\frac{2}{\mu_e}\right)^2.$$

---

[2] The Chandrasekhar limit is named after the astronomer Subrahmanyan Chandrasekhar, who famously calculated it during August 1930, on a long journey from Bombay to Venice aboard the SS *Pilsna*.

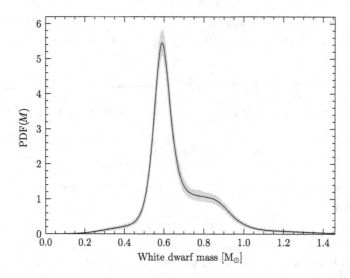

Figure 9.1 Distribution of masses for 1772 white dwarfs within 100 pc of the Sun. Gray band = 98% confidence interval on the mass function. [Data from Kilic *et al.* 2020]

The Chandrasekhar limit is very similar to (actually a bit smaller than) our estimate of the mass at which a white dwarf becomes relativistic, as given in Equation 9.11. Thus, as soon as a white dwarf becomes relativistic, it is doomed to collapse; at any mass greater than M_{Ch}, the pressure required by hydrostatic equilibrium (Equation 9.17) is higher than the pressure provided by relativistic degeneracy pressure (Equation 9.19), and the white dwarf collapses.

White dwarfs made of helium, carbon, oxygen, neon, and/or magnesium have $\mu_e = 2$, and their Chandrasekhar limit is $M_{Ch} = 1.46\,M_\odot$. However, for a white dwarf made of ^{56}Fe, $\mu_e = 56/26 = 2.154$, and the Chandrasekhar limit is lowered to $M_{Ch} = 1.26\,M_\odot$. The observed mass function for white dwarfs in the solar neighborhood is plotted in Figure 9.1. Note that the mass function peaks at $M_{WD} = 0.59\,M_\odot$, well below the Chandrasekhar limit. Only ~1% of the white dwarfs in this local sample have $M_{WD} > 1.2\,M_\odot$, with the most massive white dwarf having an estimated mass of $M_{WD} \approx 1.35\,M_\odot$. At the other end of the white dwarf mass function, about 2% of local white dwarfs have $M_{WD} < 0.3\,M_\odot$. The existence of such low-mass white dwarfs seems odd at first glance. After all, a star with $M_{init} \approx 0.9\,M_\odot$, with a lifetime comparable to the age of the universe, produces a white dwarf with $M_{WD} \sim 0.55\,M_\odot$ (Figure 8.12). A white dwarf with a significantly lower mass than this would seemingly require a lower M_{init}, implying a stellar lifetime longer than the age of the universe. The resolution of the paradox comes when we realize that extremely low-mass white dwarfs are nearly all found in short-period binary systems. In such a system, as we discuss in Chapter 12, a close companion can strip away the outer layers of a star, leaving behind, in this case, an underweight degenerate core.

As an example of a white dwarf with well-determined properties, consider Sirius B, the nearest white dwarf. It forms a binary system with a luminous main sequence star, Sirius A. The *Gaia* parallax for Sirius B, $p = 374.49 \pm 0.23$ mas, implies a distance $d = 1/p = 2.6703 \pm 0.0016$ pc. Since Sirius A and B form a visual binary, we can compute masses for both the star Sirius A ($M_A = 2.137\,M_\odot$) and the white dwarf Sirius B ($M_B = 1.054\,M_\odot$). The spectrum of Sirius B displays pressure-broadened Balmer lines from hydrogen; this implies that the photosphere of Sirius B lies within an atmosphere of hydrogen-rich, non-degenerate gas, lying atop the hydrogen-depleted, degenerate core.[3]

Fits to the spectrum of Sirius B give an effective temperature of $T_{\mathrm{eff,B}} = 25\,900$ K and a surface gravity of $\log g_B = 8.60$. This value of $\log g_B$, combined with the dynamical mass $M_B = 1.054\,M_\odot$, implies a photospheric radius $R_B \approx 0.0086\,R_\odot \approx 6000$ km. The measured effective temperature and surface gravity also imply that the atmosphere, made mostly of ionized hydrogen, has a pressure scale height

$$h_{\mathrm{atm}} \sim \frac{kT_{\mathrm{eff,B}}}{g_B(0.5m_H)} \sim 0.1 \text{ km}, \tag{9.22}$$

very much smaller than the radius R_B of the white dwarf. Combining the radius R_B with the effective temperature $T_{\mathrm{eff,B}}$ yields a luminosity

$$\frac{L_B}{L_\odot} = \left(\frac{R_B}{R_\odot}\right)^2 \left(\frac{T_{\mathrm{eff,B}}}{T_{\mathrm{eff},\odot}}\right)^4 \approx \left(\frac{0.0086}{1}\right)^2 \left(\frac{25\,900 \text{ K}}{5772 \text{ K}}\right)^4 \approx 0.030. \tag{9.23}$$

The observationally deduced radius $R_B \approx 0.0086\,R_\odot$ is $\sim$30% smaller than the radius calculated from the theoretical mass–radius relation of Equation 9.7. This is because Sirius B is massive enough that the degenerate electrons at its dense center are partially relativistic; this softens the central regions of the white dwarf and permits gravity to squeeze it to a smaller radius.

A white dwarf can thus be approximated as a spherical core of radius R_{core}, supported by electron degeneracy pressure, wrapped in an atmosphere with scale height $h_{\mathrm{atm}} \ll R_{\mathrm{core}}$, supported by ideal gas pressure. (The transition region between degeneracy and non-degeneracy is actually of finite thickness; however, for mathematical convenience, we'll treat it as infinitesimally thin.) The degenerate core contains a mass M_{core} that is only slightly less than the total mass M_{WD} of the white dwarf. Within the core, efficient conductive transport by degenerate electrons ensures that the temperature T_{core} is nearly uniform. The non-degenerate atmosphere has a mass $M_{\mathrm{atm}} \ll M_{\mathrm{WD}}$ and a temperature that decreases from T_{core} at its base, where it is in thermal contact with the core, to $T_{\mathrm{eff}} < T_{\mathrm{core}}$ at the photosphere. Although the atmosphere is physically thin, with $h_{\mathrm{atm}} \ll R_{\mathrm{core}}$, it is optically thick, and thus hides the degenerate core from direct observation.

---

[3] About 80% of white dwarfs, including Sirius B, have spectra with strong absorption lines from hydrogen. Among the remaining 20%, most have spectra dominated by absorption lines from neutral helium or, at higher effective temperatures, from singly ionized helium.

We can only indirectly deduce, for instance, what the temperature T_{core} of the degenerate core must be.

It's useful to know the temperature T_{core}, since it's key to determining how long a white dwarf can remain shining. The energy that a (non-fusing, non-shrinking) white dwarf radiates from its atmosphere comes almost entirely from the thermal energy of the atomic nuclei in its core. The degenerate electrons in the core can't radiate away their energy, and the number of non-degenerate particles in the atmosphere is small. The total thermal energy of the nuclei in the core is

$$K_{core} \approx \frac{3}{2} \frac{kT_{core}}{m_{AMU} \bar{\mu}_n} M_{WD},$$ (9.24)

where $\bar{\mu}_n$ is the mean molecular mass of the atomic nuclei. For a carbon/oxygen white dwarf, $\bar{\mu}_n \approx 14$, and the thermal energy of the nuclei is

$$K_{core} \approx 1.8 \times 10^{47} \, \mathrm{erg} \left(\frac{T_{core}}{10^7 \, \mathrm{K}} \right) \left(\frac{M_{WD}}{1 \, M_\odot} \right).$$ (9.25)

The **thermal time** for a white dwarf (or any other warm object) is the time required for it to radiate away its thermal energy, assuming constant luminosity. Scaling to properties appropriate for Sirius B, the thermal time is

$$t_{WD} \equiv \frac{K_{core}}{L_{WD}} \approx 50 \, \mathrm{Myr} \left(\frac{T_{core}}{10^7 \, \mathrm{K}} \right) \left(\frac{M_{WD}}{1 \, M_\odot} \right) \left(\frac{L_{WD}}{0.03 \, L_\odot} \right)^{-1}.$$ (9.26)

If Sirius B has an interior temperature of $10^7 \, \mathrm{K}$, then it can keep shining for $\sim 50 \, \mathrm{Myr}$. But how can we determine the actual value of T_{core}? To find the temperature of the core, we need to look more closely at the non-degenerate atmosphere that permits the thermal energy K_{core} to gradually leak out into space.

In the physically thin atmosphere, the equation of hydrostatic equilibrium can be written as

$$\frac{dP}{dr} \approx -GM_{WD} \frac{\rho}{r^2},$$ (9.27)

where $\rho(r)$ is the mass density of the atmosphere. Energy is transported through the atmosphere radiatively, so the equation of energy transport is

$$\frac{dT}{dr} \approx -\frac{3L_{WD}}{16\pi ac} \frac{\kappa \rho}{T^3 r^2},$$ (9.28)

where $\kappa(r)$ is the opacity of the atmosphere and $T(r)$ is its temperature.

From Equations 9.27 and 9.28, we find a relation among pressure, temperature, and opacity in a white dwarf atmosphere:

$$\frac{dP}{dT} \approx \frac{16\pi acG}{3} \frac{M_{WD}}{L_{WD}} \frac{T^3}{\kappa}.$$ (9.29)

Since the atmosphere is non-degenerate, its pressure is given by the ideal gas law,

$$P = \frac{k}{m_{AMU} \bar{\mu}} \rho T.$$ (9.30)

Most white dwarfs have atmospheres sufficiently dense and hot that bound–free and free–free opacity dominates. This means that the opacity κ is given by Kramers' law:

$$\kappa = \kappa_0 \rho T^{-3.5} = \frac{\kappa_0 m_{\text{AMU}} \bar{\mu}}{k} P T^{-4.5}. \tag{9.31}$$

Putting the opacity of Equation 9.31 into Equation 9.29, we find a readily integrable relation between P and T in the white dwarf's non-degenerate atmosphere:

$$P dP \approx \frac{16\pi acGk}{3\kappa_0 m_{\text{AMU}} \bar{\mu}} \frac{M_{\text{WD}}}{L_{\text{WD}}} T^{7.5} dT. \tag{9.32}$$

Using the boundary condition that $P \to 0$ and $T \to 0$ as $r \to \infty$, the pressure in the atmosphere is

$$P(T) \approx \left[\frac{64\pi acGk}{51\kappa_0 m_{\text{AMU}} \bar{\mu}} \right]^{1/2} \left(\frac{M_{\text{WD}}}{L_{\text{WD}}} \right)^{1/2} T^{4.25}. \tag{9.33}$$

Using the ideal gas law, we can write the density ρ of the atmosphere as a function of its temperature T:

$$\rho(T) \approx \left[\frac{64\pi acGm_{\text{AMU}} \bar{\mu}}{51\kappa_0 k} \right]^{1/2} \left(\frac{M_{\text{WD}}}{L_{\text{WD}}} \right)^{1/2} T^{3.25}. \tag{9.34}$$

If you dive downward by one pressure scale height in a white dwarf's atmosphere, so that P increases by a factor $e \approx 2.72$, the temperature will increase by a factor $\exp(1/4.25) \approx 1.27$ and the density by a factor $\exp(3.25/4.25) \approx 2.15$.

If you dive all the way to the base of the atmosphere at $r = R_{\text{core}}$, you reach the point where the temperature T of the atmosphere is equal to the temperature T_{core} of the isothermal degenerate core. On either side of the $r = R_{\text{core}}$ surface, the atomic nuclei have the same ideal gas pressure. However, at $r < R_{\text{core}}$ the electrons have a degeneracy pressure, and at $r > R_{\text{core}}$ the electrons have an ideal gas pressure. The electron pressures on the two sides of the surface must be equal to maintain hydrostatic equilibrium. Therefore,

$$\left[n_e k T_{\text{core}} \right]_{r=R_{\text{core}}} = \left[\frac{(3\pi^2)^{2/3}}{5} \hbar^2 m_e^{-1} n_e^{5/3} \right]_{r=R_{\text{core}}}. \tag{9.35}$$

The transition from degeneracy to non-degeneracy is not accompanied by a discontinuous leap in electron density n_e. Thus, assuming that n_e is the same at the base of the atmosphere as at the top of the core, we can solve Equation 9.35 to yield an electron number density of

$$n_e(R_{\text{core}}) = \frac{5\sqrt{5}}{3\pi^2} \left(\frac{m_e k T_{\text{core}}}{\hbar^2} \right)^{3/2}. \tag{9.36}$$

This corresponds to a mass density

$$\rho(R_{\text{core}}) = \mu_e m_{\text{AMU}} n_e \approx 0.755 \left(\frac{\mu_e}{2}\right) m_{\text{AMU}} \left(\frac{m_e k}{\hbar^2}\right)^{3/2} T_{\text{core}}^{1.5}, \tag{9.37}$$

where μ_e is the number of nucleons per electron. However, Equation 9.34 gives us another relation for the density at the base of the atmosphere:

$$\rho(R_{\text{core}}) \approx 1.99 \left[\frac{acGm_{\text{AMU}}\bar{\mu}}{\kappa_0 k}\right]^{1/2} \left(\frac{M_{\text{WD}}}{L_{\text{WD}}}\right)^{1/2} T_{\text{core}}^{3.25}. \tag{9.38}$$

Combining Equations 9.37 and 9.38, we end with an equation that tells us the luminosity of a white dwarf with a given mass M_{WD} and core temperature T_{core}:

$$L_{\text{WD}} = C_{\text{WD}} M_{\text{WD}} \frac{T_{\text{core}}^{3.5}}{\kappa_0}. \tag{9.39}$$

The constant C_{WD} depends on the composition of the white dwarf's atmosphere (via $\bar{\mu}$) and its degenerate core (via μ_e), but otherwise includes only physical constants. For a carbon/oxygen white dwarf whose atmosphere has a physically plausible opacity, the luminosity becomes

$$L_{\text{WD}} \approx 0.004 \, L_\odot \left(\frac{M_{\text{WD}}}{0.6 \, M_\odot}\right) \left(\frac{T_{\text{core}}}{10^7 \, \text{K}}\right)^{3.5}. \tag{9.40}$$

Alternatively, if you see a carbon/oxygen white dwarf with luminosity L_{WD} and mass M_{WD}, you can deduce that its internal core temperature is

$$T_{\text{core}} \approx 1.3 \times 10^7 \, \text{K} \left(\frac{M_{\text{WD}}}{0.6 \, M_\odot}\right)^{-2/7} \left(\frac{L_{\text{WD}}}{0.01 \, L_\odot}\right)^{2/7}. \tag{9.41}$$

For Sirius B, this implies a core temperature $T_{\text{core}} \approx 1.5 \times 10^7 \, \text{K}$.

We can combine Equations 9.26 and 9.41 to write the white dwarf's thermal time in terms of the white dwarf's luminosity rather than its core temperature:

$$t_{\text{WD}} \approx 110 \, \text{Myr} \left(\frac{M_{\text{WD}}}{0.6 \, M_\odot}\right)^{5/7} \left(\frac{L_{\text{WD}}}{0.01 \, L_\odot}\right)^{-5/7}. \tag{9.42}$$

The cooling of the white dwarf is temporarily paused, however, when its core temperature reaches $T_{\text{core}} \sim 4 \times 10^6 \, \text{K}$. At this temperature, the nuclei within the core start to form a crystalline lattice. The release of the latent heat of crystallization, $\epsilon_{\text{cry}} \sim 3.5 \times 10^{13} \, \text{erg g}^{-1} \sim 4 \times 10^{-8} c^2$, keeps the luminosity of a $0.6 \, M_\odot$ white dwarf at a value $L_{\text{WD}} \sim 2 \times 10^{-4} \, L_\odot$ for a time $t_{\text{cry}} \sim 2 \, \text{Gyr}$. The dimmest known white dwarfs have $L_{\text{WD}} \sim 3 \times 10^{-5} \, L_\odot$; these represent early-forming white dwarfs that have passed through the crystallization process and are now cooling again. These old dim white dwarfs have an effective temperature $T_{\text{eff}} \sim 3800 \, \text{K}$ or less, which means that it is possible to talk about a white dwarf that is red. (But not a red dwarf that is white.)

9.2 Neutron Stars and Black Holes

Stars with an initial mass greater than $\sim 10\,M_\odot$ are very rare. (For convenience, in this section we will refer to stars with $M_{init} \geq 10\,M_\odot$ as "high-mass stars.") Few high-mass stars are made in the first place. From the Chabrier initial mass function, we find that

$$\frac{N(>10\,M_\odot)}{N(>0.074\,M_\odot)} \sim 5 \times 10^{-3}. \qquad (9.43)$$

In addition, the time that a star with $M_{init} > 10\,M_\odot$ actually spends as a star is very short. Stellar models indicate that stars with $M_{init} = 10\,M_\odot$ have a main sequence lifetime $t_{MS} \approx 20\,\text{Myr}$, while the most massive stars known, with $M_{init} \sim 100\,M_\odot$, have a main sequence lifetime $t_{MS} \sim 2\,\text{Myr}$.

When we look at the evolution of high-mass stars on a Hertzsprung–Russell diagram (Figure 8.5), we find that they remain roughly constant in luminosity as they evolve off the main sequence, in contrast to low-mass stars, which are significantly brighter as red giants than as main sequence stars. The consistently high luminosity of high-mass stars, as they evolve from blue supergiants to yellow supergiants to red supergiants, is supplied by nuclear fusion. Consider, for example, a star with initial mass $M_{init} \approx 10\,M_\odot$. It remains on the main sequence for $\sim 20\,\text{Myr}$, burning hydrogen in its core by the CNO bi-cycle. Once the central hydrogen is exhausted, the core contracts and heats up, igniting helium burning. The star then spends $\sim 1\,\text{Myr}$ burning helium in its core. Once the central helium is exhausted, the core contracts and heats up, igniting carbon burning. The star then spends $\sim 0.3\,\text{kyr}$ burning carbon in its core. Affairs now move very rapidly. Oxygen burning in the core lasts for $\sim 1\,\text{yr}$, and silicon burning lasts for just $\sim 1\,\text{week}$.

Silicon burning occurs at a temperature $T \sim 3 \times 10^9\,\text{K}$. At this temperature, the critical density at which ideal gas pressure gives way to degeneracy pressure (Equation 3.78) is $\rho_{i\text{-}d} \sim 10^7\,\text{g cm}^{-3}$. As silicon burning proceeds, a core of iron peak elements is built up. This inert core is sometimes referred to as the "iron core" of the star; however, detailed modeling of silicon burning reveals that the ashes contain ^{56}Ni and ^{58}Ni as well as ^{54}Fe and ^{56}Fe. Initially, the iron/nickel core is low enough in density to be supported by thermal pressure. However, as it gravitationally contracts, it passes the threshold at which electron degeneracy pressure becomes dominant. After this point is reached, it is only a matter of time until the degenerate iron/nickel core reaches its Chandrasekhar limit.

When the degenerate core of a high-mass star exceeds the Chandrasekhar limit, it starts to contract. As it does so, its temperature rises steeply, and photons become energetic enough to photodisintegrate iron and nickel. For instance,

$$^{56}\text{Ni} + 100\,\text{MeV} \rightarrow 14\,^4\text{He}. \qquad (9.44)$$

This reaction is highly endothermic, absorbing about 2 MeV per nucleon. The loss of energy causes a drop in pressure; the contraction thus speeds up, reaching nearly freefall speeds. The temperature rises still more, and photons become energetic enough to photodisintegrate helium:

$$^4\text{He} + 28\,\text{MeV} \rightarrow 2p + 2n. \tag{9.45}$$

The core falls inward faster and faster. The density becomes high enough for protons to capture free electrons:

$$p + e^- \rightarrow n + \nu_e. \tag{9.46}$$

A core at the Chandrasekhar limit contains $\sim 10^{57}$ protons, so the production of neutrons by electron capture results in the formation of $\sim 10^{57}$ electron neutrinos.

Can the relentless collapse of this dense ball of neutrons be stopped? Under some circumstances, it can. Hydrostatic equilibrium can be restored with the intervention of **neutron degeneracy pressure**. Neutrons, like electrons, are fermions, and are also subject to degeneracy. For electrons, as we saw in Equation 3.46, the non-relativistic degeneracy pressure is

$$P_{\text{nr},e} \sim \frac{\hbar^2}{m_e} n_e^{5/3} \qquad [\text{electrons}]. \tag{9.47}$$

For neutrons, the equivalent degeneracy pressure is

$$P_{\text{nr},n} \sim \frac{\hbar^2}{m_n} n_n^{5/3} \qquad [\text{neutrons}]. \tag{9.48}$$

In a white dwarf with $\mu_e = 2$, the number density of neutrons equals the number density of electrons ($n_n = n_e$). Thus, in a white dwarf, neutron degeneracy pressure is smaller than electron degeneracy pressure by a factor $m_e/m_n \approx 1/1839$, and is therefore ignored when studying the structure of white dwarfs. Once a collapsing stellar core is converted entirely to neutrons, however, the only degeneracy pressure left is that provided by the neutrons.

A ball of neutrons supported by neutron degeneracy pressure is called a **neutron star**, even though it is not a star by our strictest definition. The term "neutron star" was introduced by Walter Baade and Fritz Zwicky in 1934, in their paper "Cosmic rays from super-novae."[4]

White dwarfs have a mass–radius relation given by Equation 9.14:

$$R_{\text{WD}} = 0.710 R_{\text{L}} \left(\frac{2}{\mu_e}\right)^{5/3} \frac{m_{\text{AMU}}}{m_e} \left(\frac{M_{\text{WD}}}{M_{\text{L}}}\right)^{-1/3}, \tag{9.49}$$

where μ_e is the nucleon-to-electron ratio, $M_{\text{L}} \approx 1.88\,M_\odot$ is the Landau mass, and $R_{\text{L}} \approx 5.55\,\text{km}$ is the Schwarzschild radius of the Landau mass. Since M_{L} is independent of the mass of the particles providing the degeneracy pressure, we

---

[4] Baade and Zwicky hyphenated the word "super-novae" because they invented that term as well, and thus had the right to spell it however they pleased.

can then make a quick and easy estimate of the mass–radius relation for neutron stars:

$$R_{NS} = 0.710 R_L \left(\frac{2}{\mu_n}\right)^{5/3} \frac{m_{AMU}}{m_n} \left(\frac{M_{NS}}{M_L}\right)^{-1/3}. \tag{9.50}$$

Since the nucleon-to-neutron ratio is $\mu_n = 1$ in a neutron star, and since $m_n \approx m_{AMU}$, we can simplify this relation to

$$R_{NS} = 0.710(2^{5/3}) R_L \left(\frac{M_{NS}}{M_L}\right)^{-1/3} \approx 2.25 R_L \left(\frac{M_{NS}}{M_L}\right)^{-1/3}. \tag{9.51}$$

These mass–radius relations imply that neutron stars will be much smaller than white dwarfs, with

$$\frac{R_{NS}}{R_{WD}} = 2^{5/3} \frac{m_e}{m_n} \left(\frac{M_{WD}}{M_{NS}}\right)^{1/3} \approx 0.0017 \left(\frac{M_{WD}}{M_{NS}}\right)^{1/3}. \tag{9.52}$$

A typical mass for a white dwarf is $M_{WD} = 0.6\,M_\odot$; this implies a radius similar to that of the Earth ($R_{WD} \sim 10\,000$ km $\sim 1.6 R_\oplus$). A typical mass for a neutron star is $M_{NS} = 1.5\,M_\odot$; from Equation 9.51, this implies a radius similar to that of a small asteroid ($R_{NS} \sim 13$ km).

Although these calculations give a valid order-of-magnitude size for a neutron star, determining the interior structure of a neutron star is more complicated than simply replacing the subscript e with the subscript n. First, Equation 9.51 tells us that the radius R_{NS} of a $1.5\,M_\odot$ neutron star is only about three times its Schwarzschild radius; in other words, its escape speed is $v_{esc} \sim c/\sqrt{3}$. These conditions are sufficiently extreme that we should be using general relativity, instead of making do with Newtonian gravity. A second problem arises when we consider that a $1.5\,M_\odot$ neutron star contains $N_n \approx 1.8 \times 10^{57}$ neutrons. If you squeeze that many neutrons into a sphere of radius $R_{NS} \approx 13$ km $\approx 1.3 \times 10^{19}$ fm, the mean number density of neutrons will be

$$n_n \approx \frac{3 N_n}{4\pi R_{NS}^3} \approx 0.2 \text{ fm}^{-3}. \tag{9.53}$$

At these high densities, the strong nuclear force, which has a range $r_{sn} \sim 1.4$ fm, comes into play.

A neutron star can be thought of as an enormous atomic nucleus with an escape speed close to the speed of light. Thus, understanding neutron stars requires combining nuclear physics with general relativity. It is not completely surprising that the equation of state for neutron stars is still a matter of debate. Plausible equations of state give a radius $R_{NS} \approx 12$ km for a mass $M_{NS} = 1.5\,M_\odot$. At a mass $M_{TOV} \sim 1.2 M_L \sim 2.3\,M_\odot$, a neutron star reaches the TOV limit, which is the neutron star equivalent of the Chandrasekhar limit for white dwarfs. (The TOV limit is the maximum mass for a non-rotating neutron star; a rapidly rotating neutron star can have a mass as much as 20% above the TOV limit.)

Isolated neutron stars can be detected by the thermal radiation from their non-degenerate atmospheres. A neutron star ~ 1 Myr old has an effective temperature $T_{\text{eff}} \sim 10^6$ K, and thus produces soft X-rays with a mean photon energy $h\bar{\nu} \sim 2.7kT_{\text{eff}} \sim 0.2$ keV. The thermal luminosity of the neutron star is

$$L_{\text{NS}} \approx 0.36 \, L_\odot \left(\frac{R_{\text{NS}}}{10 \text{ km}} \right)^2 \left(\frac{T_{\text{eff}}}{10^6 \text{ K}} \right)^4 . \tag{9.54}$$

For example, the nearest known neutron star, at a distance of $d \approx 120$ pc, is called RX J1856.5−3754; this isolated neutron star is an X-ray emitter whose spectrum indicates it has a hot patch with $T_{\text{eff}} \sim 0.7 \times 10^6$ K on a surface that otherwise has $T_{\text{eff}} \sim 0.4 \times 10^6$ K. However, some neutron stars don't emit purely thermal radiation; they increase their visibility by becoming **pulsars**. A pulsar is a rotating, magnetized neutron star that emits beams of synchrotron radiation that are not aligned with the spin axis of the pulsar. If one of the synchrotron beams sweeps across the Earth, it can be detected as a "pulse" of radiation.[5] From the time between pulses, we can deduce the rotation period $\mathcal{P}_{\text{rot}}$ of the neutron star. Some rapidly spinning pulsars, known as **millisecond pulsars**, have $\mathcal{P}_{\text{rot}} < 10$ ms.

Of the ~ 3000 pulsars that have been detected, ~ 200 are binary pulsars. That is, they are in bound binary systems; their companion is usually a white dwarf or another neutron star. For a binary pulsar, the time between observed pulses of radiation varies, even though $\mathcal{P}_{\text{rot}}$ remains constant. The observed variation is partly due to a simple Doppler shift: as the pulsar orbits the system's barycenter, the frequency with which pulses arrive is alternately redshifted and blueshifted. Observations of the time-varying Doppler shift reveal the orbital period $\mathcal{P}_{\text{orb}}$ of the binary pulsar. For some binary pulsars, the orbital period is as short as ~ 1 hr. For a short-period binary pulsar, other types of variation are seen in the arrival time of pulses. These variations are due to relativistic effects such as precession of the periapse, gravitational time dilation, and orbital shrinkage as gravitational waves carry away energy from the binary system. By fitting these relativistic effects, the masses of the two objects in the binary can be deduced. The resulting mass function for neutron stars in binary systems is shown in Figure 9.2. The function peaks at $M_{\text{NS}} = 1.34 \, M_\odot$, well below the TOV limit for neutron stars. Interestingly, neutron stars that are in a binary system with another neutron star have a strongly preferred mass; the 68% interval for this population is $M_{\text{NS}} = 1.33 \pm 0.09 \, M_\odot$. By contrast, neutron stars that are in a binary with a white dwarf have a higher mean mass and a broader distribution; the 68% interval for this population is $M_{\text{NS}} = 1.55 \pm 0.23 \, M_\odot$.

The typical mass of a neutron star is comparable to the Chandrasekhar limit for a white dwarf. This is consistent with a scenario in which the degenerate core of a high-mass star collapses to form a neutron star once it exceeds its

---

[5] This means, of course, that a pulsar does not pulsate in and out. From your encounters with the vagaries of astronomical nomenclature, this shouldn't surprise you at all.

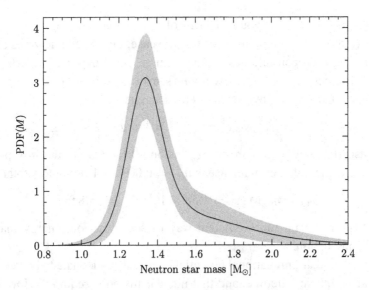

Figure 9.2 Distribution of masses for 47 pulsars in binary systems; for inclusion, pulsars must have a mass known to within ~15%. Gray band = 98% confidence interval on the mass function. [Özel and Freire 2016, and references therein]

Chandrasekhar limit; however, it requires that very little of the non-degenerate envelope falls back onto the neutron star. At the low-mass end of the neutron star mass function, the neutron star companion to the binary pulsar PSR J1756−2251 has $M_{NS} = 1.230 \pm 0.007\,M_{\odot}$; this mass is lower than that of the most massive white dwarfs. At the high-mass end, the millisecond pulsar PSR J0740+6620 has $M_{NS} = 2.08 \pm 0.07\,M_{\odot}$, and the black widow pulsar PSR J0952−0607 has $M_{NS} = 2.35 \pm 0.17\,M_{\odot}$.[6] Neutron stars this massive place a useful lower bound on the TOV limit. It is not surprising that some neutron stars are significantly more massive than the Chandrasekhar limit. Formally, the Chandrasekhar limit applies only to a zero-temperature gas; thermal support might permit more massive cores. The cores could also accrete additional mass during the collapse process or afterwards. Finally, we have omitted other processes, such as magnetism and rotation, that could provide additional support or affect the neutron star formation process in interesting ways.

9.3 Core Collapse Supernovae

Suppose that someone gave you a solar mass of low-density hydrogen and told you to extract energy from it. The energy equivalent of one solar mass is

$$E_{\odot} \equiv M_{\odot}c^2 = 1.79 \times 10^{54}\,\text{erg} = 1790\,\text{foe}, \qquad (9.55)$$

---

[6] "Black widows" are pulsars with low-mass companions ($M \leq 0.1\,M_{\odot}$) on short-period orbits. The companions are low mass because the pulsar has "eaten" their outer layers; hence the analogy to female black widow spiders, who sometimes eat the smaller male after mating.

where 1 foe $\equiv 10^{51}$ erg.[7] If you had 8 $M_\odot$ of oxygen on hand, you could burn the hydrogen (that is, "burning" in the chemical sense, otherwise known as combustion). Burning hydrogen releases energy with an efficiency $\eta_{burn} = 1.6 \times 10^{-9}$; only 1.6 parts per billion of the hydrogen's mass is converted to energy. Thus, burning one solar mass of hydrogen releases an energy

$$E_{burn} = \eta_{burn} M_\odot c^2 = 2.8 \times 10^{45} \text{ erg} = 2.8 \,\mu\text{foe}. \tag{9.56}$$

By contrast, if the hydrogen undergoes fusion all the way to the iron peak, the resulting efficiency of energy release is $\eta_{fus} = 0.008\,85$, leading to a total energy

$$E_{fus} = \eta_{fus} M_\odot c^2 = 1.58 \times 10^{52} \text{ erg} = 15.8 \text{ foe}. \tag{9.57}$$

Nuclear fusion is a far more efficient way of squeezing out energy than mere chemical combustion.

We have seen that gravitational potential energy is inadequate to power the Sun for its main sequence lifetime, and that nuclear fusion is required. However, the Sun in its current state is much larger in radius than the white dwarf that its central regions will eventually become. If 1 $M_\odot$ of hydrogen is compressed into a white dwarf of radius $R_{WD} \approx 9000$ km, the gravitational potential energy released has

$$\eta_{WD} \approx \frac{GM_\odot}{R_{WD}c^2} \approx 2 \times 10^{-4}, \tag{9.58}$$

corresponding to

$$E_{WD} = \eta_{WD} M_\odot c^2 \approx 4 \times 10^{50} \text{ erg} \approx 0.4 \text{ foe}. \tag{9.59}$$

Generally, stars that end as white dwarfs extract more energy from nuclear fusion than they do from gravitational potential energy. But what about stars that end as neutron stars? If 1 $M_\odot$ of hydrogen is squeezed into a neutron star of radius $R_{NS} \approx 14$ km, the gravitational potential energy has

$$\eta_{NS} \approx \frac{GM_\odot}{R_{NS}c^2} \approx 0.11, \tag{9.60}$$

corresponding to

$$E_{NS} = \eta_{NS} M_\odot c^2 \approx 2.0 \times 10^{53} \text{ erg} \approx 200 \text{ foe}. \tag{9.61}$$

A **supernova** can be defined as a stellar event that releases at least 10^{51} erg of energy on a very short timescale.[8] In practice, "very short timescale" means that much of the action occurs during an interval comparable to the freefall time of a white dwarf:

$$t_{ff} \approx 4 \text{ s} \left(\frac{M_{WD}}{0.6\,M_\odot} \right)^{-1}. \tag{9.62}$$

---

[7] The term "foe" is an acronym for "[ten to the] fifty-one ergs." The same unit is also known as a "bethe," in tribute to Hans Bethe.

[8] As described in Section 12.3, a mere "nova" is an event that typically releases a microfoe of energy or less.

The importance of the white dwarf's freefall time becomes apparent when we look at the two mechanisms that can release more than 10^{51} erg from a star's-worth of material. Either the supernova is powered by nuclear fusion, or it is powered by gravitational collapse to a neutron star or black hole.

A supernova powered by nuclear fusion must release its energy in a **thermonuclear runaway**, not in the slow, steady fusion reactions found within ordinary stars. During its main sequence lifetime, the Sun will release $E_{MS} \sim 10^{51}$ erg of nuclear energy. However, it does so at a rate $L_\odot \sim 4 \times 10^{33}$ erg s$^{-1} \sim 0.1$ foe Gyr^{-1}. In order to release $\sim 10^{51}$ erg of nuclear energy on a timescale of seconds rather than gigayears, a thermonuclear supernova must have nuclear fusion occur in matter supported by electron degeneracy pressure. As we saw when considering helium flashes in Section 8.4, when fusion begins in degenerate material, it triggers a positive feedback loop that causes a runaway increase in the fusion rate. The thermonuclear runaway ends only when the temperature rises to the point where ideal gas pressure exceeds electron degeneracy pressure. In Section 12.4, we will look in more detail at **thermonuclear supernovae**. In these supernovae, a carbon/oxygen white dwarf is brought toward or over the Chandrasekhar limit, either by adding mass to a single white dwarf or by merging two white dwarfs together. As the relativistic white dwarf collapses, its temperature rises until carbon burning is triggered. The fusion energy released as the white dwarf collapses is enough to blow the entire white dwarf to smithereens.

However, there's more than one way to produce $\sim 10^{51}$ erg of energy in a few seconds. Section 9.2 discusses how stars with $M_{init} > 10\,M_\odot$ develop increasingly large degenerate iron cores. When the core reaches a mass equal to the Chandrasekhar limit for iron, $M_{Ch} \approx 1.26\,M_\odot$, it starts to collapse on a timescale comparable to the white dwarf freefall time. If the core collapses to form a neutron star, the gravitational potential energy released is, from Equation 9.61, $E_{NS} \sim 2 \times 10^{53}$ erg. However, most of this energy is carried away by the $\sim 10^{57}$ electron neutrinos that are produced as the $\sim 10^{57}$ protons in the initial iron core are converted to neutrons. These neutrinos, with an average energy of $\sim 10^{-4}$ erg ~ 100 MeV, are a great boon to students of core collapse, but a neutron star surrounded by an expanding sphere of neutrinos isn't the whole story of **core collapse supernovae**. As Marvin the Martian would say, "Where's the kaboom?"[9]

As illustrated in Figure 9.3, core collapse supernovae, like thermonuclear supernovae, are characterized by a large photon luminosity. At maximum light, core collapse supernovae commonly have $M_V < -16$, corresponding to $L_{SN} > 2 \times 10^8\,L_\odot$. Some "superluminous supernovae" have $M_V < -22$, corresponding to $L_{SN} > 5 \times 10^{10}\,L_\odot$. For a brief time, a supernova in a mid-size galaxy can outshine all the stars in the galaxy combined.[10] Integrated over the complete light curve, the photon energy of a core collapse supernova is typically $E_\gamma \sim 10^{49}$ erg ~ 0.01 foe,

---

[9] *Hare-Way to the Stars*, directed by Chuck Jones. Burbank, CA: Warner Bros. Cartoons, released March 29, 1958.

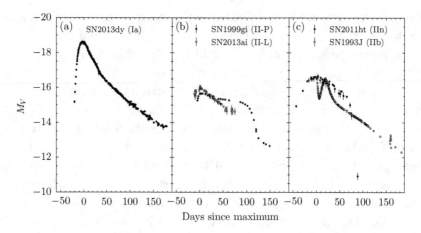

Figure 9.3 Sample light curves in the V band for different types of supernovae. (a) Thermonuclear (Type Ia) supernova. (b) and (c) Core collapse supernovae of different subtypes. [Data from the Open Supernova Catalog, sne.space]

or about one part in 10 000 of the neutrino energy. In addition to the neutrinos and photons streaming away from the supernova, core collapse supernovae also create **supernova remnants** containing hot gas expanding away from the supernova site. The supernova remnant consists of the outer layers of the supernova progenitor star along with gas from the interstellar medium that has been swept up by the expanding shock wave of the supernova explosion. For instance, the supernova remnant Cassiopeia A, seen in Figure 9.4, contains $M_{gas} \sim 10\,M_\odot$ of gas expanding at a speed $v \sim 5000$ km s^{-1}, implying a kinetic energy of expansion of

$$E_{boom} \sim \frac{1}{2}M_{gas}v^2 \sim 2 \text{ foe}. \tag{9.63}$$

Although core collapse supernovae are not uniform in energy, a useful back-of-envelope accounting is that the neutrino energy of a core collapse supernova is $E_\nu \sim 100$ foe, the kinetic energy of the expanding ejecta is $E_{boom} \sim 1$ foe, and the energy of emitted photons is $E_\gamma \sim 0.01$ foe. The production of photons is just a side effect of core collapse. The truly puzzling question is how the inward collapse of the core leads to the outward expansion of the progenitor star's outer layers. Once the iron core exceeds its Chandrasekhar limit, it collapses inward with a speed $v \sim R_{WD}/t_{ff,WD} \sim 3000$ km s^{-1}. The 10^{57} neutrinos rush outward with a speed nearly equal to that of light. How can you use the energy of the collapsing core or the expanding neutrino cloud to prompt several solar masses of gas (the outer envelope of the progenitor star) to move outward at $v \sim 10^4$ km s^{-1}?

---

[10] Note that although the core collapse itself takes only a few seconds, it can take many months for most of the generated light to escape from the expanding remnant.

(a)   (b)

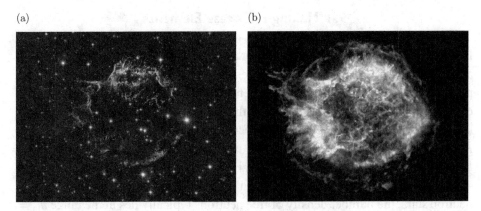

Figure 9.4 (a) Supernova remnant Cassiopeia A ($d \approx 3.33\,\text{kpc}$) seen at visible wavelengths. [NASA, ESA, and the Hubble Heritage Collaboration; R. Fesen (Dartmouth) and J. Long (ESA/Hubble)] (b) Cas A seen in X-rays with 0.5–6 keV photon energy. [NASA/CXC/SAO] Each panel is 8.0×6.0 arcmin, corresponding to $7.7 \times 5.8\,\text{pc}$ at the distance of Cas A.

Numerical simulations indicate that **core bounce** can transfer energy to the outer envelope. As the iron core collapses to form a neutron star, it initially overcompresses, then springs back as the short-range repulsive nature of the strong nuclear force comes into play. The core bounce sends a shock wave through the outer envelope of the star. Within $\sim$0.04 s, the amount of gas swept up by the shock wave equals the amount of matter in the shock itself. This causes the shock to stall. In the earliest generation of numerical supernova simulations, this caused a failure to explode. However, the earliest numerical simulations in the 1960s were purely one dimensional, due to a lack of computing power. By the 1980s, two-dimensional simulations were the state of the art. In these, **neutrino opacity** caused the onset of convection. Although the cross section for the interaction of a neutrino with baryonic matter is small (even at a neutrino energy $E \sim 100\,\text{MeV}$), the density in and near the collapsing star is sufficiently large that the material is partially opaque to neutrinos. The trapped neutrinos heat the gas and lead to violent convection just above the core. In two-dimensional simulations, this convection carries energy outward and restarts the expansion of the shock wave after a breathless pause of $\sim$0.3 s. The regenerated shock wave heats and accelerates the outer envelope of the star.

By the twenty-first century, Moore's law brought us to the point where three-dimensional simulations of core collapse supernovae are possible. Unfortunately, this is a case where more dimensions equals more problems. Complications that are introduced in fully three-dimensional simulations include (but are not limited to) asymmetric explosions, neutron star kicks, magnetic fields, and convection instabilities.

9.4 Making r-Process Elements

During a core collapse supernova, a great many neutrons are present (about 10^{57}), and the energy per particle is high. These are the conditions necessary for the production of neutron-rich heavy elements, beyond the iron peak. In fact, core collapse supernovae are locations where the **rapid neutron-capture process**, also known as the **r-process**, is thought to occur. During the s-process, as it occurs in AGB stars, the number density of free neutrons is typically $n_{\text{fn}} \leq 10^{10} \text{ cm}^{-3}$; from Equation 8.49, this leads to a typical time between neutron captures of $t_{\text{nc}} \geq 40 \text{ d}$. During the r-process, as it occurs in core collapse supernovae and colliding neutron stars, the number density of free neutrons typically lies in the range $n_{\text{fn}} = 10^{20}$–10^{24} cm^{-3}, leading to a time between neutron captures of $t_{\text{nc}} = 0.03$–300 μs. These times are short compared to a typical beta decay time, permitting extremely neutron-rich nuclei to be built.

Generally, for a nucleus of given proton number Z, its lifetime τ against beta decay decreases as it accumulates more neutrons. For instance, although the beta-decay lifetime of ^{60}Fe is a lengthy $\tau_{60} = 3.8 \text{ Myr}$, that of ^{66}Fe is $\tau_{66} \approx 0.6 \text{ s}$, and that of ^{72}Fe is $\tau_{72} \sim 30 \text{ ms}$. Eventually, a nucleus that has been capturing neutrons will reach a neutron number $N = A - Z$ for which the beta decay time τ is less than the mean time t_{nc} between neutron captures. The nucleus will then decay to a nucleus of the next higher atomic number, which is likely to capture additional neutrons before undergoing a beta decay. When the high flux of free neutrons eventually ends, and t_{nc} becomes much longer, the extremely neutron-rich nuclei that have been formed will undergo successive beta decays until they become stable. For example, ^{72}Fe will decay to ^{72}Ge.

One important aspect of the r-process is that it makes a different mix of elements than the s-process. Figure 9.5, for example, shows the abundances of heavy elements present in the Sun at the time of its formation. The abundances are subdivided into the amount produced by the s-process (dotted line) and by the r-process (solid line). Some elements, such as strontium ($Z = 38$) and barium ($Z = 56$), are more readily made by the s-process; thus, barium abundance (as plotted in Figure 1.9) is a useful tracer of gas that has undergone the s-process. By contrast, other elements, most strikingly in the vicinity of platinum ($Z = 78$), are more readily made by the r-process. In addition, thorium ($Z = 90$) and uranium ($Z = 92$) can be made *only* by the r-process, since the methodical s-process path stops short at lead ($Z = 82$) and bismuth ($Z = 83$).

While the r-process is ongoing, the production of nuclei with a "magic number" of neutrons is preferred. When nuclei have $N = 2, 8, 20, 28, 50, 82,$ or 126 neutrons, they have particularly long lifetimes τ against beta decay, and particularly small cross sections for neutron capture. Nuclei built up by neutron capture are thus disproportionately likely to have a magic number of neutrons. For instance,

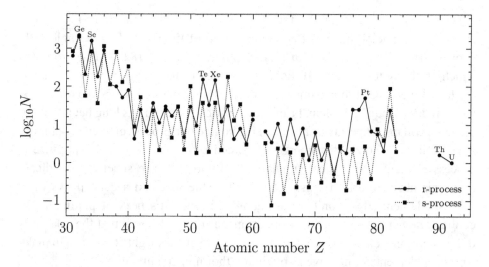

Figure 9.5 Abundances (by number) of heavy elements present in the Sun at its formation. Solid line + circles = amount of each element made by the r-process. Dotted line + squares = amount made by the s-process. Abundances are normalized to $N(\mathrm{H}) = 10^{12}$ hydrogen atoms. [Data from Cowan *et al.* 2006]

the r-process makes relatively large amounts of the isotope ^{78}Ni, which is "doubly magic," with 28 protons and 50 neutrons. When the ^{78}Ni ultimately decays, it ends as stable ^{78}Se, with 34 protons and 44 neutrons. In addition, the decay of "singly magic" ^{74}Cr, ^{76}Fe, and ^{80}Zn, all of which have 50 neutrons, produces the isotopes ^{74}Ge, ^{76}Ge, and ^{80}Se. The preference for $N = 50$ neutrons thus results in a peak near Ge and Se on the r-process abundance plot (Figure 9.5).

Similarly, the r-process makes relatively large amounts of the doubly magic isotope ^{132}Sn, with 50 protons and 82 neutrons. When the ^{132}Sn decays, it produces stable ^{132}Xe, with 54 protons and 78 neutrons. In addition, the decay of singly magic ^{128}Pd and ^{130}Cd, both of which have 82 neutrons, produces the isotopes ^{128}Te and ^{130}Te. The preference for $N = 82$ neutrons thus results in a second peak near Te and Xe on the r-process abundance plot (Figure 9.5). The third peak on the r-process abundance plot, near platinum (Pt), results from the preference for the magic number of $N = 126$ neutrons. Although the doubly magic isotope ^{208}Pb, with 82 protons and 126 neutrons, is stable against beta decay, the singly magic isotopes ^{193}Ho, ^{194}Er, ^{195}Tm, ^{196}Yb, and ^{197}Lu, all of which have 126 neutrons, can be made in significant quantities during the r-process. These isotopes ultimately decay to ^{193}Ir, ^{194}Pt, ^{195}Pt, ^{196}Pt, and ^{197}Au, respectively. The third r-process abundance peak is thus sometimes called the "jeweller's peak," since it contains the elements platinum and gold.[11]

[11] Iridium is sometimes used in jewelry as well, but its high melting point makes it difficult to work.

Building heavy elements by the r-process requires a high ratio of free neutrons to seed nuclei, but not for a long period of time. Building a single ^{238}U nucleus from a seed nucleus of ^{56}Fe requires capturing 182 free neutrons (of which 66 decay to protons). If the time between neutron captures is $t_{nc} \sim 3\,\mu s$, the total time needed to accumulate the necessary neutrons is $t \sim 0.5\,ms$. Thus, the r-process occurs in violent, brief, neutron-rich events. It has long been thought that core collapse supernovae are a leading source of r-process elements, but more recently it was realized that mergers between neutron stars can also produce r-process elements. Numerical simulations of core collapse supernovae indicate that their neutrino-driven winds can produce the first and second peaks of r-process elements, but don't go as far as the jeweller's peak at platinum. By contrast, numerical simulations of neutron star mergers reveal that the tidal tails of hot, neutron-rich material made in the merger have the right conditions to make r-process elements as massive as platinum, thorium, and uranium.

Exercises

9.1 A star needs to develop a carbon/oxygen core of order $1\,M_\odot$ in order to ignite carbon burning; this requires an initial stellar mass $M_\star \approx 8\,M_\odot$. Why does the initial mass need to be so large compared to the required C/O core mass?

9.2 Suppose that Betelgeuse is at a distance $d = 800$ light-years, and that it underwent core collapse 799 years ago.

(a) How many neutrinos from the assumed core collapse supernova of Betelgeuse will pass through your body? [In general, this will depend on the orientation of your body relative to Betelgeuse; for simplicity, you may assume you are spherical.]

(b) At the typical energy of a supernova neutrino, a nucleon's cross section for interacting with a neutrino through the weak force is $\sigma \sim 10^{-40}\,cm^2$. About how many neutrinos from the assumed core collapse supernova of Betelgeuse will interact with a nucleon in your body?

(c) At the lower energy of solar neutrinos, a nucleon's cross section for interacting with a neutrino is $\sigma \sim 10^{-42}\,cm^2$. Given the flux of solar neutrinos at 1 au from the Sun (see Section 4.1), what is the rate at which interactions between solar neutrinos and nucleons occur in your body?

10

Rotating Stars

Darn the wheel of the world!
Why must it continually turn over?
Where is the reverse gear?
Fondly, Uncle Jack.

Jack London (1876–1916)
Book inscription [1916]

Stars rotate. In previous chapters, we tended to ignore this fact. Our justification, given in Section 2.1, was that the Sun's equatorial rotation speed, $v_{rot,\odot} \sim 2 \, \mathrm{km \, s^{-1}}$, is very small compared to the Sun's escape speed, $v_{esc,\odot} \sim 600 \, \mathrm{km \, s^{-1}}$, and thus cannot cause any significant flattening of the Sun's shape. Although this argument is true for the Sun, some other stars rotate more rapidly. For instance, Altair's measured rotation speed, $v_{rot} \sin i = 240 \, \mathrm{km \, s^{-1}}$, has a profound effect on its shape, its internal structure, and its lifetime. In addition, even modest rotation rates can cause significant alterations to the internal structure of stars. Finally, knowing how stars are spun up and spun down is important if we are to study a star's complete history, from a slowly whirling gas cloud to a swiftly spinning white dwarf or a millisecond pulsar.

10.1 Effects of Rotation on Structure

To gain insight on how rotation changes the structure of a star, it helps to start with a simple case: an object with uniform angular speed Ω. Such an object is referred to as having "solid-body rotation" (even if it happens to be fluid). We might expect that a rotating star could become an extremely flattened object, like a spinning disk of pizza dough, in the limit of very large Ω. However, the impact of rotation depends on the mass distribution, and stars (unlike lumps of dough) are strongly centrally concentrated. To develop our physical intuition, we therefore consider the extreme case in which the dense core of the star is approximated as a point of mass $M_\star$ and the extended envelope of the star has negligible mass. In this case, the effective potential in a frame of reference co-rotating with the star can be written as

$$\Phi_{\text{eff}}(r,\theta) = -\frac{GM_\star}{r} - \frac{1}{2}\Omega^2 r^2 \sin^2\theta, \tag{10.1}$$

where θ is the polar angle, measured from the rotation axis. The acceleration as measured in the rotating frame is then

$$\vec{g}_{\text{eff}} = -\vec{\nabla}\Phi_{\text{eff}} = \left(-\frac{GM_\star}{r^2} + \Omega^2 r \sin^2\theta\right)\hat{r} + (\Omega^2 r \sin\theta\cos\theta)\hat{\theta}. \tag{10.2}$$

In the limit of small r, the magnitude of g_{eff} will be smaller than the gravitational acceleration $g = GM_\star/r^2$ in the absence of rotation; this implies that a rotating star behaves as if it has an effective mass smaller than $M_\star$. In addition, the reduction of g_{eff} depends on Ω; this implies that stars with the same mass and composition, but with different rotation rates, will evolve in a different manner. The classical Vogt–Russell theorem can therefore be expanded to read "the structure of a rotating star is uniquely determined by its total mass, its composition, and its angular momentum distribution."

In the equatorial plane of the star ($\theta = \pi/2$), the effective acceleration equals zero at a critical radius,

$$r_{\text{crit}} = \left(\frac{GM_\star}{\Omega^2}\right)^{1/3}. \tag{10.3}$$

Beyond this radius, $\vec{g}_{\text{eff}}$ points away from the star's center. Thus, if the low-density stellar envelope stretched past r_{crit}, its pressure would have to *increase* with radius in order to preserve hydrostatic equilibrium. Rapidly rotating stars are not, in reality, confined by a convenient high-pressure circumstellar medium; thus, gas at $r > r_{\text{crit}}$ in the equatorial plane will be accelerated away from the star. Suppose that a centrally concentrated star of mass $M_\star$ rotates with a non-zero angular speed Ω. The star's envelope has an equatorial radius $R_{\text{eq},\star}$; thanks to the effects of rotation, this is somewhat larger than its polar radius. The equatorial radius will be exactly equal to the critical radius r_{crit}, and the star will thus be on the verge of shedding gas from its equator, if the angular speed Ω equals the **breakup speed**,

$$\Omega_{\text{BU}} = \left(\frac{GM_\star}{R_{\text{eq},\star}^3}\right)^{1/2}. \tag{10.4}$$

Expressed as a physical speed for material at the star's equator, the breakup speed is

$$v_{\text{BU}} = \left(\frac{GM_\star}{R_{\text{eq},\star}}\right)^{1/2} = 437\,\text{km}\,\text{s}^{-1}\left(\frac{M_\star}{1\,M_\odot}\right)^{1/2}\left(\frac{R_{\text{eq},\star}}{1\,R_\odot}\right)^{-1/2}. \tag{10.5}$$

(Notice that the technical definition of the breakup speed is actually equal to $v_{esc}/\sqrt{2}$.) The rotation period corresponding to the breakup speed is

$$P_{BU} = \frac{2\pi}{\Omega_{BU}} = 2.78 \,\text{hr} \left(\frac{M_\star}{1\,M_\odot}\right)^{-1/2} \left(\frac{R_{eq,\star}}{1\,R_\odot}\right)^{3/2}. \tag{10.6}$$

This minimum possible rotation period is comparable to the freefall time of a star. For a $0.6\,M_\odot$ white dwarf, we expect P_{BU} to be half a second, while for a $1.5\,M_\odot$ neutron star, P_{BU} is about half a millisecond.

The speed limit on stellar rotation is not of purely theoretical interest. For instance, there is a subclass of stars with spectral type B that have strong Balmer emission lines in their spectrum. (The first star found to have such a spectrum was γ Cassiopeiae: in 1866, Angelo Secchi discovered that it had a "*très belle*" Balmer Hβ emission line where other stars had an absorption line.) The emission lines of these "Be stars," as they are called, come from a circumstellar disk of hot gas that has been ejected from the star due to its rapid rotation. As an example, the prototype Be star γ Cas has a measured rotation speed of $v_{rot} \sin i = 432\,\text{km s}^{-1}$.

For a rotating, centrally concentrated star, a surface of constant effective potential Φ_{eff} can be used to infer a figure of merit for the departure from spherical symmetry. Consider an equipotential surface with polar radius r_{pol} (at $\theta = 0$) and equatorial radius r_{eq} (at $\theta = \pi/2$). From Equation 10.1, we find

$$\frac{GM_\star}{r_{pol}} = \frac{GM_\star}{r_{eq}} + \frac{1}{2}\Omega^2 r_{eq}^2. \tag{10.7}$$

The relation between the polar and equatorial radii is therefore

$$\frac{r_{eq}}{r_{pol}} = 1 + \frac{1}{2}\frac{\Omega^2 r_{eq}^3}{GM_\star}. \tag{10.8}$$

Thus, we can view the dimensionless number $0.5\Omega^2 r_{eq}^3/GM_\star$ as a measure of the departure from spherical symmetry. Note that even in the limit that the equatorial radius equals the critical radius from Equation 10.3, the equipotential surface has $r_{eq}/r_{pol} = 1.5$, and thus is not very highly flattened.

Because stars are centrally concentrated, their equipotential surfaces resemble those of the "point mass core" approximation. Thus, we never expect stars to be flat as a pancake,[1] even if they undergo maximal rotation with $v_{rot} = v_{BU}$. Moreover, since the departure from spherical symmetry scales as r_{eq}^3, even a star that is maximally rotating at its surface will be nearly spherical in its deep interior. When we relax our assumption that the star's core is a point mass, the departure from spherical symmetry still scales as $\Omega^2/\bar{\rho}$, where $\bar{\rho}$ is the mean density interior to an equipotential surface. The Sun, for example, has a central density ~ 100

---

[1] It is possible for compact objects to have flat-as-a-pancake gaseous accretion disks. However, these disks violate the assumption of uniform Ω; instead, they are in nearly Keplerian rotation, with $\Omega \propto r_{eq}^{-3/2}$.

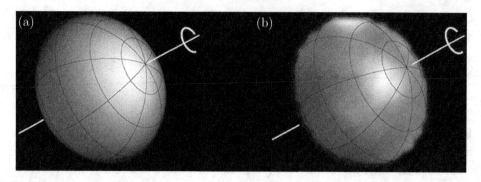

Figure 10.1 (a) Model of a rapidly rotating star. (b) Image of Altair made using the CHARA interferometer; dark gray = lower effective temperature, white = higher effective temperature. [Monnier *et al.* 2007]

times its mean density; thus, the departure from spherical symmetry in its core will be only $\sim 1/100$ as large as that of its surface layers.

We can test our ideas about the shapes of rotating stars by looking at massive main sequence stars that rotate rapidly. For instance, Altair is a main sequence star of spectral class A7V. Its mass is $M_\star = 1.8\,M_\odot$; this would imply a main sequence lifetime $t_{\mathrm{MS}} \approx 1.5\,\mathrm{Gyr}$ for a non-rotating star. As mentioned in Section 1.2, the rotational broadening of Altair's absorption lines indicates that $v_{\mathrm{rot}} \sin i = 240\,\mathrm{km\,s}^{-1}$. At a distance $d = 5.13\,\mathrm{pc}$, Altair is close enough for its apparent shape to have been determined by interferometry. As seen in Figure 10.1, the apparent shape of Altair is decidedly non-circular. The observed properties of Altair are well described by an oblate model in which the star's intrinsic flattening is $R_{\mathrm{eq}}/R_{\mathrm{pol}} = 1.23$ and the inclination of the rotation axis relative to the line of sight is $\sin i = 0.89$. Given the observed value of $v_{\mathrm{rot}} \sin i$, this implies that Altair has an equatorial rotation speed of $v_{\mathrm{rot}} = 270\,\mathrm{km\,s}^{-1}$, about 70% of the star's breakup speed, $v_{\mathrm{BU}} \approx 400\,\mathrm{km\,s}^{-1}$.

10.2 Meridional Circulation

Notice in Figure 10.1 that the effective temperature of Altair is higher near its pole than near its equator. At the poles, $T_{\mathrm{eff}} = 8500\,\mathrm{K}$; at the equator, $T_{\mathrm{eff}} = 6900\,\mathrm{K}$. This **gravity darkening** of the equatorial regions[2] is an effect that is detected in other rapidly rotating oblate stars such as Regulus ($M_\star \sim 3.8\,M_\odot$), α Ophiuchi ($M_\star \sim 2.4\,M_\odot$), and α Cephei ($M_\star \sim 1.7\,M_\odot$). All these flattened stars have $\mathcal{P}_{\mathrm{rot}} \leq 16\,\mathrm{hr}$ and $R_{\mathrm{eq}}/R_{\mathrm{pol}} \geq 1.2$.

---

[2] Gravity darkening of a star's equatorial regions is sometimes referred to as gravity brightening of its polar regions. The term you choose presumably depends on whether you always look on the bright side of life.

To understand why rapidly rotating stars have hot poles and cooler equators, we must look more closely at the impact of rotation on stellar structure. In this context, it is useful to adopt cylindrical coordinates (R, z, ϕ), with the $R = 0$ axis being the rotational axis of the star and the $z = 0$ plane being its equatorial plane. To keep the problem simple, we assume that the star is in solid-body rotation, with angular speed Ω uniform throughout the star. However, we relax our assumption of a point-mass core, and simply require that the mass distribution $\rho(R, z)$ has rotational symmetry around the axis of stellar rotation.

For a spherical star, the equation of hydrostatic equilibrium is

$$\frac{dP}{dr} = -\rho \frac{GM}{r^2} \quad \text{[spherical]}. \tag{10.9}$$

For a rotating, non-spherical star, we must use the more general equation

$$\vec{\nabla} P = -\rho \vec{\nabla} \Phi_{\text{eff}}, \tag{10.10}$$

where the effective potential is

$$\Phi_{\text{eff}}(R, z) = \Phi(R, z) - \frac{1}{2}\Omega^2 R^2 \tag{10.11}$$

and the gravitational potential $\Phi(R, z)$ is given by the Poisson equation,

$$\nabla^2 \Phi = 4\pi G \rho. \tag{10.12}$$

In a frame of reference co-rotating with the star, the effective acceleration is $\vec{g}_{\text{eff}} = -\vec{\nabla} \Phi_{\text{eff}}$. Mathematically, this implies that at any given point, $\vec{g}_{\text{eff}}$ must be perpendicular to the equipotential surface passing through that point (keeping in mind that the potential in question is the effective potential Φ_{eff}). Since, from Equation 10.10, $\vec{\nabla} P$ is parallel to $\vec{g}_{\text{eff}}$, surfaces of constant Φ_{eff} must also be surfaces of constant P for a rotating star in hydrostatic equilibrium. To think of it another way, if P were *not* constant on an equipotential surface, there would be a component of the pressure gradient $\vec{\nabla} P$ parallel to the equipotential surface. However, since the gravitational acceleration must be exactly perpendicular to the equipotential surface, any component of $\vec{\nabla} P$ parallel to the equipotential surface would violate hydrostatic equilibrium.

Since every equipotential surface corresponds to a single value of Φ_{eff} and a single value of P, we can make the mapping $P = P(\Phi_{\text{eff}})$. Since Equation 10.10 also implies

$$\rho = -\frac{dP}{d\Phi_{\text{eff}}}, \tag{10.13}$$

we can write $\rho = \rho(\Phi_{\text{eff}})$. Thus, the equipotential surfaces are surfaces of constant ρ, P, and Φ_{eff}. Also, if the pressure is given by an ideal gas law, and if $\bar{\mu}$ is constant throughout the star, we can write $T \propto P/\rho = T(\Phi_{\text{eff}})$, and the equipotential surfaces will be surfaces of constant temperature, as well. One property of

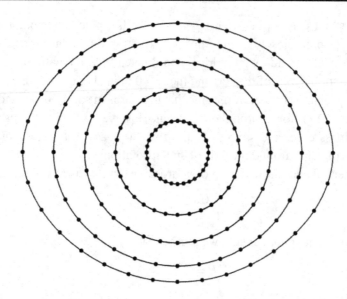

Figure 10.2 Equipotential contours of a stellar model with $M_\star = 2\,M_\odot$ and $\Omega = 2.2 \times 10^{-4}\,\mathrm{s}^{-1}$, corresponding to $\mathcal{P}_{\mathrm{rot}} \approx 8\,\mathrm{hr}$. [Roxburgh 2004]

the star that is not constant on equipotential surfaces is the effective gravitational acceleration, g_{eff}. Figure 10.2 shows equipotential surfaces for an Altair-like rotating star with a physically realistic density profile; it illustrates that equipotential surfaces are more closely spaced in the polar direction than in the equatorial direction. Thus, g_{eff}, which is given by the gradient of Φ_{eff}, is larger near the poles than near the equator on a given equipotential surface.

In the radiative zone of a spherical star, the equation of energy transfer can be written in the form

$$\frac{L}{4\pi r^2} = -\frac{4acT^3}{3\kappa\rho}\frac{dT}{dr} \qquad \text{[spherical]}. \qquad (10.14)$$

For a rotating, non-spherical star, the generalized equation of radiative energy transfer is

$$\vec{F} = -\frac{4acT^3}{3\kappa\rho}\frac{dT}{d\Phi_{\mathrm{eff}}}\vec{\nabla}\Phi_{\mathrm{eff}}, \qquad (10.15)$$

where $\vec{F}$ is the radiative energy flux. Equation 10.15 tells us that the flux $\vec{F}$ points in the direction opposite to $\vec{g}_{\mathrm{eff}}$, climbing out of the effective potential well. Since T and ρ are both functions of Φ_{eff}, and the opacity κ is a function of T and ρ, we can write Equation 10.15 as

$$\vec{F} = f\vec{\nabla}\Phi_{\mathrm{eff}} = -f\vec{g}_{\mathrm{eff}}, \qquad (10.16)$$

where the function

$$f(\Phi_{\text{eff}}) \equiv -\frac{4acT^3}{3\kappa\rho}\frac{dT}{d\Phi_{\text{eff}}} \tag{10.17}$$

is constant on an equipotential surface. Taking the divergence of the flux in Equation 10.16, we find that

$$\vec{\nabla}\cdot\vec{F} = \frac{df}{d\Phi_{\text{eff}}}\left|\vec{\nabla}\Phi_{\text{eff}}\right|^2 + f\nabla^2\Phi_{\text{eff}}. \tag{10.18}$$

Making use of the definition of Φ_{eff}, and substituting from the Poisson equation, we find

$$\vec{\nabla}\cdot\vec{F} = \frac{df}{d\Phi_{\text{eff}}}g_{\text{eff}}^2 + f(4\pi G\rho - 2\Omega^2). \tag{10.19}$$

We can find another relation for the divergence of the flux by looking at the equation of thermal equilibrium. In a spherical star,

$$\frac{1}{4\pi r^2}\frac{dL}{dr} = \rho\epsilon \quad \text{[spherical]}. \tag{10.20}$$

For a non-spherical star, the generalized equation of thermal equilibrium is

$$\vec{\nabla}\cdot\vec{F} = \rho\epsilon. \tag{10.21}$$

Outside the fusion powerhouse at the core of the star, this is equivalent to stating that the divergence of the flux vanishes:

$$\vec{\nabla}\cdot\vec{F} = 0. \tag{10.22}$$

By combining Equations 10.19 and 10.22, we have a relation that must hold in the radiative zone of a star undergoing solid-body rotation:

$$\frac{df}{d\Phi_{\text{eff}}}g_{\text{eff}}^2 + f(4\pi G\rho - 2\Omega^2) = 0. \tag{10.23}$$

Everything that goes into Equation 10.23 is constant on an equipotential surface, except for g_{eff}. Thus, for Equation 10.23 to hold true everywhere on an equipotential surface, we require $df/d\Phi_{\text{eff}} = 0$ and also $4\pi G\rho = 2\Omega^2$. That is, we have come to the absurd conclusion that rotating stars with radiative transport must be of uniform density, and the more slowly they rotate, the lower that density must be! The absurdity implied by Equation 10.23 is known as von Zeipel's paradox, after the astronomer Hugo von Zeipel.

The escape from von Zeipel's paradox is simply to acknowledge that a rotating star cannot transport energy purely by radiation. On an equipotential surface, the value of $\vec{\nabla}\cdot\vec{F}$ averages to zero, but the polar regions, with their higher values of g_{eff}, receive an excess of radiative heating, while the equatorial regions receive a deficiency of radiative heating. This is the cause of the gravity darkening of the equatorial regions at the star's surface. In the interior of the star, the hotter gas near the poles rises while the cooler gas near the equator sinks; this sets up

meridional circulation. This circulation can be effective at mixing pristine gas from the outer layers of the star with processed material from the star's core. However, the characteristic time to complete one circuit of a meridional cell is

$$t_{mer} \sim t_{KH} \left(\frac{v_{BU}}{v_{rot}}\right)^2. \tag{10.24}$$

Thus, for a star like the Sun, with $v_{BU} \approx 210 v_{rot}$, the time for meridional circulation is $t_{mer} \sim 4 \times 10^4 t_{KH} \sim 1000\,\text{Gyr}$, too long to have a significant effect on the Sun's internal structure. However, for a star rotating rapidly enough to have $R_{eq}/R_{pol} \approx 1.2$, like Altair, the time for meridional mixing will be a few times the Kelvin–Helmholtz time. For Altair, whose Kelvin–Helmholtz time is $t_{KH} \sim 5\,\text{Myr}$ and whose main sequence lifetime is $t_{MS} \sim 1.5\,\text{Gyr}$, meridional mixing has a significant effect on the internal composition of the star.

In a rapidly rotating star, we might expect efficient meridional mixing to keep the hydrogen fraction X uniform throughout the star; this would be a radical departure from the composition gradients that are set up in non-rotating main sequence stars (for instance, see Figure 8.4). In reality, we do find that rapidly rotating stars evolve off the main sequence and develop hydrogen-burning shells. Thus, there must be some effect that prevents full mixing in rapid rotators. In part, meridional mixing is suppressed in the deep interior by the fact that the departure from spherical symmetry there is reduced by the high degree of central mass concentration. Meridional mixing can also be suppressed by the development of gradients in the mean molecular mass $\bar{\mu}$. Earlier, we assumed that $\bar{\mu}$ was constant throughout the star; this assumption led us, via von Zeipel's paradox, to the conclusion that rotating stars must be hotter near the poles and cooler near the equator, thus setting up meridional circulation. However, by permitting a difference in $\bar{\mu}$ between the poles and the equator, we can reduce the difference in T and thus suppress meridional circulation.

10.3 Angular Momentum Transport

Our assumption of uniform angular speed Ω is computationally useful, but is not necessarily physically correct. There is no particular reason to assume that stars are born in a state of uniform rotation. Even if they were, stellar evolution naturally generates differential rotation within a star. Suppose, for instance, that a Sun-like star is uniformly rotating while on the main sequence, with a rotation period of approximately a month. As the star evolves up the red giant branch, its envelope swells to ~ 200 times its main sequence radius, while its core contracts to $\sim 1/20$ its main sequence radius. Angular momentum conservation would then predict an envelope rotation period of a few millennia and a core rotation period of an hour or two. However, even if the star's total angular momentum $J_\star$ is conserved, there are physical processes that transport angular momentum within the star. These processes can work to reduce gradients in the angular speed. (The situation is further complicated by the fact that the total angular momentum $J_\star$ is not necessarily constant; as we discuss later, stars can lose angular momentum in stellar winds and gain angular momentum from accretion. For the moment, though, let's consider an isolated star of constant $J_\star$.)

Internal angular momentum transport is a complicated subject; we can, however, outline some of the transport mechanisms. First, moving blobs of gas carry angular momentum, just as they carry thermal energy. The transport of angular momentum by the bulk motion of gas is called **hydrodynamic** angular momentum transport. Stars in differential rotation, where Ω varies either with depth or with latitude, are subject to a wealth of instabilities that can transport angular momentum. Since these instabilities usually result in turbulence, hydrodynamic angular momentum transport, like convective energy transport, is difficult to model numerically.

Angular momentum can also be transported by waves, without bulk motion of gas. This **wave-driven** angular momentum transport is mediated by two families of waves created by convective turbulence. First are sound waves (known as p-modes, in this context), where the restoring force is provided by pressure gradients. Second are gravity waves (known as g-modes), where the restoring force is provided by perturbations in gravitational potential.[3] The g-modes propagate in radiative stellar interiors; in the presence of differential rotation, there can be a net exchange of angular momentum between incoming and outgoing g-waves.

Finally, angular momentum can be transported by the interaction of a star's magnetic field with the ionized gas in its interior. This **magnetic** angular momentum transport can be highly effective even if the magnetic field is fairly weak. The solar sunspot cycle is an example of magnetic angular momentum transport in action. At the base of the Sun's convection zone, differential rotation amplifies the local magnetic field; when the field becomes large enough, magnetic field lines rise to the surface and appear as sunspots. The characteristic time for this process is the length of the solar activity cycle, $t \approx 11$ yr; this is ~ 200 times the Sun's rotation period, but is much shorter than other solar timescales of interest. In regions where it operates effectively, magnetic angular momentum transport can strongly suppress differential rotation as a function of depth. Because the three types of angular momentum transport – hydrodynamic, wave-driven, and magnetic – operate differently and have different effects, understanding the balance among them is important for determining the internal structure of rotating stars.

How can we determine whether stars are in differential rotation or not? As is often the case, our best knowledge comes from the Sun. Observations of

---

[3] Waves on the Earth's oceans are a form of gravity wave, if their wavelength is long enough for surface tension to be negligible.

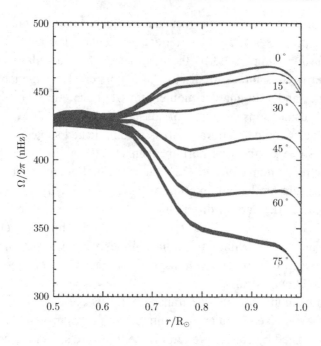

Figure 10.3 Angular rotation rate within the Sun as a function of $r/R_\odot$ at different latitudes, indicating the 1σ error bounds. Data are from the Global Oscillation Network Group. [Howe *et al.* 2000]

sunspots reveal that the Sun's photosphere is in differential rotation as a function of latitude, with $\Omega_\odot = 2.97 \times 10^{-6}\,\mathrm{s}^{-1}$ at the equator, but $\Omega \approx 2.6 \times 10^{-6}\,\mathrm{s}^{-1}$ at latitude $\ell = \pm 45°$. The differential rotation of the Sun also extends to its interior. The angular speed of rotation below the Sun's photosphere can be determined by helioseismology, the study of waves propagating through the Sun. As discussed in Section 11.3, helioseismology yields information about the temperature and chemical composition of the Sun's interior, since these factors determine the speed of p-modes (that is, the speed of sound). However, sound propagation is also affected by whether the sound is going "upstream" against the Sun's rotation, or "downstream" with the Sun's rotation. Figure 10.3 shows the Sun's internal rotation frequency (plotted as $\nu \equiv \Omega/2\pi$) for different latitudes, as found by helioseismology. In the Sun's convective zone, at $r > 0.71\,R_\odot$, the rotation is differential, with higher angular speed Ω near the equator than near the poles. However, in the radiative zone, the rotation is nearly solid-body, with $\Omega_c \approx 2.7 \times 10^{-6}\,\mathrm{s}^{-1}$, or $\mathcal{P}_c \approx 27\,\mathrm{d}$. Since the radiative zone contains 97.6% of the Sun's mass, this means that our initial assumption of uniform Ω is not a wretchedly bad one, at least in the Sun's case.

The **tachocline** is the relatively thin layer between the uniform rotation of the radiative zone and the differential rotation of the convective zone. The word "tachocline" comes from the ancient Greek roots *tachos* (meaning "speed") and *klino* (meaning "slope" or "slant"), referring to the gradients in angular speed that exist within the tachocline. At solar latitude $\ell = \pm 60°$, for instance, the middle of the tachocline is at $r_t = 0.71\,R_\odot$, where the angular speed is $\Omega_t = 2.5 \times 10^{-6}\,s^{-1}$; however, the angular speed decreases by a fractional amount $\delta\Omega / \Omega_t \approx -0.17$ moving radially outward through a tachocline of thickness $\delta r / r_t \approx 0.06$. (The tachocline is the layer of the Sun where the magnetic field is amplified by differential rotation, ultimately giving rise to sunspots on the surface.)

Measuring differential rotation in stars other than the Sun is difficult; however, enough is known to reveal that the Sun's pattern of internal rotation is not universal. Differential rotation in a star's photosphere as a function of latitude can be deduced from Zeeman Doppler imaging. As a simple example of how this technique works, consider a star whose rotation axis is perpendicular to the line of sight ($\sin i = 1$); the rotation is sufficiently slow that we can approximate the star as being spherical. A starspot is at latitude ℓ, where the angular speed is $\Omega(\ell)$. Since starspots are highly magnetized, absorption or emission lines from the spot show circular polarization from the Zeeman effect. Using a spectropolarimeter, we can measure the Doppler shift specifically of the polarized light from the starspot. The radial velocity of the spot, relative to the star's mean radial velocity, is seen to vary from $-\Omega(\ell)R_* \cos \ell$, when the star's rotation first brings it into view, to $+\Omega(\ell)R_* \cos \ell$ when rotation carries it out of sight. The time from maximum blueshift to maximum redshift is $\pi / \Omega(\ell)$; that is, half the rotation period at the spot's latitude. If we know the star's radius R_*, the observed properties allow us to find both $\Omega(\ell)$ and $\cos \ell$. For stars with multiple starspot groups, a more sophisticated analysis permits an image to be made of the star's magnetic field (this is the "imaging" of Zeeman Doppler imaging), as well as providing an estimate of Ω as a function of latitude. Main sequence stars similar in mass to the Sun are found to have differential rotation in their photospheres; usually, as with the Sun, Ω decreases going from the equator toward the poles. Massive, rapidly rotating stars, by contrast, are found to have nearly uniform Ω in their photospheres. This is because the strong Coriolis forces in rapidly rotating stars act to reduce gradients in the angular speed.

Since the Sun offers us photons in overflowing abundance, helioseismology lets us reconstruct $\Omega(r)$ within the Sun in fine detail. For stars other than the Sun, given the relative scarcity of photons, asteroseismology is usually limited to finding the difference (if any) between the angular speed of a star's core and that of its outer envelope. Figure 10.4, for instance, shows the rotation period of the core and of the envelope for apparently bright stars in the later stages of their evolution. Note that the rotation periods are plotted as a function of the surface gravity $\log g$ of the star; as a star's envelope expands, it moves rightward on the plot (toward lower $\log g$). As a star evolves along the subgiant bridge, the rotation period of its core tends to decrease, while that of its envelope increases. However, by the time a star becomes a red giant, the contrast in rotation period is much

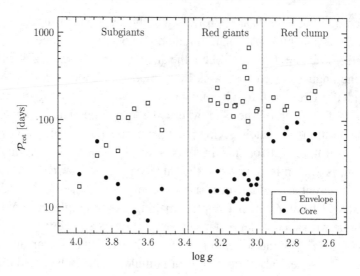

Figure 10.4 Rotation period for cores (filled circles) and envelopes (open squares) of evolved stars. Approximate boundaries in $\log g$ between subgiants, red giants, and red clump stars are given as vertical lines. [Data from Aerts *et al.* 2019 and Deheuvels *et al.* 2020]

smaller than you would expect if angular momentum were conserved for core and envelope separately. Instead of cores swiftly pirouetting once per hour while envelopes slowly rotate once every few millennia, actual red giants, as shown in Figure 10.4, have $\mathcal{P}_{rot} \sim 20$ d for the core and ~ 200 d for the envelope. By the time a star develops core helium burning and settles into the red clump on an H–R diagram, the discrepancy between its core and envelope rotation period has decreased to a factor of ~ 2. Thus, as stars evolve off the main sequence, angular momentum transport must occur from their core to their envelope.

Making an accurate numerical model of a star, difficult enough when the star is not rotating, becomes still harder when stars are permitted to rotate. The simplifying assumption of spherical symmetry is lost, and we must come to grips with angular momentum transport as well as energy transport. In addition, meridional mixing can significantly alter the composition profile of a star. As an example of how rotation affects a star's composition, consider the isotope ^{14}N. In a relatively massive main sequence star, we have seen how the CN cycle enhances the mass fraction X_{14} of ^{14}N in a star's core. In the modern Sun, for instance, X_{14} increases from a photospheric value of 9×10^{-4} to a central value of 5×10^{-3}, as shown in Figure 5.9. However, in a rapidly rotating star, meridional mixing can bring some of the central store of ^{14}N up to the surface. Figure 10.5 shows how the surface mass fraction of ^{14}N evolves with time in models of massive stars. For these models, an initial mass fraction $X_{14} = 7.4 \times 10^{-4}$ was assumed. In the non-rotating models (shown by the solid lines), the surface ^{14}N remains nearly constant with

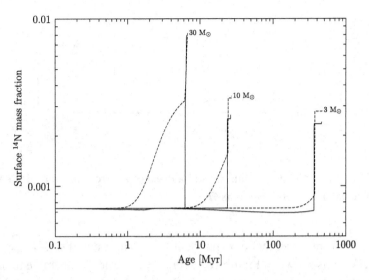

Figure 10.5 Surface mass fraction of ^{14}N as a function of age for non-rotating stars (solid lines) and rotating stars (dashed lines).

time until the star leaves the main sequence and undergoes the first dredge-up. However, rapid-rotating models (shown by the dashed lines) have surface ^{14}N enhanced by meridional mixing while the star is still on the main sequence. Thus, the main sequence stellar wind from a $30\,M_\odot$ star will inject more ^{14}N into the interstellar medium than a non-rotating stellar model would predict. Thus, rapid stellar rotation affects not merely the structure of the rotating star, but has implications for the composition of the next generation of stars as well.

10.4 Rotation and Star Formation

The saga of a star's angular momentum evolution actually begins before the star even forms. To become a star at all, a region of interstellar gas must get rid of most of its angular momentum. Consider a sphere of interstellar gas that is destined to become a star of mass $M_\star = 1\,M_\odot$. The gas is not yet a protostar; it's just a denser than average region of interstellar space, with a number density $n_{\rm H}$ of hydrogen nuclei. For concreteness, let's take $n_{\rm H} = 100\,{\rm cm}^{-3}$ as a density typical of molecular clouds. At a density $n_{\rm H}$, the radius of a sphere containing $1\,M_\odot$ of gas, assuming solar abundance, is

$$dr \approx 1.8 \times 10^7 \, {\rm R}_\odot \left(\frac{n_{\rm H}}{100\,{\rm cm}^{-3}}\right)^{-1/3} \approx 0.41 \, {\rm pc} \left(\frac{n_{\rm H}}{100\,{\rm cm}^{-3}}\right)^{-1/3}. \qquad (10.25)$$

We are writing this radius as a small quantity dr because, although it is vastly larger than the final size of the star, it is tiny compared to the size of the Milky Way Galaxy.

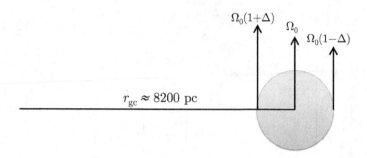

Figure 10.6 A gas sphere of radius dr is at a distance r_{gc} from the galactic center. The angular speed at the sphere's center is Ω_0.

Suppose, for instance, that our spherical gas cloud is at a distance r_{gc} from the galactic center, as shown in Figure 10.6. If $r_{gc} \approx 8200\,\mathrm{pc}$, comparable to the Sun's current distance from the galactic center, then

$$\Delta \equiv \frac{dr}{r_{gc}} \approx 5.0 \times 10^{-5} \left(\frac{n_H}{100\,\mathrm{cm}^{-3}}\right)^{-1/3}. \tag{10.26}$$

Near the Sun's location in the Galaxy, the rotation curve is quite flat. That is, the orbital speed v_c of gas around the galactic center is $v_c \approx 240\,\mathrm{km\,s}^{-1}$ regardless of location. This means that the Galaxy is not in solid-body rotation; instead, it has $\Omega(r) = v_c/r \propto r^{-1}$. At the center of the gas cloud, the angular speed around the galactic center is

$$\Omega_0 = \frac{v_c}{r_{gc}} \approx 0.029\,\mathrm{km\,s}^{-1}\,\mathrm{pc}^{-1} \approx 9.5 \times 10^{-16}\,\mathrm{s}^{-1}. \tag{10.27}$$

On the side of the cloud closest to the galactic center, as shown in Figure 10.6, the angular speed is $\Omega \approx \Omega_0[1 + \Delta]$. On the side farthest from the galactic center, it is $\Omega \approx \Omega_0[1 - \Delta]$.

The difference in Ω across the spherical gas cloud means that an observer at the sphere's center will see the side closest to the galactic center pull ahead at a speed

$$v_{shear} = \Omega_0 dr = v_c \Delta \approx 12\,\mathrm{m\,s}^{-1} \left(\frac{n_H}{100\,\mathrm{cm}^{-3}}\right)^{-1/3}, \tag{10.28}$$

about the maximum speed of a world champion sprinter. The observer at the cloud's center will see a point at the opposite side of the cloud fall behind at the same speed. The angular momentum of the motion about the cloud center is of order

$$J_{cloud} \sim 0.4 M v_{shear} dr \sim 0.4 M \Omega_0 (dr)^2$$

$$\sim 1.2 \times 10^{53}\,\mathrm{g\,cm}^2\,\mathrm{s}^{-1} \left(\frac{M}{1\,M_\odot}\right)^{5/3} \left(\frac{n_H}{100\,\mathrm{cm}^{-3}}\right)^{-2/3}. \tag{10.29}$$

For a $1\,M_\odot$ cloud, this is $\sim$60 000 times the current angular momentum of the Sun's rotation, $J_\odot = 2.0 \times 10^{48}\,\mathrm{g\,cm^2\,s^{-1}}$. For this calculation, we assumed quiescently rotating gas; in practice, turbulence in star-forming regions can create even larger birth angular momenta than we have calculated. Thus, even if the ZAMS Sun had been a rapid rotator with $v_{rot} \sim 0.9 v_{BU}$, the gas cloud from which it formed somehow had to dispose of more than 99.99% of its initial angular momentum.

The case of the lost angular momentum becomes slightly less puzzling when we look at binary systems. Suppose that the primary in a binary system has mass M_1, while the secondary (the lower mass object) has mass $M_2 = qM_1$, where $q \le 1$. The orbital angular momentum of the system is then

$$
\begin{aligned}
J_{orb} &= \frac{M_1 M_2}{(M_1 + M_2)^{1/2}} \left[Ga(1 - e^2) \right]^{1/2} \\
&= \frac{q}{(1 + q)^2} M_{tot}^{3/2} \left[Ga(1 - e^2) \right]^{1/2},
\end{aligned}
\tag{10.30}
$$

where $M_{tot} = M_1 + M_2$ is the total mass of the binary system, and a and e are the semimajor axis and eccentricity of the objects' relative orbit. In the case of a circular orbit, the angular momentum is

$$
J_{orb} = 8.86 \times 10^{52}\,\mathrm{g\,cm^2\,s^{-1}}\,\frac{q}{(1 + q)^2} \left(\frac{M_{tot}}{1\,M_\odot} \right)^{3/2} \left(\frac{a}{1\,\mathrm{au}} \right)^{1/2}.
\tag{10.31}
$$

Suppose that a gas cloud starts in quiescent rotation, with J_{cloud} given by Equation 10.29. If the cloud evolves into a binary system, then the orbital angular momentum from Equation 10.31 will equal the initial angular momentum J_{cloud} if the final binary separation is

$$
a \approx 1.8\,\mathrm{au}\,\frac{(1 + q)^4}{q^2} \left(\frac{M}{1\,M_\odot} \right)^{1/3},
\tag{10.32}
$$

assuming an initial cloud density $n_H \approx 100\,\mathrm{cm^{-3}}$. Thus, a $1\,M_\odot$ gas cloud can deposit its initial angular momentum into an equal-mass binary ($q = 1$) with a separation $a \sim 30\,\mathrm{au}$ between the two M1V stars in the system. However, if it splits into a G2V star and a brown dwarf ($q \approx 0.015$), the brown dwarf will have to be at a distance $a \sim 10\,000\,\mathrm{au}$ to conserve the initial angular momentum.

Although a small amount of mass can act as an angular momentum "scapegoat," carrying a large amount of angular momentum on its Keplerian orbit, we can be quite confident that the Sun does not have a brown dwarf companion at $a \sim 10\,000\,\mathrm{au}$. In addition to perturbing the inner solar system at a measurable level, it would have been detected by all-sky infrared surveys. What, then, is the mechanism by which the Sun and other single stars lost the bulk of their initial angular momentum? To answer that question, let's look back at the protostellar and pre-main sequence phases of star formation, now paying closer attention to angular momentum transport.

Consider, for instance, a newly unveiled PMS star with mass $M_\star = 1\,M_\odot$. It is on the deuterium-burning birthline, with a luminosity $L_\star \sim 10\,L_\odot$, an effective temperature $T_{\text{eff}} \approx 4200\,\text{K}$, and thus a radius $R_\star \sim 6\,R_\odot$. The breakup speed of this young PMS star is $v_{\text{BU}} \sim 200\,\text{km s}^{-1}$. Observations of young solar-mass PMS stars show that they have a typical rotation period of $\mathcal{P}_{\text{rot}} \sim 8\,\text{d}$; this implies an equatorial rotation speed $v_{\text{rot}} \sim 40\,\text{km s}^{-1}$, about 20% of the breakup speed. It is also seen that almost every young PMS star is surrounded by a protoplanetary disk. The final stages of accretion onto a protostar, as well as the last trickles of accretion onto a young PMS star, come from this dusty gas disk. If the disk is in Keplerian rotation, then the specific angular momentum $j_{\text{disk}} = (GM_\star r)^{1/2}$ of matter in the disk must be larger than that of matter in the central protostar or young PMS star; this follows from the fact that the disk is outside the protostar ($r > R_\star$) and the protostar is rotating slower than breakup ($v_{\text{rot}} < v_{\text{BU}}$). Accretion from a Keplerian disk rotating in the same sense as a protostar should thus spin up the protostar until it reaches its breakup speed. The observed fact that young PMS stars rotate at only $\sim$20% of their breakup speed is therefore puzzling.

The solution to the puzzle of slow rotation is the presence of magnetic **star–disk coupling**. In this scenario, the inner edge of the protoplanetary disk is magnetically coupled to the outer layers of the protostar, enforcing co-rotation at an angular speed $\Omega = (GM_\star/r_{\text{in}}^3)^{1/2}$, where r_{in} is the distance from the inner edge of the disk to the protostar's center. As gas accretes onto the star along magnetic field lines, its specific angular momentum $j = \Omega r^2$ decreases, thus avoiding excessive spin-up of protostars. Later, as the PMS star contracts on its Kelvin–Helmholtz time, star–disk coupling keeps its rotation period $\mathcal{P}_{\text{rot}}$ roughly constant as it develops into a zero age main sequence star, as opposed to decreasing as $\mathcal{P}_{\text{rot}} \propto R_\star^2$, as you would expect in homologous contraction with no star–disk coupling. Observationally, it is found that solar-mass stars in young clusters like the Pleiades (Figure 1.14) have a median rotation period $\mathcal{P}_{\text{rot}} \sim 4\,\text{d}$, about half that of a typical solar-mass PMS star. However, ZAMS stars have a range of rotation periods at any given mass, indicating that the effectiveness of star–disk coupling varies from one star to another.

10.5 Rotation on the Main Sequence

Measuring differential rotation of stars, whether by Zeeman Doppler imaging or by asteroseismology, can only be done for apparently bright stars (what astronomers call "stars with names," as opposed to fainter stars that have only catalog numbers). For fainter stars, the best we can do is measure $v_{\text{rot}} \sin i$ from the rotational broadening of absorption lines, or alternatively, try to determine the mean rotation period $\mathcal{P}_{\text{rot}}$ at the star's photosphere. As mentioned in Section 1.4, $\mathcal{P}_{\text{rot}}$ can be most readily determined for stars that have starspots. The fact

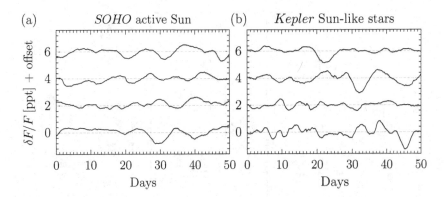

Figure 10.7 (a) Flux of the Sun measured by the *SOHO* satellite around the time of solar maximum. Flux variation is shown as parts per thousand (ppt) of the mean flux. (b) Flux of stars with activity levels similar to that of the Sun, measured by *Kepler*. [Following Basri *et al.* 2010]

that starspots change their size and shape while drifting about in the photosphere makes measuring the exact rotation period somewhat challenging.

To illustrate the difficulty of finding $\mathcal{P}_{rot}$, Figure 10.7(a) shows the solar irradiance measured by the *SOHO* satellite over four different one-month intervals. These observations were taken near the time of solar maximum, so there is plenty of variability in the measured flux; however, the 24.5 d rotation period at the Sun's equator is not immediately obvious from the data. Figure 10.7(b) shows the Sun's flux at the top and the flux of three different Sun-like stars observed by *Kepler* below. These stars show flux variations comparable in amplitude to those of the Sun, but finding a rotation period requires more sophisticated statistical analysis than simply staring at the light curves. One method of analysis involves monitoring the star's flux $f(t)$ for a time τ much longer than the expected rotation period. The autocorrelation function

$$R(\Delta t) = \int_0^{\tau - \Delta t} f(t + \Delta t) f(t) dt \qquad (10.33)$$

will have a peak at $\Delta t = \mathcal{P}_{rot}$ (as well as peaks at $\Delta t = 2\mathcal{P}_{rot}$, $3\mathcal{P}_{rot}$, and so forth).

Studies of the rotation periods of stars in open clusters reveal that young stars ($t < 250\,\mathrm{Myr}$) all rotate rapidly compared to the Sun. Figure 10.8(a) shows the rotation period of stars in the Upper Scorpius association. The stars in this cluster are extremely young ($t \approx 10\,\mathrm{Myr}$) and still retain dust disks, a lingering reminder of the gaseous protoplanetary disks that once regulated their rotation. In the mass range $M_\star = 0.4$–$1.1\,M_\odot$, the young stars of Upper Sco have a median rotation period of $\widetilde{\mathcal{P}}_{rot} \approx 5\,\mathrm{d}$. Note, however, that there is a wide spread in $\mathcal{P}_{rot}$ in each mass bin; for instance, stars in Upper Sco with $M_\star \approx 1\,M_\odot$ have measured rotation periods ranging from 0.8 d to 12 d. This range in rotation periods reflects the range

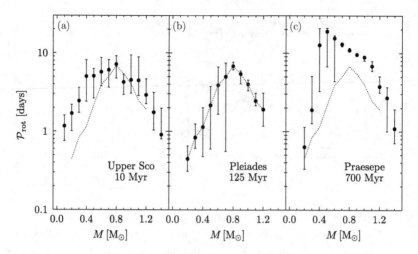

Figure 10.8 Rotation period versus mass for stars in (a) Upper Sco, (b) the Pleiades, and (c) Praesepe. In each panel, dots = median $\mathcal{P}_{\text{rot}}$ for each mass bin; error bars = interquartile range. The relation for the Pleiades is repeated as the dotted line in each panel. [Data from García *et al.* 2014, Godoy-Rivera *et al.* 2021]

in effectiveness of magnetic star–disk coupling, with weaker coupling and shorter disk lifetimes leading to shorter rotation periods.

Figure 10.8(b) shows the rotation period of stars in the Pleiades, which has $t \approx 125$ Myr. This panel is similar to Figure 1.14(b); however, instead of sorting main sequence stars by color (with bluer, and hence more massive, stars on the left), it sorts them by mass (with more massive, and hence bluer, stars on the right). In the Pleiades, stars with $M_\star \approx 0.8\,\text{M}_\odot$ have the longest median rotation period, with $\widetilde{\mathcal{P}}_{\text{rot}} \approx 7$ d. Lower-mass stars in the Pleiades have a shorter median rotation period, but also have a larger spread in $\mathcal{P}_{\text{rot}}$.

Finally, Figure 10.8(c) shows the rotation periods of stars in the older cluster Praesepe, with $t \approx 700$ Myr. At a given stellar mass, stars in Praesepe have a longer median rotation period than stars in the Pleiades. In Praesepe, stars with $M_\star \approx 0.5\,\text{M}_\odot$ have the longest median rotation period, with $\widetilde{\mathcal{P}}_{\text{rot}} \approx 19$ d. Strikingly, for Praesepe stars in the mass range $M_\star = 0.7 \rightarrow 1.1\,\text{M}_\odot$, there is a very tight correlation between stellar mass and rotation period, with $\widetilde{\mathcal{P}}_{\text{rot}} \propto M_\star^{-1.3}$; within this mass range, knowing a Praesepe star's mass lets you deduce its rotation period with a fractional uncertainty of order 10%.

Note that in Praesepe, stars with $M_\star > 1.2\,\text{M}_\odot$ remain rapidly rotating, while their less massive neighbors have had their rotation slowed down. In populations of mature main sequence stars, there is generally a fairly abrupt transition, known as the **Kraft break**, between rapidly rotating high-mass stars and slowly rotating low-mass stars. The transition between rapid rotators and slow rotators occurs in the mass range $M_\star \approx 1.5$–$1.2\,\text{M}_\odot$. Mature main sequence stars with $M_\star \approx 1.5\,\text{M}_\odot$, corresponding to $T_{\text{eff}} \approx 7000$ K and spectral class F2V, typically rotate

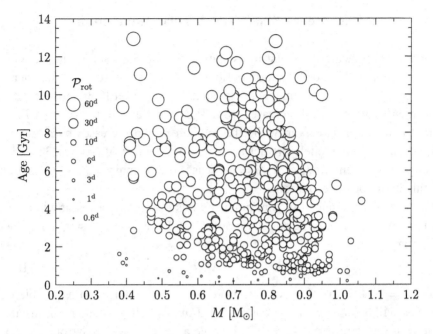

Figure 10.9 Stars of known mass and rotation period, with ages deduced from gyrochronology. The area of each open circle is proportional to $\mathcal{P}_{\rm rot}$. [Data from Claytor *et al.* 2020]

with $v_{\rm rot} \sin i \sim 80 \, \mathrm{km \, s^{-1}}$; by contrast, stars with $M_\star \approx 1.2 \, \mathrm{M_\odot}$, corresponding to $T_{\rm eff} \approx 6200 \, \mathrm{K}$ and spectral class F8V, typically have $v_{\rm rot} \sin i \sim 10 \, \mathrm{km \, s^{-1}}$.

Figure 10.9 shows values of $\mathcal{P}_{\rm rot}$ for main sequence stars with $T_{\rm eff} < 5500 \, \mathrm{K}$ whose rotation period has been determined from *Kepler* data. For the stars shown, the rotation period ranges from $\mathcal{P}_{\rm rot} = 0.56 \, \mathrm{d}$ (at an age $t = 0.16 \, \mathrm{Gyr}$) to $\mathcal{P}_{\rm rot} = 59.6 \, \mathrm{d}$ (at $t = 12.9 \, \mathrm{Gyr}$). For solar-mass stars, the period–age relation is approximately

$$\mathcal{P}_{\rm rot} \approx 24.5 \, \mathrm{d} \left(\frac{t}{4.57 \, \mathrm{Gyr}} \right)^{1/2} \approx 11 \, \mathrm{d} \left(\frac{t}{1 \, \mathrm{Gyr}} \right)^{1/2}. \tag{10.34}$$

At a given stellar mass below the Kraft break, the relation $\mathcal{P}_{\rm rot} \propto t^{1/2}$ generally gives a good fit. The normalization depends on the stellar mass, with more massive stars having a shorter rotation period at any given age. A reasonably tight period–age relation is the basis of **gyrochronology**; that is, the determination of a star's age from its rotation period or rotation speed. Be aware, however, that a tight period–age relation is found only for isolated main sequence stars below the Kraft break. It breaks down for very young stars ($t < 250 \, \mathrm{Myr}$), for evolved stars, and for stars in close binaries whose rotational angular momentum is affected by tidal braking and mass transfer.

10.6 Stellar Winds and Angular Momentum

Observations indicate that solar-mass stars arrive on the main sequence with a range of (relatively short) rotation periods, then converge to a unique rotation period that increases with time. For solar analogs, Equation 10.34 tells us that the converged rotation period increases from $\mathcal{P}_{\rm rot} \approx 6\,\mathrm{d}$ at $t = 0.25\,\mathrm{Gyr}$ to $\mathcal{P}_{\rm rot} \approx 30\,\mathrm{d}$ at $t = 8\,\mathrm{Gyr}$. However, while the rotation period increases by a factor of ~ 5, standard solar models predict an increase in radius of only $\sim 33\%$. Thus, the angular momentum $J_\star \propto R_\star^2/\mathcal{P}_{\rm rot}$ decreases on the main sequence, assuming roughly homologous evolution.

One way a star can lose angular momentum is by emitting a stellar wind of gas with relatively high specific angular momentum. For a spherical star, the moment of inertia is

$$I_\star = k_{\rm I} M_\star R_\star^2, \tag{10.35}$$

where $k_{\rm I} = 0.4$ for a sphere of uniform density. For a star of compressible gas, the mass will be more centrally concentrated, and $k_{\rm I}$ will be smaller. For instance, an $n = 1.5$ polytrope has $k_{\rm I} = 0.205$, while a softer $n = 3$ polytrope has $k_{\rm I} = 0.0754$. If the star is in solid-body rotation with angular speed Ω, the total angular momentum is $J_\star = I_\star \Omega$ and the mean specific angular momentum is

$$j_\star = J_\star/M_\star = k_{\rm I} \Omega R_\star^2. \tag{10.36}$$

The gas at the star's equator, being farthest from the rotation axis, is the stellar material with the highest specific angular momentum,

$$j_{\rm eq} = \Omega R_\star^2. \tag{10.37}$$

Thus, a stellar wind preferentially from the star's equator will drive down the star's mean specific angular momentum. Given a mass loss rate $\dot{M}_{\rm wind}$, a wind from the equatorial regions will translate to an angular momentum loss rate of

$$\dot{J}_{\rm wind} \approx \dot{M}_{\rm wind} \Omega R_\star^2. \tag{10.38}$$

The time over which the star's angular momentum is lost, also known as the "spindown time," is then

$$t_J \approx \frac{J_\star}{\dot{J}_{\rm wind}} \approx k_{\rm I} \frac{M_\star}{\dot{M}_{\rm wind}} \approx k_{\rm I} t_M, \tag{10.39}$$

where t_M is the mass loss time for the star. Notice that the spindown time, in this approximation, is independent of Ω. The Sun's mean mass loss rate is $\dot{M}_\odot = 2.0 \times 10^{-8}\,M_\odot\,\mathrm{Myr}^{-1}$ and its dimensionless moment of inertia is $k_{\rm I} = 0.07$, close to that of an $n = 3$ polytrope. Scaling to the solar values, a star's spindown time is then

$$t_J = 3500\,\mathrm{Gyr} \left(\frac{k_{\rm I}}{0.07}\right) \left(\frac{M_\star}{1\,M_\odot}\right) \left(\frac{\dot{M}_{\rm wind}}{\dot{M}_\odot}\right)^{-1}. \tag{10.40}$$

Thus, although the calculated spindown time for the Sun is only 7% of its mass loss time, it is still much longer than the Sun's main sequence lifetime.

One way in which a stellar wind can effectively spin down a star is by being torqued so that its specific angular momentum is much greater than ΩR_*^2. In the case of the Sun and other magnetically active stars, this can be done by the star's **magnetic field**. The magnetic field in the Sun's photosphere is notably complicated. We have seen, for instance, that the photosphere develops highly magnetized sunspots every 11 years, during the maximum of the solar activity cycle. Even during solar minimum, the magnetic field in the photosphere varies with time and with location. In our discussion of magnetic stellar winds, we will simply scale the magnetic field strength in the photosphere to a value $B_{\text{phot}} = 3 \, \text{G}$, typical for the Sun during the last few solar minima. At a distance greater than $r \sim 3 \, R_\odot$ from the Sun's center, the field lines of the Sun's magnetic field are mostly radial. If we assume the radial component dominates in other stars as well, the magnetic field strength outside a star has the form

$$B(r) \approx B_{\text{phot}} \left(\frac{r}{R_*} \right)^{-2}.$$ (10.41)

The magnetic energy density associated with this field is

$$U_{\text{mag}}(r) = \frac{B(r)^2}{8\pi} \approx 220 \, \text{GeV cm}^{-3} \left(\frac{B_{\text{phot}}}{3 \, \text{G}} \right)^2 \left(\frac{r}{R_*} \right)^{-4}.$$ (10.42)

Because of the steep dependence of magnetic energy density on distance, the Sun's magnetic energy density at $r = 1 \, \text{au} = 215 \, R_\odot$ is only $U_{\text{mag}} \sim 100 \, \text{eV cm}^{-3}$.

The solar wind is an outward flow of charged particles: protons, helium nuclei, and a few heavier ions, balanced with enough electrons to keep things electrically neutral. It is driven by the pressure gradient within the Sun's hot ($T \sim 2 \times 10^6 \, \text{K}$) corona. Once the ionized gas of the solar wind has reached $r \sim 20 \, R_\odot$, it has been accelerated to a speed comparable to the escape speed from the Sun's photosphere: $v_{\text{wind}} \sim 600 \, \text{km s}^{-1}$. The mass continuity equation implies that a spherically symmetric wind must have a density

$$\rho_{\text{wind}} = \frac{\dot{M}_{\text{wind}}}{4\pi r^2 v_{\text{wind}}}.$$ (10.43)

The kinetic energy density of the outflowing wind is then

$$U_{\text{wind}}(r) = \frac{1}{2} \rho_{\text{wind}} v_{\text{wind}}^2 = \frac{\dot{M}_{\text{wind}} v_{\text{wind}}}{8\pi R_*^2} \left(\frac{r}{R_*} \right)^{-2}.$$ (10.44)

Scaling to values appropriate for the Sun (and adopting $v_\odot = 600 \, \text{km s}^{-1}$ and $\dot{M}_\odot = 2 \times 10^{-8} \, M_\odot \, \text{Myr}^{-1}$) the kinetic energy density of the wind is

$$U_{\text{wind}} = 0.39 \, \text{GeV cm}^{-3} \left(\frac{\dot{M}_{\text{wind}}}{\dot{M}_\odot} \right) \left(\frac{v_{\text{wind}}}{v_\odot} \right) \left(\frac{1 \, R_\odot}{R_*} \right)^2 \left(\frac{r}{R_*} \right)^{-2}.$$ (10.45)

Comparison of the magnetic energy density (Equation 10.42) with the wind's kinetic energy density (Equation 10.45) tells us that for Sun-like stars the magnetic energy density dominates near the star. However, the wind's kinetic energy density, thanks to its shallower dependence on r, dominates at large radii. The radius at which $U_{mag} = U_{wind}$ is called the **Alfvén radius**. Expressed as a multiple of the star's photospheric radius $R_\star$, the Alfvén radius is

$$\frac{r_A}{R_\star} = \frac{B_{phot}R_\star}{(\dot{M}_{wind}v_{wind})^{1/2}}$$

$$= 24 \left(\frac{B_{phot}}{3\,G}\right)\left(\frac{R_\star}{1\,R_\odot}\right)\left(\frac{\dot{M}_{wind}}{\dot{M}_\odot}\right)^{-1/2}\left(\frac{v_{wind}}{v_\odot}\right)^{-1/2}. \tag{10.46}$$

Consider a magnetized star with $r_A > R_\star$. Since the material of its stellar wind is highly ionized, it is pinned to the magnetic field lines. At $r < r_A$, the magnetic field lines dominate the dynamics. They rotate with the same angular speed Ω as the star; thus, as the ionized gas of the stellar wind is constrained to move along the magnetic field lines, it is torqued to higher angular momentum. Outside the Alfvén radius, by contrast, the gas in the stellar wind dominates the dynamics; instead of exerting torque on the gas, the magnetic field lines are now tugged tamely along by the gas. Since the material in the gas shares the angular speed Ω of the magnetic field lines as far as r_A, the specific angular momentum that it carries away to infinity is not $j_{eq} = \Omega R_\star^2$ but $j_{eq} = \Omega r_A^2$. This reduces the star's spindown time to a value (compare with Equation 10.39) of

$$t_J = k_I \frac{M_\star}{\dot{M}_{wind}}\left(\frac{r_A}{R_\star}\right)^{-2}. \tag{10.47}$$

Scaling to solar values, and using Equation 10.46 for the Alfvén radius, this becomes

$$t_J = 6.1\,\text{Gyr}\left(\frac{k_I}{0.07}\right)\left(\frac{M_\star/R_\star^2}{1\,M_\odot/R_\odot^2}\right)\left(\frac{B_{phot}}{3\,G}\right)^{-2}\left(\frac{v_{wind}}{v_\odot}\right). \tag{10.48}$$

Thus, spindown times are shorter for centrally concentrated stars with low surface gravity, possessing a strong magnetic field and a slow wind. Note that the mass loss rate $\dot{M}_{wind}$ cancels out of the final value for the magnetized spindown time t_J.

If we specifically look at stars in the mass range $0.4\,M_\odot < M_\star < 2\,M_\odot$, we can adopt the mass–radius relation $R_\star/R_\odot \approx M_\star/M_\odot$ and assume a main sequence lifetime of

$$t_{MS} \approx 10\,\text{Gyr}\left(\frac{M_\star}{1\,M_\odot}\right)^{-3.5}. \tag{10.49}$$

Using Equation 10.48 for the spindown time, the requirement that a star be effectively spun down on the main sequence, $t_J < t_{MS}$, becomes the mass requirement

$$M_\star < 1.2\,M_\odot \left(\frac{k_{\rm I}}{0.07}\right)^{-0.4} \left(\frac{v_{\rm wind}}{v_\odot}\right)^{-0.4} \left(\frac{B_{\rm phot}}{3\,{\rm G}}\right)^{0.8}. \qquad (10.50)$$

Stars in the mass range $M_\star = 1.2\text{--}2\,M_\odot$ have density profiles nearly homologous to the Sun, and thus have similar values of $k_{\rm I}$. Their stellar wind speeds are likely to be similar to the solar wind speed (just as their escape speeds $v_{\rm esc}$ are similar to the Sun's escape speed). Thus, for stars with $M_\star > 1.2\,M_\odot$ to be effectively spun down, they would have to have magnetic fields stronger than $B_\odot \sim 3$ G. However, the stellar mass $M_\star \approx 1.2\,M_\odot$ is where the Kraft break begins, and mature main sequence stars are seen to have *not* been spun down. Thus, stars in this mass range must have relatively weak magnetic fields.

In standard dynamo theory, the Sun's magnetic field is generated by an interplay between the shearing motions in the tachocline, which wind up magnetic field lines, and the turbulent motions in the convective zone, which stretch field lines in the vertical direction. In this scenario, the net efficiency of the dynamo is parameterized by the Rossby number, defined as the ratio of the star's rotation period $\mathcal{P}_{\rm rot}$ to its convective overturn timescale $\tau_{\rm co}$.[4] In the Sun, for example, with a convective zone of thickness $r_{\rm cz} = 0.29\,R_\odot = 2.0 \times 10^5$ km, we found that convective motion has the leisurely subsonic speed $v \sim 0.02\,{\rm km\,s}^{-1}$ (Equation 4.83). This leads to a convective overturn timescale of $\tau_{\rm co} \sim r_{\rm cz}/v \sim 10^7$ s ~ 100 d. Given the Sun's rotation period, $\mathcal{P}_\odot = 24.5$ d, this implies a solar Rossby number ${\rm Ro} = \mathcal{P}_\odot/\tau_{\rm co} \sim 0.2$.

A low Rossby number (${\rm Ro} \le 1$) implies that the star's rotation has a strong effect on the properties of its convection; the resulting interplay between rotation and convection leads to a strong magnetic field. A high Rossby number (${\rm Ro} > 1$), by contrast, corresponds to the limit of a weak magnetic field. Stars with $M_\star > 1.2\,M_\odot$ have shallow surface convective zones, which leads to a short convective overturn timescale $\tau_{\rm co}$ and thus to a large Rossby number. With weak magnetic fields, these relatively massive stars are unable to spin down by a magnetized wind during their main sequence lifetime.

Although the spindown time t_J of a magnetized star does not explicitly depend on its rotational angular speed Ω (Equation 10.48), dynamo theory leads us to expect that more rapidly rotating stars have higher magnetic field strength B. As Sun-like stars rotate more rapidly, they have larger shears in their tachoclines, and build stronger equilibrium magnetic fields. Let us assume, for simplicity, that $B_{\rm phot} \propto \Omega$. Equation 10.48 then tells us, assuming the star's moment of inertia remains constant, that

$$t_J = \Omega/\dot\Omega \propto B_{\rm phot}^{-2} \propto \Omega^{-2}. \qquad (10.51)$$

The resulting relation $\dot\Omega \propto \Omega^3$ has the solution $\Omega \propto t^{-1/2}$. This implies that the rotation period increases with time at the rate $\mathcal{P}_{\rm rot} \propto t^{1/2}$. Since this is the same

---

[4] The Rossby number is named after the meteorologist Carl-Gustaf Rossby, who studied the effect of the Earth's rotation on instabilities in the jet stream.

time dependence as the empirical gyrochronology relation of Equation 10.34, our simplifying assumption that $B_{phot} \propto \Omega$ turns out to be justifiable.

Given $t_J \propto \Omega^{-2}$, if two stars with the same $M_\star$ and $R_\star$ have different values of Ω, the more rapidly rotating star will have the shorter spindown time t_J. This leads to convergence to a unique rotation rate for a given value of $M_\star$ at late times, in agreement with what we see in Praesepe (Figure 10.8) and other relatively old clusters. However, although the assumption $B_{phot} \propto \Omega$ enables us to explain some observed properties of rotating stars, it doesn't apply to the most rapid rotators. Above a critical angular speed Ω_{crit}, the photospheric magnetic field strength saturates at a maximum value B_{max}. For these rapid-rotating stars, the spindown time has the value

$$t_J = \Omega/\dot{\Omega} \propto B_{max}^{-2} \propto \Omega_{crit}^{-2}. \tag{10.52}$$

Observations indicate that this minimum spindown time for magnetized stars is $t_J \sim 40\,\text{Myr}$. From Equation 10.48, this implies that, for Sun-like stars, the magnetic field strength saturates at $B_{max} \sim 40\,\text{G}$ for rotation periods shorter than $\mathcal{P}_{crit} = 2\pi/\Omega_{crit} \sim 2\,\text{d}$.

Thus, although magnetized stellar winds can be effective at spinning down main sequence stars, the dynamo mechanism within stars cannot produce arbitrarily strong magnetic fields. Stars with $M_\star > 1.2\,\text{M}_\odot$, with their shallow outer convection zones, always have weak magnetic fields. At lower masses, although faster rotation tends to produce stronger magnetic fields, there is an upper limit on B_{phot}; even the fastest-spinning main sequence stars take at least $\sim 40\,\text{Myr}$ for their magnetic winds to spin them down significantly.

Exercises

10.1 Suppose that Maxwell's daemon (as mischievous as ever) squeezes the Sun into a neutron star while conserving its mass $M_\odot$ and angular momentum $J_\odot$. What would be the rotation period $\mathcal{P}_{rot}$ of the resulting neutron star, assuming solid-body rotation? How does the rotation speed of the neutron star compare to its breakup speed?

10.2 Assume that the magnetic field strength B_{phot} of a Sun-like star is proportional to its rotation rate Ω, while the mass loss rate $\dot{M}_{wind}$ is held to a fixed value. With these assumptions, what will be the asymptotic time dependence $\Omega(t)$ for the rotation rate of the star?

10.3 Consider an O star of initial mass $M_\star$ and with no magnetic field. For what mass loss rate $\dot{M}$ is the spindown time t_J equal to the lifetime of the star? State explicitly the assumptions that you make in your calculation.

11

Pulsating Stars

I'd rather learn from one bird how to sing
than teach ten thousand stars how not to dance.

e e cummings (1894–1962)
"you shall above all things be glad and young" [1938]

The life cycles of stars play out across vast oceans of time. The Sun's current age, for instance, is 14 orders of magnitude greater than the Sun's freefall time of 30 min. However, although stars are generally in hydrostatic equilibrium, all stars are also variable. We can distinguish broad families of variation. The most spectacular variation involves catastrophic changes such as supernova explosions. Some stellar variability arises from external influences, such as accretion of mass from a disk or a companion. Other variability is stochastic, involving mass loss from gusty stellar winds, for instance, or changes in brightness as hot and cold convective elements rise and sink near a star's photosphere. However, some stars change their brightness regularly with time as a result of pulsations. Massive time-domain surveys, which monitor the flux of large numbers of stars over an extended length of time, have uncovered a wealth of stellar pulsations, in stars of different masses at different stages of their lifetime. The study of stellar pulsations, known as **asteroseismology**, is transforming our understanding of stellar structure and evolution. It is therefore right and proper for us to discuss stellar oscillations and their observable consequences.

The topic of asteroseismology is vast, and entry to the study is made somewhat difficult by the daunting number of empirical classes of pulsating variable stars. As a simplified introduction, Figure 11.1 shows some of the more famous classes of pulsating variables. We have already encountered the Mira variable stars; these are AGB stars that have pulsation periods $P > 100$ d.

Classical Cepheid variable stars (known as Cepheids, for short) are evolved massive stars, with masses in the range $M_\star = 4$–$20\,M_\odot$. Their spectral type generally varies from F at maximum luminosity to G at minimum. The pulsation period of Cepheids usually lies in the range $P \approx 0.3$–30 d, with the more

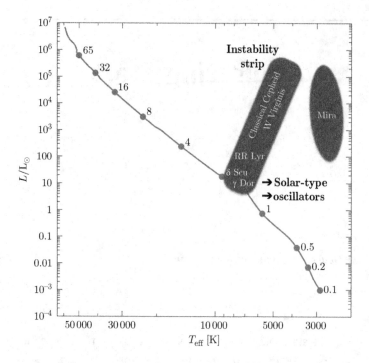

Figure 11.1 Approximate location of some classes of pulsating variable stars on the Hertzsprung–Russell diagram.

luminous Cepheids having longer pulsation periods. Figure 11.2, for example, shows the mean apparent magnitude of Cepheid stars in the Large Magellanic Cloud as a function of their pulsation period. As we discuss in Section 11.1, a star can pulsate at its fundamental frequency or at higher-order overtones. In Figure 11.2, the black points indicate the properties of stars pulsating at their fundamental frequency. For these fundamental-mode Cepheids, assuming the Large Magellanic Cloud is at a distance $d = 50\,\text{kpc}$ (corresponding to $m - M = 18.49$), the period–luminosity relation is

$$\langle M_I \rangle \approx -2.96 \log \left(\frac{\mathcal{P}}{1\,\text{d}} \right) - 1.94. \tag{11.1}$$

This relation is known as Leavitt's law, after Henrietta Leavitt, who first discovered it for Cepheids in the Magellanic Clouds. Leavitt's law is empirically useful in determining distances to relatively nearby galaxies.

W Virginis variable stars lie close to Cepheids on the Hertzsprung–Russell diagram; a W Virginis variable is ~1.5 magnitudes dimmer than a Cepheid with the same color. W Virginis stars are AGB stars with masses $M_\star < 1\,M_\odot$; they are low-metallicity, relatively old stars ($t \sim 10\,\text{Gyr}$). The pulsation period of W Virginis stars usually lies in the range $\mathcal{P} \approx 4$–$20\,\text{d}$.

Other types of variable stars tend to lie in compact regions of the H–R diagram. For instance, **RR Lyrae** variable stars have $L_\star \sim 50\,L_\odot$ and $T_{\text{eff}} \sim 6100\,\text{K}$. They

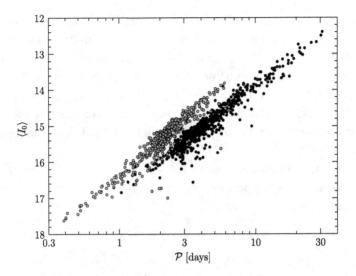

Figure 11.2 Mean I band apparent magnitude as a function of pulsation period $\mathcal{P}$ for a sample of Cepheids in the Large Magellanic Cloud ($d \approx 50\,\mathrm{kpc}$). Black points: Cepheids pulsating at their fundamental mode. Gray points: Cepheids pulsating at their first overtone. [Data from Udalski *et al.* 1999]

are horizontal branch stars, frequently found in globular clusters, with masses $M_\star \sim 0.6\,\mathrm{M_\odot}$ and pulsation periods usually lying in the range $\mathcal{P} \approx 7\text{–}20\,\mathrm{hr}$. **Delta Scuti** variable stars have $L_\star \sim 40\,\mathrm{L_\odot}$ and $T_{\mathrm{eff}} \sim 8000\,\mathrm{K}$. They are fairly young stars; although the prototype of this class, δ Scuti itself, is a giant star, the main sequence star Vega is also a low-amplitude Delta Scuti variable. Delta Scuti stars have masses in the range $M_\star \sim 1\text{–}2\,\mathrm{M_\odot}$ and pulsation periods that are typically $\mathcal{P} \approx 0.2\text{–}5\,\mathrm{hr}$. **Gamma Doradus** variable stars are dimmer and cooler, on average, than Delta Scuti variables, with $L_\star \sim 10\,\mathrm{L_\odot}$ and $T_{\mathrm{eff}} \sim 7000\,\mathrm{K}$. Gamma Doradus stars have masses comparable to those of Delta Scuti variables, but are distinguished by having longer pulsation periods, with $\mathcal{P} \approx 7\text{–}70\,\mathrm{hr}$.

Each listed class of variable star – Gamma Dor, Delta Scu, RR Lyr, W Vir, and classical Cepheids – represents a different type of star at a different stage of its evolution. What these classes have in common, however, is that they fall in the same region of the H–R diagram: a diagonal band, roughly perpendicular to the main sequence, called the **instability strip**. Why stars in the instability strip are subject to pulsation is a question that we will grapple with in Section 11.2. Let's start, however, by looking at the low-amplitude pulsations that are typical of main sequence stars in hydrostatic equilibrium.

11.1 Adiabatic Radial Pulsations

Start with a spherical star in hydrostatic equilibrium, then squeeze it very slightly, being careful to preserve its spherical shape. When you stop squeezing, the star will expand outward, then oscillate about its equilibrium configuration. That is, we are considering **radial perturbations** of small amplitude. In the case of pure radial oscillations, the enclosed mass M is a useful choice of radial coordinate, since if we divide up the star into thin spherical gas shells, the mass M enclosed by a given shell will remain constant as the shell's radius r increases and decreases. In other words, by taking M as our radial coordinate, we are adopting the viewpoint of a Lagrangian observer who rides along with a gas shell as the star oscillates in and out.

If we write the equilibrium state of the star as $r_0(M)$, $\rho_0(M)$, and $P_0(M)$, we can break down the radial perturbations into Fourier components, each with a different frequency ω:

$$r(M, t) = r_0(M) + r_1(M)e^{i\omega t}, \tag{11.2}$$

$$\rho(M, t) = \rho_0(M) + \rho_1(M)e^{i\omega t}, \tag{11.3}$$

$$P(M, t) = P_0(M) + P_1(M)e^{i\omega t}. \tag{11.4}$$

The perturbations to the equilibrium state are assumed to be small, with $|r_1| \ll r_0$, $|\rho_1| \ll \rho_0$, and $|P_1| \ll P_0$. The frequency ω is a constant, but is not required to be a real number. If it is real, corresponding to sinusoidal variations, we expect $\omega \propto t_{\mathrm{ff}}^{-1}$. However, a more sophisticated analysis is required to determine under what circumstances ω is real, and what numerical values of ω correspond to permitted stable oscillations of a star with known equilibrium structure r_0, ρ_0, and P_0.

Since a perturbed star is no longer in equilibrium, we must replace the equation of hydrostatic equilibrium (Equation 2.13) with the momentum equation

$$\frac{\partial^2 r}{\partial t^2} = -4\pi r^2 \left(\frac{\partial P}{\partial M}\right) - \frac{GM}{r^2}, \tag{11.5}$$

where the two terms on the right-hand side represent the competing pressure gradient force and gravitational force. Pulsations don't create or destroy mass, so we can still embrace the mass conservation equation

$$\frac{\partial r}{\partial M} = \frac{1}{4\pi r^2 \rho} \tag{11.6}$$

at any given moment t. Now we use the standard linear perturbation technique of inserting the perturbed values of r, ρ, and P into our equations and keeping only those terms linear in r_1, ρ_1, and P_1. For instance, the linearized version of the momentum equation (Equation 11.5) is

$$-\omega^2 r_1 = -4\pi r_0^2 \frac{dP_1}{dM} - 8\pi r_0 \frac{dP_0}{dM} r_1 + 2\frac{GM}{r_0^3} r_1. \tag{11.7}$$

We can use the hydrostatic equilibrium relation for the unperturbed star,

$$\frac{dP_0}{dM} = -\frac{GM}{4\pi r_0^4}, \tag{11.8}$$

to rewrite Equation 11.7 as

$$4\pi r_0 \frac{dP_1}{dM} = \left(4\frac{GM}{r_0^3} + \omega^2\right) r_1.$$ (11.9)

Similarly, the linearized version of the mass conservation equation (Equation 11.6) becomes

$$\frac{dr_1}{dM} = -\frac{1}{4\pi r_0^2 \rho_0}\left(2\frac{r_1}{r_0} + \frac{\rho_1}{\rho_0}\right).$$ (11.10)

We expect that small-amplitude oscillations will be adiabatic, so that we can assume $P \propto \rho^\gamma$, and thus $P_1/P_0 = \gamma(\rho_1/\rho_0)$. For adiabatic oscillations, we can then rewrite Equation 11.10 as

$$\frac{dr_1}{dM} = -\frac{1}{4\pi r_0^2 \rho_0}\left(2\frac{r_1}{r_0} + \frac{1}{\gamma}\frac{P_1}{P_0}\right).$$ (11.11)

In the general case, we can combine the momentum equation (Equation 11.9) and the mass conservation equation (Equation 11.11) to find a single **linear adiabatic wave equation**, giving the relation between the wave amplitude $r_1(M)$ and frequency ω in a star of given equilibrium structure. In general, though, the linear adiabatic wave equation is rather complicated. Thus, for purposes of illustration, let's take a particularly simple case, and assume a **homologous** displacement; that is, one where $r_1 = \varepsilon r_0$, with ε being constant throughout the star. In this simple case, the mass conservation relation of Equation 11.11 tells us that

$$\varepsilon\frac{dr_0}{dM} = -\frac{1}{4\pi r_0^2 \rho_0}\left(2\varepsilon + \frac{1}{\gamma}\frac{P_1}{P_0}\right).$$ (11.12)

However, since mass conservation in the unperturbed star (Equation 11.6) requires

$$\frac{dr_0}{dM} = \frac{1}{4\pi r_0^2 \rho_0},$$ (11.13)

this reduces to

$$P_1 = -3\gamma\varepsilon P_0.$$ (11.14)

Thus, homologous contraction or expansion requires a homologous increase or decrease in the pressure profile of the star. Inserting the result of Equation 11.14 into the momentum equation as given in Equation 11.9, we find a simple dispersion relation,

$$\omega^2 = (3\gamma - 4)\frac{GM}{r_0^3} \qquad [r_1 = \varepsilon r_0].$$ (11.15)

If the adiabatic index is $\gamma > 4/3$, then ω is real, and the star pulsates stably with a period

$$P = \frac{2\pi}{\sqrt{3\gamma - 4}} \left(\frac{r_0^3}{GM} \right)^{1/2} = \frac{4\sqrt{2}}{\sqrt{3\gamma - 4}} t_{\mathrm{ff}}. \tag{11.16}$$

Consider, for example, a population of pulsating stars which all have a similar effective temperature (which we expect to be $T_{\mathrm{eff}} \sim 6500\,\mathrm{K}$ for stars in the instability strip). These stars will have a luminosity–radius relation $R_\star \propto L_\star^{1/2}$. If, in addition, they have a mass–radius relation of the form $R_\star \propto M_\star^\alpha$, then their period–luminosity relation will be, from Equation 11.16,

$$\log L_\star = \frac{3\alpha - 1}{4\alpha} \log P + K. \tag{11.17}$$

Thus, as long as the mean density ρ decreases with increasing stellar mass (implying $\alpha > 1/3$), the more luminous stars will have a longer pulsation period, as in Leavitt's law.

If the adiabatic index is $\gamma < 4/3$, then ω is imaginary, and the star is unstable. The e-folding time for growth (or decay) of the pulsations is

$$\tau = \frac{1}{\sqrt{-\omega^2}} = \frac{2\sqrt{2}}{\pi} \frac{1}{\sqrt{4 - 3\gamma}} t_{\mathrm{ff}}. \tag{11.18}$$

In general, stars with $\gamma > 4/3$ everywhere are dynamically stable; stars with $\gamma < 4/3$ everywhere are dynamically unstable. (A gas of photons has $\gamma = 4/3$, and thus is on the brink of instability. This precarious state helps to explain the extreme mass loss that we see in the most massive main sequence stars, where radiation pressure starts to dominate.) Stars with $\gamma < 4/3$ in part of their interior have to be studied in more detail to see whether they are stable or not. Regions of partial ionization, as we saw in Section 4.2, can have $\gamma \approx 1.2$; however, these regions are usually sufficiently small compared to the entire star that they don't lead to dynamical instabilities.

Notice from Equation 11.15 that since ω^2 is a constant, adiabatic radial pulsations with $r_1 \propto r_0$ can occur only in a uniform density star, with

$$\rho_0 = \rho_\star \equiv \frac{3M_\star}{4\pi R_\star^3}. \tag{11.19}$$

In a real star, adiabatic radial pulsations must be non-homologous; that is, r_1/r_0 must vary throughout the star. Suppose that you have a stellar model whose equilibrium state is known; it may be a polytrope, or it may be a fancier stellar model. The linear adiabatic wave equation, with appropriate boundary conditions at $M = 0$ and $M = M_\star$, then has an infinite number of eigenvalues ω_i, with $i = 0, 1, 2, \ldots$, and related eigenfunctions $r_1 \exp(i\omega_i t)$. If $\gamma > 4/3$, then all the eigenvalues ω_i are real, and all the radial modes are stable.

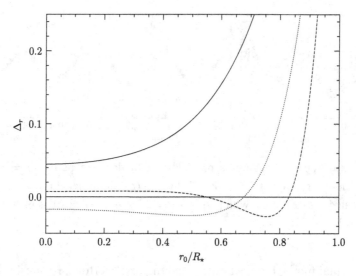

Figure 11.3 Dimensionless radial displacement $\Delta_r \equiv r_1/r_0$, arbitrarily normalized to $\Delta_r = 1$ at $r_0 = R_\star$, for the first three modes in an $n = 3$ polytrope with $\gamma = 5/3$. Solid line: fundamental mode; dotted line: first overtone; dashed line: second overtone. [Data from Schwarzschild 1941]

If we order the eigenvalues so that $\omega_0 < \omega_1 < \omega_2 < \cdots$, then the solution with $\omega = \omega_0$ is the **fundamental mode** of the star, the solution with $\omega = \omega_1$ is the first overtone, the solution with $\omega = \omega_2$ is the second overtone, and so forth. For example, Figure 11.3 shows the first three radial pulsation modes of an $n = 3$ polytrope consisting of a gas with $\gamma = 5/3$. (As shown in Section 6.4, this provides a good, but not perfect, model of the Sun's radiative zone, at $r < 0.71\,R_\odot$.) The fundamental mode of this polytrope has period $\mathcal{P}_0 = 2\pi/\omega_0 \approx 1.86 t_{\mathrm{ff}}$; assuming a freefall time equal to that of the Sun, this corresponds to $\mathcal{P}_0 \approx 55\,\mathrm{min}$ or $\omega_0 \approx 0.0019\,\mathrm{s}^{-1}$. The fractional displacement r_1/r_0 for the different modes is plotted in Figure 11.3; notice that r_1/r_0 for the fundamental mode, the solid line, decreases toward the center, where the density is higher. The first overtone, with $\mathcal{P}_1 = 0.738\mathcal{P}_0 \sim 40\,\mathrm{min}$, has r_1/r_0 plotted as the dotted line. The second overtone, with $\mathcal{P}_2 = 0.570\mathcal{P}_0 \sim 31\,\mathrm{min}$, is plotted as the dashed line. Note that the first overtone has one node, where $r_1 = 0$, while the second overtone has two nodes. This is a general result; the eigenvalue ω_i corresponds to the displacement r_1 having i nodes. In this way, a radially pulsating star has been compared to an organ pipe, where higher harmonics correspond to standing waves with larger numbers of nodes.

Many Cepheids pulsate at their fundamental mode; however, some are first-overtone pulsators, as shown in Figure 11.2. It is also possible for a star to pulsate in two or more modes simultaneously. (It is thought that Polaris A, the closest Cepheid to the Earth, is a double-mode Cepheid.) Figure 11.4, for instance, shows

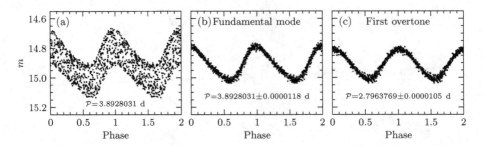

Figure 11.4 (a) Apparent magnitude, phased by the fundamental period $\mathcal{P}_0 = 3.893$ d, of the double-mode Cepheid OGLE-LMC-CEP-0832. (b) The isolated fundamental mode. (c) The isolated first overtone, with period $\mathcal{P}_1 = 2.796$ d. [OGLE Atlas of Variable Star Light Curves]

the fundamental mode, Figure 11.4(b), and the first overtone, Figure 11.4(c), of a double-mode Cepheid observed by OGLE. The ratio of the first overtone period to the fundamental period for this double-mode Cepheid is $\mathcal{P}_1/\mathcal{P}_0 = 0.718$, slightly smaller than the corresponding ratio $\mathcal{P}_1/\mathcal{P}_0 = 0.738$ for the simple $n = 3$ polytrope shown in Figure 11.3.

11.2 Non-adiabatic Radial Pulsations

In the previous section, we assumed that pulsations are adiabatic, and that after one complete pulsation period, the star returns to the exact same state in which it began. Pulsating stars aren't perfectly adiabatic, of course. During one pulsation period, heat continues to flow through the star, and photons continue to escape into space. This is undeniably a process that increases entropy.

However, we can approximate a star's pulsations as being adiabatic if the fundamental pulsation period $\mathcal{P}_0$ is short compared to the thermal time t_{th}. The total thermal energy of a star is

$$K_\star = \frac{3}{2} \int \frac{kT}{m_{\text{AMU}}\bar{\mu}} dM, \tag{11.20}$$

so the thermal time (the time required to radiate away its thermal energy at constant $L_\star$) is

$$t_{\text{th}} = \frac{K_\star}{L_\star}. \tag{11.21}$$

However, the virial theorem tells us that for a star in equilibrium, thermal energy is simply related to gravitational potential energy, with $K_\star = -W_\star/2$. Thus, the thermal time for a star in virial equilibrium is

$$t_{\text{th}} = \frac{K_\star}{L_\star} = -\frac{1}{2}\frac{W_\star}{L_\star} = \frac{1}{2}t_{\text{KH}}, \tag{11.22}$$

where t_{KH} is the Kelvin–Helmholtz time defined in Section 2.3 as the time for a star to radiate away its gravitational potential energy.[1]

For the Sun, we computed $t_{KH} \sim 30$ Myr, and thus $t_{th} = t_{KH}/2 \sim 15$ Myr. By contrast, the fundamental pulsation period of the Sun is $\mathcal{P}_0 \sim 1$ h $\sim 10^{-10}$ Myr, 11 orders of magnitude shorter than the thermal time. Within an equilibrium star with no pulsations, heat flows from the hot core to the cool photosphere. This increases entropy, but on a timescale $t_{th} \gg \mathcal{P}_0$. During one pulsation period, a Sun-like star can increase its entropy by only one part in 100 billion, which is a good approximation to being adiabatic.

Let's see if a strongly pulsating star, such as a Cepheid, can increase its entropy during one pulsation period. Pulsations cause changes in ρ and P, and thus also in the temperature T. The energy generation rate $\epsilon_{nuc}(\rho, T)$ and the opacity $\kappa(\rho, T)$ will therefore also undergo changes. As we have seen, a small change in T can lead to a large change in ϵ_{nuc}, especially for the triple alpha process, with its strong sensitivity to temperature. In addition, there are regions of density–temperature space where the opacity κ is strongly dependent on temperature. As we saw in Section 4.1, for instance, H^- opacity has the dependence $\kappa \propto T^9$.

Let's suppose that pulsations in a star cause the energy generation rate in its core to fluctuate wildly in amplitude with a period $\mathcal{P}_0 \ll t_{th}$. The large fluctuations in luminosity that result will be smeared out during the time t_{th} that it takes the energy to reach the photosphere. Thus, as far as the outer layers of the star are concerned, a star whose energy generation pulsates with a period $\mathcal{P}_0 \ll t_{th}$ will be identical to a star steadily generating energy with the same time-averaged rate. Similarly, if the opacity of a layer within the star varies greatly with a period $\mathcal{P}_0 \ll t_{th}$, it will act as a "shutter" that alternately traps and releases light. However, these fluctuations in radiative energy transport will also be smeared out during the time it takes the energy to reach the photosphere. Thus, periodic changes to ϵ_{nuc} in the star's core and to κ in the star's interior will be smeared out as long as their period is short compared to the thermal time. To an outside observer, the star will look like an ordinary, boring, equilibrium star, with no sign of the rapid swings in ϵ_{nuc} or κ happening deep within the star.

We do, however, see stars whose observable properties vary quite strongly on a timescale $t \sim \mathcal{P}_0$. These are the pulsating variable stars such as Cepheids and RR Lyrae stars. What's happening with these stars? Instead of looking at an entire star, focus just on its cooler, lower-density, outer envelope. Consider an outer envelope with mass $M_{env} \ll M_\star$ and average temperature $T_{env} < T_c$. The thermal energy of this envelope is

$$K_{env} = \frac{3k}{2m_{AMU}\bar{\mu}} T_{env} M_{env}. \tag{11.23}$$

---

[1] Because of the close link between the thermal time and the Kelvin–Helmholtz time for a star in virial equilibrium, the terms are sometimes used interchangeably.

Since the outer envelope is well outside the energy-generating core, we can write $L_{\mathrm{env}} = L_\star$, and thus find that the thermal time for the envelope alone is

$$t_{\mathrm{th,env}} = \frac{3k}{2m_{\mathrm{AMU}}\bar{\mu}L_\star} T_{\mathrm{env}} M_{\mathrm{env}}. \tag{11.24}$$

If we choose a thin enough outer envelope, this will be shorter than the fundamental pulsation period of the star, $\mathcal{P}_0 \sim t_{\mathrm{ff}}$. Thus, for every star, there is a critical envelope mass M_{crit} for which the time to radiate away the envelope's thermal energy is equal to the fundamental pulsation period of the star. This critical envelope mass is determined by the criterion that

$$K_{\mathrm{crit}} \sim \frac{k}{m_{\mathrm{AMU}}\bar{\mu}} T_{\mathrm{crit}} M_{\mathrm{crit}} \sim \mathcal{P}_0 L_\star, \tag{11.25}$$

where T_{crit} is the temperature at the base of the critical envelope. That is, the thermal energy of the critical envelope must be comparable to the photon energy radiated during one pulsation of the star. For the Sun, with $\mathcal{P}_0 \sim 1\,\mathrm{hr}$, this energy is $\mathcal{P}_0 L_\odot \sim 1.4 \times 10^{37}$ erg. The thermal energy of the solar atmosphere, from the base of the photosphere upward, is $K_{\mathrm{atm}} \sim 2 \times 10^{35}$ erg. Thus, the base of the Sun's critical envelope lies in the upper region of the convection zone, $\sim 1000\,\mathrm{km}$ below the photosphere, where the temperature is $T_{\mathrm{crit}} \sim 20\,000\,\mathrm{K}$ and the density is $\rho_{\mathrm{crit}} \sim 10^{-6}\,\mathrm{g\,cm}^{-3}$. The mass of the Sun's critical envelope is $M_{\mathrm{crit}} \sim 4 \times 10^{-9}\,M_\odot$.

Pulsating variable stars such as Cepheids are stars for which some mechanism is driving non-adiabatic pulsations at the base of the critical envelope. If the driving mechanism is at work well below the base of the critical envelope, then the long thermal times will smear out the pulsations. If the driving mechanism is at work well above the base of the critical envelope, then the short thermal times mean that temperature fluctuations and the resulting opacity fluctuations can't be sustained. However, driving mechanisms that are triggered at the conditions prevailing at the base of the critical envelope (T_{crit}, ρ_{crit}, and so forth) hit the sweet spot for driving high-amplitude non-adiabatic pulsations of the outer envelope of the star.

Non-adiabatic radial pulsations of the outer regions can be approximated as a **Carnot heat engine**, named after the nineteenth-century engineer Nicolas Léonard Sadi Carnot. A Carnot heat engine is a theoretical device that has the most efficient cycle for converting a given amount of heat energy into work; for instance, the work of lifting the outer envelope of a star against the force of gravity. The four basic steps of the Carnot cycle are shown in Figure 11.5. For concreteness, the steps are shown for a Carnot heat engine consisting of a piston (rather than a massive stellar envelope) and a cylinder of gas (rather than a gas layer at the base of the stellar envelope). The steps are:

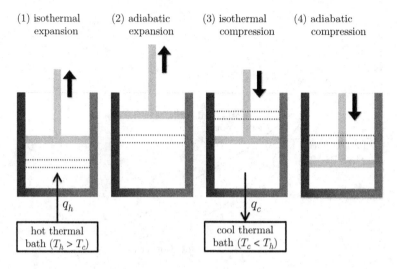

(1) isothermal expansion (2) adiabatic expansion (3) isothermal compression (4) adiabatic compression

Figure 11.5 The four steps of the Carnot cycle, illustrated for a piston rather than a stellar envelope.

(1) Gas expands isothermally at a temperature T_h. To do this, it absorbs an amount of heat q_h from its surroundings.
(2) Gas expands adiabatically. As it does this, its temperature drops from T_h to T_c.
(3) Gas is compressed isothermally at a temperature T_c. To do this, it dumps an amount of heat q_c into its surroundings.
(4) Gas is compressed adiabatically. As it does this, its temperature rises from T_c back to T_h.

Figure 11.6 shows how the properties of the Carnot heat engine evolve during one complete cycle (or pulsation). The pressure within the cylinder is plotted as a function of the volume for the successive steps of isothermal expansion, adiabatic expansion, isothermal compression, and adiabatic compression. The net work done on the piston by the gas during one complete cycle is

$$\Delta W = \oint_{abcda} P dV, \tag{11.26}$$

which is just the area enclosed by the closed curve in the P–V plane. Since the first law of thermodynamics tells us that the change in internal energy is $dE = TdS - PdV$, and since $\Delta E = 0$ over a complete cycle of a Carnot heat engine, we can also write

$$\Delta W = \oint_{abcda} T dS. \tag{11.27}$$

Obviously, a pulsating star is not an ideal Carnot heat engine. However, since $\Delta W > 0$ for any clockwise closed curve in the P–V plane, as long as the gas of

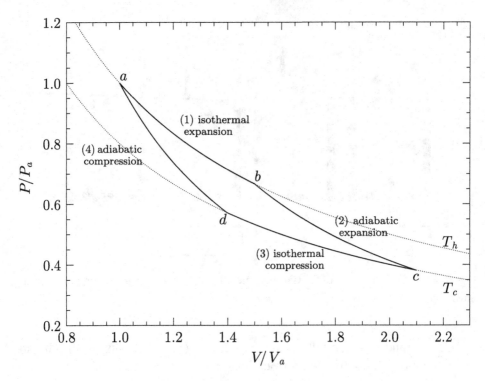

Figure 11.6 Pressure–volume diagram for a Carnot heat engine with $T_c = 0.8T_h$ and $\gamma = 5/3$.

the star's outer envelope expands at a higher temperature and is compressed at a lower temperature, the star will be an engine of sorts, and ΔW will be positive.

The driving mechanism for Cepheids and similar pulsating variable stars is called the **kappa mechanism**. It has been given this name because it involves changes in the opacity $\kappa(\rho, T)$ at the base of the critical envelope. (Although it is theoretically possible for pulsations to be driven by changes in the specific energy generation rate ϵ_{nuc}, the critical envelope for pulsating stars tends to lie far above the region where nuclear fusion takes place.) Consider the layer of gas at the base of the critical envelope, in the case when energy transport through the layer is radiative. If the opacity of the layer increases when it is compressed, then it will trap radiation which will then be released when the built-up pressure causes the layer to expand again.

If we write the opacity κ in a generalized Kramers' law form,

$$\kappa \propto \rho^\alpha T^{-\beta},\tag{11.28}$$

then if the compression is nearly adiabatic, with $T \propto P/\rho \propto \rho^{\gamma-1}$, the dependence of κ on the layer's density will be

$$\kappa \propto \rho^{\alpha-\beta(\gamma-1)}.\tag{11.29}$$

For free–free and bound–free opacity, with $\alpha = 1$ and $\beta = 3.5$, this leads to a density dependence of

$$\kappa_{\text{ff,bf}} \propto \rho^{1-3.5(\gamma-1)}. \tag{11.30}$$

If we assume $\gamma = 5/3$, the "standard" value for a monatomic gas, this would imply $\kappa \propto \rho^{-1.3}$; the opacity would *decrease* as the layer was compressed, and the kappa mechanism would not work. This explains why most stars are not high-amplitude pulsators. However, there are domains where the opacity can increase as gas is compressed.

Consider, for instance, the case of classical Cepheid pulsators. In Cepheids, the base of the critical envelope is at a temperature $T_{\text{crit}} \sim 40\,000\,\text{K}$, implying a critical envelope mass (from Equation 11.25) of

$$M_{\text{crit}} \approx 3 \times 10^{-4}\,M_{\odot} \left(\frac{P_0}{5\,\text{d}}\right)\left(\frac{L_{\star}}{2000\,L_{\odot}}\right), \tag{11.31}$$

scaling to the properties of Delta Cephei. At the relevant temperature $T \sim 40\,000\,\text{K}$, free–free and bound–free transitions dominate the opacity. However, this is also the temperature at which helium starts to become doubly ionized, as shown in Figure 4.5(a). Thus, as a layer at $T \sim 40\,000\,\text{K}$ is compressed, part of the increase in the specific internal energy,

$$du = -Pd(1/\rho), \tag{11.32}$$

goes to ionizing He^+ rather than to increasing the temperature T. As shown in Figure 4.5(b), the second ionization of helium can drive the adiabatic index as low as $\gamma \approx 1.23$ in a low-density gas, resulting in an opacity

$$\kappa \propto \rho^{1-3.5(\gamma-1)} \propto \rho^{1-3.5(0.23)} \propto \rho^{0.2}. \tag{11.33}$$

Thus, as the layer is compressed and helium becomes more highly ionized, the opacity increases with density, as required by the kappa mechanism.

Stars with $T_{\text{eff}} > 7500\,\text{K}$ have partial ionization zones that are located above the base of the critical envelope. Stars with $T_{\text{eff}} < 5500\,\text{K}$ have partial ionization zones that are located below the base of the critical envelope. It is only in the narrow range of temperatures corresponding to the instability strip of Figure 11.1 that the partial ionization zone corresponds to the base of the critical envelope. (Note that although the Sun, with $T_{\text{eff}} = 5772\,\text{K}$, lies within the temperature range in question, the base of its critical envelope is well inside the convective zone. Thus, the opacity is irrelevant, as long as it remains sufficiently high that convection is the most efficient means of energy transport.)

11.3 Adiabatic Non-radial Pulsations

Even if a star is not a flamboyantly pulsating variable like a Cepheid or RR Lyrae star, it will usually quiver a bit. These low-amplitude stellar pulsations provide both an elegant test of stellar theory and a window into the interiors of stars. By analyzing the observable consequences of pulsation, we can deduce properties of a star's interior, such as the sound speed, temperature, chemical composition, and (as briefly discussed in Section 10.3) the interior rotation.

Why do stars pulsate at all? A gong will oscillate if it is struck by a mallet, but otherwise will sit in silent equilibrium. We need to identify the "mallet" that causes a star to oscillate. In a relatively cool star (cooler than the instability strip on the H–R diagram), the source of pulsations lies in the star's deep surface convection zone. The chaotic, turbulent convection results in random collisions between convective blobs; this generates waves with a broad spectrum of frequencies. Some frequencies produce standing waves, which cause the star's surface to oscillate and produce observable periodic fluctuations in the star's flux. Pulsations triggered by turbulent convection are known as **solar-like oscillations**, since the Sun's observed pulsations are of this type. Solar-like oscillations have been detected in numerous other stars, and are likely to be a nearly universal phenomenon in the cooler portion of the H–R diagram.

Solar-like oscillations have important differences from the radial pulsations seen in Cepheids and other classical pulsators. The radial pulsations in a Cepheid arise from a feedback loop among opacity, expansion, and contraction. Once a radial mode is excited, it retains phase coherence. By contrast, solar-like oscillations are stochastically excited by turbulence. Only certain frequencies produce a standing wave pattern, and there is no requirement that all the turbulent collisions produce those frequencies at the same phase. Occasionally, by the luck of the draw, positive interference will occur for one frequency or another. In general, however, pulsation amplitudes are far lower for solar-like oscillations than for classical pulsators. Among stars with solar-like oscillations, observed variations in flux range from about 1% for luminous red giants to one part in 10^5 for Sun-like stars.

Cepheid stars, as we have seen, typically oscillate at only one or two of their lowest-order radial modes. By contrast, we observe a rich spectrum of pulsation frequencies for solar-like oscillators. The many standing waves within a Sun-like star will not necessarily be radial. Thus, to describe solar-like oscillations, we need a more sophisticated mathematical framework than we used for purely radial pulsations in Section 11.1. In a non-spherical, non-equilibrium star, the equation of hydrostatic equilibrium must be replaced with the general momentum equation (which we have already encountered as Equation 7.9):

$$\rho \frac{\partial \vec{v}}{\partial t} + \rho(\vec{v} \cdot \vec{\nabla})\vec{v} = -\vec{\nabla}P - \rho\vec{\nabla}\Phi. \qquad (11.34)$$

Here, $\vec{v}$ is the bulk velocity of gas in the star's interior, and the gravitational potential Φ is given by the Poisson equation,

$$\Phi(\vec{r}) = -G \int \frac{\rho(\vec{x})d^3x}{|\vec{x} - \vec{r}|}. \qquad (11.35)$$

As the gas moves, it must obey the mass continuity equation

$$\frac{\partial \rho}{\partial t} + \vec{\nabla} \cdot (\rho \vec{v}) = 0. \qquad (11.36)$$

Since we are dealing with low-amplitude oscillations, we can assume a spherical equilibrium state[2] with pressure $P_0(r)$, density $\rho_0(r)$, potential $\Phi_0(r)$, and so forth. Starting with the equilibrium state, we take the gas at each position $\vec{r}$ and displace it through a small distance $\vec{\xi}$. Since $\vec{\xi}$ is not constrained to be in the radial direction, we can write it in spherical coordinates as $\vec{\xi} = \xi_r \hat{r} + \xi_\theta \hat{\theta} + \xi_\phi \hat{\phi}$. The small displacement is associated with small perturbations in the pressure P_1, density ρ_1, and potential Φ_1. Keeping only those terms linear in $\vec{\xi}$, P_1, ρ_1, and Φ_1, the momentum equation (Equation 11.34) takes the linearized form

$$\rho_0 \frac{d^2\vec{\xi}}{dt^2} = -\vec{\nabla}P_1 - \rho_1 \vec{\nabla}\Phi_0 - \rho_0 \vec{\nabla}\Phi_1. \qquad (11.37)$$

In Equation 11.37, the acceleration on the left-hand side is tied to three driving terms on the right-hand side; these correspond to fluctuations in the pressure, density, and gravitational potential respectively. If the dominant driving term is the pressure term, then the resulting wave is a **p-mode**, or sound wave. If the dominant driving terms are the density and potential terms, then the resulting wave is a **g-mode**, or gravity wave.

We can simplify Equation 11.37 by using the Cowling approximation, which ignores the effects of the potential perturbation Φ_1. When Thomas Cowling introduced this approximation in 1941, he acknowledged that "the oscillations of a heterogeneous compressible fluid sphere present a problem which is, in any case, of great mathematical complexity." Thus, he made the most physically reasonable approximation that would reduce the complexity. Cowling's reason for ignoring Φ_1 (rather than P_1 or ρ_1) becomes more apparent when we look at the linearized version of the Poisson equation:

$$\Phi_1(\vec{r}) = -G \int \frac{\rho_1(\vec{x})d^3x}{|\vec{x} - \vec{r}|}. \qquad (11.38)$$

If the density perturbation ρ_1 switches frequently between negative and positive within a star (for instance, if it represents a high-order radial overtone), the broad $1/x$ smoothing kernel in Equation 11.38 makes the perturbed potential Φ_1

---

[2] Although rotating stars also undergo oscillations, complicated coupling exists between rotation and pulsation in the limit $v_{\text{rot}} \rightarrow v_{\text{BU}}$. To keep life simple, we will look only at oscillations in non-rotating stars.

very nearly uniform. Using the Cowling approximation, the radial portion of the momentum equation (Equation 11.37) becomes

$$\rho_0 \frac{d^2 \xi_r}{dt^2} = -\frac{dP_1}{dr} - \rho_1 g_0, \tag{11.39}$$

where $g_0 = GM_0/r^2$ is the gravitational acceleration in the unperturbed star. However, when we look at the displacement in the horizontal direction, $\vec{\xi}_h \equiv \xi_\theta \hat{\theta} + \xi_\phi \hat{\phi}$, the momentum equation yields

$$\rho_0 \frac{d^2 \vec{\xi}_h}{dt^2} = \frac{1}{r} \frac{dP_1}{d\theta} \hat{\theta} + \frac{1}{r \sin \theta} \frac{dP_1}{d\phi} \hat{\phi}, \tag{11.40}$$

independent of ρ_1 as well as Φ_1. Thus, in the Cowling approximation, displacements perpendicular to the radial direction are associated only with p-modes.

The linearized momentum equations (Equations 11.39 and 11.40) are supplemented by the mass continuity equation, which in its linearized form is

$$\begin{aligned} \rho_1 &= -\vec{\nabla} \cdot (\rho_0 \vec{\xi}) \\ &= -\frac{1}{r^2} \frac{d}{dr} (r^2 \rho_0 \xi_r) - \rho_0 \vec{\nabla} \cdot \vec{\xi}_h. \end{aligned} \tag{11.41}$$

Solar-like oscillations are nearly adiabatic. This implies that the pressure perturbation P_1, the density perturbation ρ_1, and the radial displacement ξ_r are linked by the adiabatic condition

$$\frac{\rho_1}{\rho_0} + \frac{\xi_r}{\rho_0} \frac{d\rho_0}{dr} = \frac{1}{\gamma} \left[\frac{P_1}{P_0} + \frac{\xi_r}{P_0} \frac{dP_0}{dr} \right]. \tag{11.42}$$

Equations 11.39–11.42 are equations that must be satisfied for any low-amplitude perturbations within an otherwise spherical star. When we look specifically at standing waves, we realize that they will have solutions with an integral number of nodes in radius, latitude, and longitude. Thus, the perturbations associated with standing waves are usefully expressed in terms of spherical harmonics. Consider, for instance, a mode for which the pressure perturbation can be written as

$$P_1 = P_{n\ell}(r) Y_{\ell m}(\theta, \phi) e^{i\omega t}, \tag{11.43}$$

with a similar form for ρ_1, ξ_r, and ξ_h. Here, the number $n = 0, 1, 2, \ldots$ is the radial order, which is simply the number of nodes in the eigenfunction $P_{n\ell}$. The parameters $\ell = 0, 1, 2, \ldots$ and $m = -\ell, \ldots, \ell$ are the usual angular degree ℓ and azimuthal order m that enter into spherical harmonics. (The $\ell = 0$, $m = 0$ modes are thus the purely radial modes that we have already considered in Section 11.1.) In a non-rotating star with a spherical equilibrium state, the frequency ω will differ for modes with different n and ℓ, but is independent of m.

For modes of the form given in Equation 11.43, the momentum equation in the radial direction (Equation 11.39) is

$$-\omega^2 \rho_0 \xi_r = -\frac{dP_1}{dr} - \rho_1 g_0. \tag{11.44}$$

Using the spherical harmonic identities that make introductory quantum mechanics such a joy, we find that the momentum equation in the horizontal direction (Equation 11.40) reduces to

$$-\omega^2 \rho_0 \xi_h = -\frac{P_1}{r}. \tag{11.45}$$

The mass continuity equation (Equation 11.41) becomes

$$\rho_1 = -\frac{1}{r^2}\frac{d}{dr}(r^2 \rho_0 \xi_r) + k_h^2 r \rho_0 \xi_h, \tag{11.46}$$

where the **horizontal wavenumber** k_h is related to the angular degree ℓ through the relation

$$k_h^2 = \frac{\ell(\ell+1)}{r^2}. \tag{11.47}$$

For non-radial perturbations ($\ell > 0$), the wavelength corresponding to the horizontal wavenumber is

$$\lambda_h = \frac{2\pi}{k_h} = \frac{2\pi r}{\sqrt{\ell(\ell+1)}}. \tag{11.48}$$

In the limit $\ell \gg 1$, this is nearly equal to the circumference at r divided by the angular degree ℓ.

Finally, the adiabatic equation (Equation 11.42) can be usefully rewritten as

$$P_1 = P_0 \left[\gamma \frac{\rho_1}{\rho_0} - \left(\frac{d \ln P_0}{dr} - \gamma \frac{d \ln \rho_0}{dr} \right) \xi_r \right]. \tag{11.49}$$

Starting with Equations 11.44, 11.45, 11.46, and 11.49, it is possible to eliminate the two variables P_1 and ρ_1, leaving behind a pair of equations that tell us the radial and horizontal displacements as a function of the equilibrium properties of the star. The first equation is

$$\frac{d\xi_r}{dr} = \left[\frac{g_0}{c_s^2} - \frac{2}{r} \right] \xi_r + r \frac{\omega^2}{c_s^2} \left[\frac{S_\ell^2}{\omega^2} - 1 \right] \xi_h, \tag{11.50}$$

where c_s is the sound speed, given by the relation $c_s^2 = \gamma(P_0/\rho_0)$, and S_ℓ is the **Lamb frequency**, defined by the relation

$$S_\ell^2 = k_h^2 c_s^2 = \frac{\ell(\ell+1)}{r^2} c_s^2. \tag{11.51}$$

The Lamb frequency, named after the applied mathematician Horace Lamb, is also called the **critical acoustic frequency**. Note that all non-radial modes ($\ell > 0$) have a Lamb frequency that becomes arbitrarily large in the limit $r \to 0$.

Figure 11.7(b) shows the Lamb frequency (dashed lines) for a few values of ℓ inside a standard solar model.

The second equation involving ξ_r and ξ_h is

$$\frac{d\xi_h}{dr} = \frac{1}{r}\left[1 - \frac{N_{BV}^2}{\omega^2}\right]\xi_r + \left[\frac{N_{BV}^2}{g_0} - \frac{1}{r}\right]\xi_h, \tag{11.52}$$

where N_{BV} is the **Brunt–Väisälä frequency**, defined by the relation

$$N_{BV}^2 = g_0\left[\frac{1}{\gamma}\frac{d\ln P_0}{dr} - \frac{d\ln\rho_0}{dr}\right]. \tag{11.53}$$

The Brunt–Väisälä frequency is also called the **buoyancy frequency**, since it is analogous to the frequency with which a vertically displaced buoy bobs up and down in the ocean.[3] Note that the Schwarzschild criterion for instability against convection (Equation 4.51) can be written as

$$\frac{1}{\gamma}d\ln P_0 < d\ln\rho_0. \tag{11.54}$$

Thus, within the Sun's convection zone, $N_{BV}^2 < 0$, and buoyancy-driven oscillations cannot exist. As shown in Figure 11.7(b), N_{BV} is real within the Sun's radiative zone, but goes to zero both at the base of the convective zone and at the Sun's center, where $g_0 = 0$.

We now want to solve Equations 11.50 and 11.52 for a star of known equilibrium state. That is, we know the gravitational acceleration g_0, sound speed c_s, and Brunt–Väisälä frequency N_{BV} everywhere inside the equilibrium star. If we

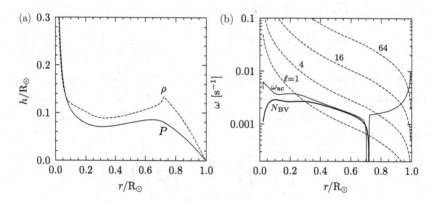

Figure 11.7 (a) Pressure scale height h (solid line) and density scale height h_ρ (dashed line) for a standard solar model. (b) Lamb frequency S_ℓ for $\ell = 1, 4, 16$, and 64 (dashed lines), Brunt–Väisälä frequency N_{BV} (heavy solid line), and acoustic cutoff frequency ω_{ac} (light solid line) for the same solar model. All frequencies are displayed as angular frequency ω rather than circular frequency $\nu = \omega/(2\pi)$.

---

[3] Brunt and Väisälä (like the Rossby who gave his name to the Rossby number) were meteorologists; the concept of buoyant oscillations is important to the study of the Earth's atmosphere.

choose an angular degree $\ell > 0$, we can then compute the relevant Lamb frequency S_ℓ from Equation 11.51. Thus, for any frequency ω, all we need to solve Equations 11.50 and 11.52 are appropriate boundary conditions. It is relevant to note, when attempting to find boundary conditions, that waves will be refracted in the presence of steep density gradients. We can quantify the steepness of a star's density gradient by computing the density scale height,

$$h_\rho(r) = \left| \frac{d \ln \rho_0}{dr} \right|^{-1}. \tag{11.55}$$

The density scale height h_ρ for the Sun is shown as the dashed line in Figure 11.7(a); since the Sun is not isothermal, h_ρ is generally larger than the pressure scale height h, shown as the solid line. Within the convection zone of the Sun, and of other Sun-like stars, the density scale height h_ρ decreases with increasing r. The amount of refraction depends on the angular degree ℓ of a mode, through the Lamb frequency S_ℓ. Given the presence of refraction, we anticipate the possibility that standing waves will be trapped in a cavity. If the top of the cavity reaches the photosphere, the standing waves will be directly visible there. To identify the trapping regime, we can combine Equations 11.50 and 11.52. This can be done most easily by introducing the function

$$\Psi(r) = c_s^2 \rho_0^{1/2} \vec{\nabla} \cdot \vec{\xi} = c_s^2 \rho_0^{1/2} \left(\frac{1}{r^2} \frac{d}{dr} (r^2 \xi_r) + k_h^2 r \xi_h \right). \tag{11.56}$$

The function Ψ then satisfies the relation

$$\frac{d^2 \Psi}{dr^2} = -K(r)^2 \Psi, \tag{11.57}$$

where the wavenumber $K(r)$ is given by

$$K^2 = \frac{1}{c_s^2} \left[\omega^2 - \omega_{ac}^2 - S_\ell^2 \left(1 - \frac{N_{BV}^2}{\omega^2} \right) \right]. \tag{11.58}$$

In Equation 11.58, we have introduced yet another frequency. This is the **acoustic cutoff frequency**, defined by

$$\omega_{ac}^2 = \frac{c_s^2}{4h_\rho^2} \left(1 - 2 \frac{dh_\rho}{dr} \right). \tag{11.59}$$

In regions where h_ρ is nearly constant (for the Sun, that would be in the middle of the radiative zone), the acoustic cutoff frequency is the inverse of the time it takes sound to travel two scale heights. Figure 11.7(b) shows ω_{ac} within the Sun as the light solid lines. In the middle of the Sun's radiative zone, the acoustic cutoff frequency is $\omega_{ac} \sim 0.003 \, \text{s}^{-1}$, reflecting the fact that sound, traveling at $c_s \sim 360 \, \text{km s}^{-1}$, takes just under three minutes to travel a distance $h_\rho \sim 0.09 \, R_\odot \sim 60\,000$ km. Notice, however, that at the top of the radiative zone, where $r \sim 0.7 \, R_\odot$, the acoustic cutoff frequency is imaginary, since the value of h_ρ increases steeply there, with $dh_\rho/dr > 1/2$. Within the convective zone at $r > 0.71\,\mathrm{R}_\odot$, the acoustic cutoff frequency is real, and increases as $r \rightarrow 1\,\mathrm{R}_\odot$. This is because h_ρ decreases with radius, and does so more steeply than the sound speed c_s decreases.[4]

In general, modes with a given value of $\vec{\xi}$, and hence of the function Ψ, will be oscillatory if K^2 is positive, and will be damped if K^2 is negative. For purely radial modes, with $S_\ell = 0$, the criterion for stable oscillation reduces to $\omega > \omega_{\mathrm{ac}}$; radial modes with a frequency less than the acoustic cutoff frequency will be damped. For non-radial modes, the situation is more complicated; if we know the values of S_ℓ, N_{BV}, and ω_{ac} as a function of r, we can deduce in what regions of a star a standing wave will be oscillatory rather than damped. For instance, in the deep interior of a star, Figure 11.7 indicates that it is generally true that $S_\ell^2 \gg N_{\mathrm{BV}}^2$ and $S_\ell^2 \gg \omega_{\mathrm{ac}}^2$ when $\ell > 0$. In this regime, there are two families of oscillatory modes. One family consists of high-frequency modes with $\omega > S_\ell$; the other consists of low-frequency modes with $\omega < N_{\mathrm{BV}}$. Notice that in the high-frequency limit, Equation 11.58 reduces to $\omega^2 \approx c_s^2 K^2$; this is the dispersion relation for a sound wave (or p-mode). In the low-frequency limit, however, Equation 11.58 reduces to

$$\omega^2 \approx \frac{S_\ell^2 N_{\mathrm{BV}}^2}{c_s^2 K^2} \approx \frac{k_h^2}{K^2} N_{\mathrm{BV}}^2. \tag{11.60}$$

This is the dispersion relation for a g-mode, which has a frequency ω independent of the sound speed c_s, but closely tied to the buoyancy frequency N_{BV}. In short, the physical conditions inside a Sun-like star give rise to p-modes at high frequency and g-modes at low frequency.

The Sun's g-modes exist only in its radiative zone, where $N_{\mathrm{BV}}^2 > 0$. Since g-modes are required to have $\omega < N_{\mathrm{BV}}$, and since the maximum value of the Brunt–Väisälä frequency in the Sun is $N_{\mathrm{BV}} = 0.0028\,\mathrm{s}^{-1}$ (see Figure 11.7), g-modes in the Sun are required to have $\omega < 0.0028\,\mathrm{s}^{-1}$, corresponding to oscillation periods $\mathcal{P} = 2\pi/\omega > 37\,\mathrm{min}$. Since the g-modes are buried deep in the interior of the Sun (and other Sun-like stars) they cannot be directly detected using telescopes. The observable oscillations are the p-modes that can travel through the Sun's convection zone. Consider a sound wave moving upward; as $r \rightarrow 1\,\mathrm{R}_\odot$, the acoustic cutoff frequency ω_{ac} becomes larger. In the regime just below the photosphere, where $\omega_{\mathrm{ac}}^2 \gg S_\ell^2$ and $\omega_{\mathrm{ac}}^2 \gg |N_{\mathrm{BV}}^2|$, only high-frequency modes with $\omega > \omega_{\mathrm{ac}}$ are able to propagate. (This is the same as the simple criterion for radial oscillations.) If waves propagate upward to a level where ω_{ac} is greater than ω, then they are reflected downward.

We can use Figure 11.7 to identify the trapping regions where p-modes are able to propagate. (Similar diagrams can be made for other stars, but let's stick to the Sun for our example.) Consider a p-mode with frequency $\omega \sim 0.01\,\mathrm{s}^{-1}$

---

[4] Within the photosphere itself, the relatively short scale height $h_\rho \sim 110\,\mathrm{km}$, combined with a local sound speed $c_s \sim 7\,\mathrm{km\,s}^{-1}$, leads to $\omega_{\mathrm{ac}} \sim 0.03\,\mathrm{s}^{-1}$, an order of magnitude larger than in the radiative zone.

and angular degree $\ell > 0$. The inner boundary of the trapping region is defined where $\omega = S_\ell$. If we look at the cases where $\ell = 1, 4, 16$, and 64, we note that the inner bound moves closer to the surface as ℓ increases. However, the upper bound, defined where $\omega = \omega_{ac}$, is independent of ℓ. For a fixed ω, modes of higher ℓ will be trapped in narrower cavities. Note that Figure 11.7 indicates over which range of ω it is possible for modes of given ℓ to exist; it doesn't indicate over which range of ω modes are excited, or over which range a standing wave pattern with an integer number of nodes is possible.

11.4 Observational Asteroseismology

In general, the frequency ω of an oscillating mode can be labeled with its value of n, ℓ, and m. In non-rotating stars, however, ω is independent of m. Although our analysis in the previous section described oscillations in terms of the angular frequency $\omega_{n\ell}$, observers often prefer to use the **circular frequency** $\nu_{n\ell} \equiv \omega_{n\ell}/(2\pi)$. Since asteroseismology is, at last, an observationally driven field, we'll stick with ν rather than ω for the rest of this chapter. Pulsations with frequency $\nu_{n\ell}$ cause the temperature of the photosphere to oscillate with the same frequency, and thus cause small fluctuations with frequency $\nu_{n\ell}$ in the star's flux. (There will also be small fluctuations in the star's color, but flux variations are easier to observe.) Fluctuations with $\ell \gg 1$ produce a large number of both hot and cool patches on the hemisphere of a star facing us; this cancels out the observable effect of fluctuations with large values of ℓ. Asteroseismology thus gives information only about modes with relatively low values of ℓ. In *Kepler* data for bright stars, for instance, the $\ell = 0, 1, 2$, and 3 modes were detected. In the special case of the Sun, the solar disk is very well resolved; thus, helioseismology can give us information about high-ℓ modes for the Sun. For other stars, though, digging out information about high-ℓ modes will be challenging.

In general, as we have seen, the frequencies $\nu_{n\ell}$ depend on a star's interior structure in a complicated way. However, there is a useful asymptotic approximation in the limit of very high overtone p-modes, with $n \gg \ell$. In this limit,

$$\nu_{n\ell} \approx \Delta\nu_\star \left[n + \frac{\ell}{2} + \epsilon \right] - (\Delta\nu_\star)^2 \left[\frac{A\ell(\ell+1) - B}{\nu_{n\ell}} \right]. \tag{11.61}$$

In Equation 11.61, the frequency $\Delta\nu_\star$, known as the **large separation**, is simply the inverse of the sound travel time along the star's diameter:

$$\Delta\nu_\star = \left(2 \int_0^{R_\star} \frac{dr}{c_s} \right)^{-1}. \tag{11.62}$$

The dimensionless parameters ϵ and B in Equation 11.61 depend on the surface boundary conditions of the star, while the dimensionless parameter A depends on the gradient of the sound speed within the star, with

$$A \approx -\frac{1}{4\pi^2 \Delta v_\star} \int_0^{R_0} \frac{dc_s}{dr} \frac{dr}{r}. \tag{11.63}$$

For large n, the leading term on the right-hand side of Equation 11.61 is much larger than the second term. This implies that two modes with the same value of ℓ and with adjacent values of n have

$$\delta v_{n\ell} \equiv v_{n\ell} - v_{n-1,\ell} \approx \Delta v_\star. \tag{11.64}$$

What does this mean from an observational point of view? In any particular star, some values of $v_{n\ell}$ correspond to standing waves that produce variations in the star's flux; these variations in flux have a period $\mathcal{P}_{n\ell} = 1/v_{n\ell}$. If you monitor a star's flux $F(t)$ for a sufficiently long time, then take its Fourier transform, the resulting power spectrum will have peaks at those values of $v_{n\ell}$ that produce detectable flux variations. According to Equation 11.64, if all the detectable modes of the star had the same value of ℓ, its power spectrum would look like a comb, with spikes in the power separated by a frequency $\Delta v_\star$.

The power spectrum of a real star is more complicated. One well-monitored star is the solar analog 16 Cygni A, a main sequence star of spectral class G1.5V, very slightly hotter than the Sun. Since 16 Cyg A lies in the field of view observed by the *Kepler* satellite, its flux was monitored with high precision for many months. Figure 11.8 shows the resulting power spectrum of 16 Cyg A.[5] The peaks in the spectrum are labeled with symbols indicating ℓ for the associated mode. The detected modes have radial orders ranging from $n = 13$ to $n = 26$. The highest peak in the data is at $v_{max} \approx 2110\,\mu\mathrm{Hz}$, corresponding to $\mathcal{P} = 1/v_{max} \approx 7.9\,\mathrm{min}$. This peak represents the $n = 18$, $\ell = 0$ oscillation mode. The value of $\Delta v_\star$ can be found from the distance between adjacent peaks with the same ℓ; for 16 Cyg A, it's $\Delta v_\star = 104\,\mu\mathrm{Hz}$, implying a sound travel time of $t \approx 2.7\,\mathrm{hr}$ across the star's diameter.

From Equation 11.61, we note that the leading term on the right-hand side is the same for the (n, ℓ) and $(n - 1, \ell + 2)$ modes. Thus, these two modes have a frequency separated by the **small separation**

$$\delta v_{\ell,\ell+2}(n) \equiv v_{n\ell} - v_{n-1,\ell+2} \approx (4\ell + 6)\frac{(\Delta v_\star)^2}{v_{n\ell}}A$$

$$\approx -\frac{1}{\pi^2}\frac{2\ell + 3}{2n + \ell + 2\epsilon}\int_0^{R_\star} \frac{dc_s}{dr}\frac{dr}{r}. \tag{11.65}$$

For example, 16 Cyg A has $\delta v_{0,2} \approx 5.8\,\mu\mathrm{Hz}$. While the large separation $\Delta v_\star$ tells us the sound crossing time for the star, the small separation tells us about the gradient in the sound speed weighted by $1/r$; thus, it gives us valuable information

---

[5] Like the Sun, 16 Cyg A has convective granulation (see Figure 4.4). The seething granulation causes flux variations, but without a strongly preferred frequency v. The effects of granulation have been subtracted from Figure 11.8.

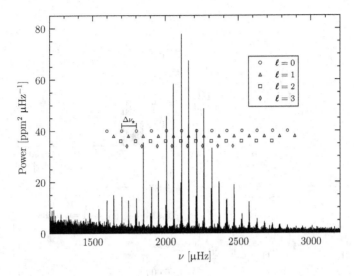

Figure 11.8 Power spectrum of the observed flux variations of the Sun-like star 16 Cyg A. The units used for flux variation are "parts per million," or "ppm." Symbols indicate the value of ℓ for each spectral peak found by Metcalfe *et al.* 2012.

about conditions near the star's center. By comparing the observed power spectrum of 16 Cyg A to that computed for numerical stellar models, the properties of 16 Cyg A can be pinned down quite well. Asteroseismology gives $M_\star = 1.11\,M_\odot$ and $R_\star = 1.24\,R_\odot$ for 16 Cyg A, as well as revealing that its age is a relatively mature $t = 6.9\,\mathrm{Gyr}$.

The frequency $\nu_{\max}$ at which the power spikes are highest (for instance, $\nu_{\max} \approx 2110\,\mu\mathrm{Hz}$ for 16 Cyg A) tells us about conditions near the surface of a star. It is found that the power is greatest when the pulsation period is comparable to the sound travel time through the stellar photosphere. More precisely,

$$\nu_{\max} \approx \frac{c_s}{4\pi h}, \tag{11.66}$$

where h is the pressure scale height of the photosphere and c_s is its sound speed. The scale height and sound speed are linked by the relation $h = c_s^2/g$, where $g = GM_\star/R_\star^2$. Therefore, we may also write

$$\nu_{\max} \propto \frac{g}{c_s} \propto \frac{g}{T_{\mathrm{eff}}^{1/2}}. \tag{11.67}$$

Thus, the observed value of $\nu_{\max}$ is a diagnostic of the surface gravity g.

Figure 11.9, for instance, shows flux variations of two red giant stars observed with *Kepler*. The upper panels show the results for KIC 2425375, a $1.5\,M_\odot$ star that has evolved up the red giant branch to have $R = 11\,R_\odot$ and $\log g = 2.53$. The lower panels show the results for KIC 5858947, a $1.05\,M_\odot$ star that is still near the base of the red giant branch, with $R = 4.5\,R_\odot$ and $\log g = 3.16$. Since

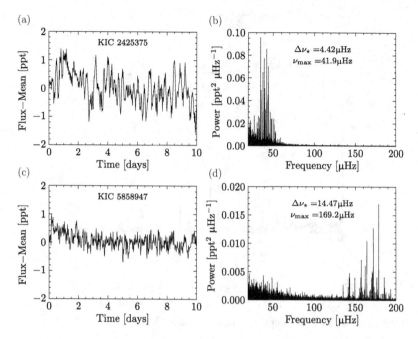

Figure 11.9 Upper left: (a) Light curve of the red giant KIC 2425375 ($\log g = 2.53$); flux variation is shown as parts per thousand [ppt] of the mean flux. (b) Derived power spectrum of KIC 2425375. (c) Light curve of the red giant KIC 5858947 ($\log g = 3.16$). (d) Derived power spectrum of KIC 5858947. [Data from *Kepler*]

the surface gravity g for the less evolved red giant is larger by a factor of four, so is its measured value of ν_{max}. Although the spectra of the red giants are noisy, modeling indicates $\nu_{max} \approx 169\,\mu$Hz for KIC 5858947 versus $\nu_{max} \approx 42\,\mu$Hz for the fluffier, more evolved red giant KIC 2425375. (For comparison, the much denser and slightly hotter main sequence star 16 Cyg A has $\log g = 4.92$ and $\nu_{max} \approx 2110\,\mu$Hz.)

As an example of the power of asteroseismology in studying stellar interiors, consider the difficult task of distinguishing red clump (RC) stars from red giant branch (RGB) stars. As shown in Figure 1.10, the red clump of core helium-burning stars inconveniently overlaps the red giant branch. If a star has $T_{eff} \approx 5000\,$K and $L_\star \approx 60\,L_\odot$, it might be either a red clump star or a red giant: you cannot tell which solely from the luminosity and color. In a plot of radius versus mass, as shown in Figure 11.10, the distribution of RC stars also shows an overlap with RGB stars. A star with $M_\star = 1.2\,M_\odot$ and $R_\star = 11\,R_\odot$ might be either a red clump star or a red giant: you cannot tell which solely from the mass and radius.

However, it is possible to distinguish red clump stars from red giant stars (which have denser degenerate cores) by their different asteroseismic frequency patterns. In Figure 11.10, stars are identified as shell-burning giants (open circles)

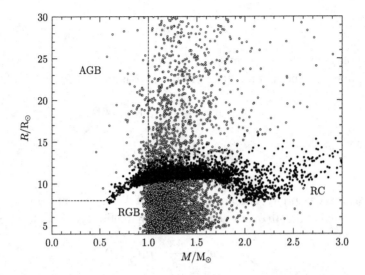

Figure 11.10 Radius and mass of stars observed by *Kepler*. Open circles: shell-burning stars with degenerate cores. Dashed line indicates the separation between red giant branch (RGB) stars and asymptotic giant branch (AGB) stars. Filled circles: core helium-burning stars, also known as red clump (RC) stars. [Data from Pinsonneault *et al.* 2023]

or as core helium-burning red clump stars (filled circles); this identification is based purely on asteroseismology. Stars powered only by shell hydrogen burning (labeled "RGB") evolve upward in Figure 11.10, from small to large radii. At small radii, the abrupt edge at $M_\star \approx 1\,M_\odot$ is a consequence of our galaxy's finite age; lower-mass stars have not yet had time to evolve up the giant branch. Core helium-burning stars (filled circles labeled "RC") have a small range in radius over a large range in mass; note that red clump stars can have $M_\star < 1\,M_\odot$ because of mass loss during their earlier RGB phase. The radius of red clump stars has a local minimum at a stellar mass $M_\star \approx 2.2\,M_\odot$, corresponding to the transition between lower-mass stars that ignite helium with a helium flash and higher-mass stars that ignite helium more sedately. The shell-burning stars with $M_\star < 1\,M_\odot$, labeled "AGB" in Figure 11.10, are asymptotic giant branch stars with two shell sources (hydrogen burning and helium burning). The AGB stars have a frequency pattern similar to RGB stars, but are lower in mass because of mass loss through stellar winds on both the red giant branch and the asymptotic giant branch.

Exercises

11.1 A uniform-density star with density $\rho_\star$, radius $R_\star$, and adiabatic index $\gamma > 4/3$ undergoes adiabatic radial pulsations. In the text, we showed that the fundamental mode for this star has frequency (Equation 11.15)

$$\omega_0^2 = \frac{4\pi}{3} G\rho_\star(3\gamma - 4) \tag{11.68}$$

and homologous displacement $r_1/r_0 = \varepsilon$.

(a) Show that the first overtone of this star has frequency

$$\omega_1^2 = \frac{4\pi}{3} G\rho_\star(10\gamma - 4) \tag{11.69}$$

and displacement

$$r_1/r_0 = \frac{\varepsilon}{2}(7x^2 - 5), \tag{11.70}$$

where $x \equiv r_0/R_\star$, and ε is the fractional displacement at $r_0 = R_\star$.

(b) Show that the jth overtone of this star, in the limit $j \to \infty$, has frequency

$$\omega_j^2 = \frac{4\pi}{3} G\rho_\star(2j^2\gamma). \tag{11.71}$$

11.2 Starting from the empirical Leavitt's law (Equation 11.1), derive the relation between mean density $\rho_\star$ and luminosity $L_\star$ for Cepheid variable stars. Assuming that all Cepheids have the same effective temperature, $T_{\text{eff}} = 6000\,\text{K}$, find the relation between the radius $R_\star$ and luminosity $L_\star$ for Cepheids. Finally, find the relation between mass $M_\star$ and luminosity $L_\star$ for Cepheids; compare it quantitatively to the mass–luminosity relation for main sequence stars.

11.3 The black hole at the center of the Milky Way has mass $M_{\text{bh}} = 4.1 \times 10^6\,M_\odot$ and is at a distance $d = 8200\,\text{pc}$ from the Earth. Suppose that the black hole is accreting pure ionized hydrogen.

(a) What is the Eddington luminosity of the black hole, assuming that scattering from free electrons dominates the opacity?

(b) What would be the flux of light (in $\text{erg}\,\text{s}^{-1}\,\text{cm}^{-2}$) detected at Earth from the black hole, if it were accreting at its Eddington luminosity? How does this compare to the flux of light we receive from the star Sirius A?

(c) Suppose that gas falling from a large distance to the black hole's Schwarzschild radius, $r_s = 2GM_{\text{bh}}/c^2$, converts gravitational potential energy to photon energy with an efficiency $\xi = 0.1$. At what rate $\dot{M}$ must the black hole accrete in order to maintain its Eddington luminosity?

Binary Stars

It's so much more friendly with two.

A. A. Milne (1882–1956)
Winnie-the-Pooh, Chapter 9 [1926]

Up to this point, we have mainly been treating stars as if they exist in splendid isolation. However, stars are frequently found in binary systems, with two stars orbiting their barycenter. If the stars are sufficiently close to each other, then they will be tidally distorted, destroying the spherical symmetry of the standard equations of stellar structure. Even if the binary separation is too large for strong tidal effects on the main sequence, the natural course of evolution swells stars to large sizes; consider, for instance, the ∼4 au radius of the evolved red supergiant Betelgeuse. If two stars are very close to each other, then gas can be transferred from one to another on relatively short timescales, destroying the time independence of the standard equations of stellar structure. Mass transfer can lead to observable consequences such as ultraviolet and X-ray emission, to say nothing of cataclysmic variable outbursts. Sometimes, the two stars in a close binary system can merge, forming a single "blue straggler" star. If two white dwarfs in a close binary system merge, they can trigger a thermonuclear, or type Ia, supernova by the double degenerate scenario. Mergers of neutron stars and white dwarfs also produce gravitational wave signals. Since binary systems can lead to physically interesting outcomes (including explosions of varying energy) it pays to take a closer look at them.

In this chapter we will use theorists' language, and refer to the more massive star in a binary as the primary and the less massive star as the secondary. (In observers' language, the more luminous star is the primary.) The mass ratio of the two stars in a binary is $q \equiv M_2/M_1$, where M_1 is the mass of the primary and M_2 the mass of the secondary. Given this definition, $q \leq 1$. In some binary systems, the two stars are "twins," with $q \sim 1$. For instance, we have seen that α Centauri A and B have $q = 0.84$; they are not identical twins, but are fairly close in mass. However, if the mass ratio is much smaller than one, the lifetimes of the two stars in the binary will be quite different, with interesting consequences, as we see later, for the evolution of both stars.

12.1 Observed Properties of Binaries

Many stars are in gravitationally bound binary, triple, or quadruple systems; quintuple, sextuple, and septuple star systems are also known to exist, but they are rare. To make a quantitative statement, consider the 100 closest known stellar systems to the Earth; to embrace this many systems, we have to reach as far as $d = 6.3$ pc. In our census, we will use a broad definition of "star" that includes brown dwarfs and white dwarfs.[1] Of the 100 closest stellar systems, 71 are single stars (stars without a bound stellar companion – although they may have planets), 22 are binaries, and 7 are triples, including the nearby α Centauri system. Thus, 71% of local stellar systems contain single stars; the solar system, by this accounting, is part of a comfortable majority. However, when we look at individual stars, 71 out of 136 (52%) are single stars; the Sun, by this alternate accounting, is part of a far more slender majority.

The probability that a star is part of a binary or higher-multiple system is dependent on its mass. To compute the prevalence of multiple-star systems as a function of stellar properties, start with a list of target stars. (For instance, you might have a target list of young K dwarfs, or of low-metallicity M dwarfs; it all depends on what type of star interests you.) You then search for stellar companions to the target stars. A target star might be part of an *eclipsing* binary, in which the two stars periodically pass in front of each other as they orbit. It might be part of a *spectroscopic* binary, in which the absorption lines of the stars show periodic Doppler shifts; if one star is much brighter than the other, it can result in a *single-lined* spectroscopic binary, with only the absorption lines of the brighter star being detected. It might be part of an *astrometric* binary, in which one star is seen to move on an elliptical path relative to the other. It might be part of a *common proper motion* binary, in which two widely separated stars in a binary don't move appreciably along their orbit during the time of observation, but do show the same proper motion as they move across the sky. (The association of Proxima Centauri with the close binary α Centauri AB was realized in 1915, when Robert Innes found that an otherwise undistinguished M dwarf had the same proper motion as α Centauri.)

A search for stellar companions to a set of N target stars results in a list of N stellar systems. A fraction f_1 of those systems are single stars (target stars without a bound stellar companion), a fraction f_2 are binary systems, a fraction f_3 are triple

---

[1] The star formation process doesn't know about the boundary between stars and brown dwarfs, and we can think of a white dwarf as a "retired star."

systems, and so forth. The **multiplicity fraction** F_m of the sample is the fraction of stellar systems that are binaries or higher-order systems. That is,

$$F_m = \sum_{i=2}^{7} f_i = 1 - f_1. \tag{12.1}$$

Here, we're assuming that septuple systems are the highest-order systems; if octuple or nonuple systems are discovered, the extension is straightforward. The **companion frequency** F_c of the sample is the mean number of bound stellar companions per target star in the sample. Since a fraction f_1 of the target stars have no companions, a fraction f_2 of the target stars have one companion, and so forth, the companion frequency is

$$F_c = \sum_{i=2}^{7} (i-1) f_i = F_m + \sum_{i=3}^{7} (i-2) f_i \geq F_m. \tag{12.2}$$

The companion frequency can be greater than one for a sample that is rich in triples and higher-order systems.[2]

As an example, when we look within 6.3 pc of the Earth, and take as our target stars objects that are the primary in their system, we find $f_1 = 0.71, f_2 = 0.22$, and $f_3 = 0.07$. This results in a multiplicity fraction $F_m = 0.29$ and a companion frequency $F_c = 0.36$. It must be kept in mind, however, that dim companions to luminous, massive primaries are always difficult to detect. Often this difficulty is acknowledged by considering systems with a mass ratio q that is bigger than a minimum value: in Figure 12.1, for instance, only systems with $q \geq 0.1$ are included. Figure 12.1 shows the estimated multiplicity fraction (open circles) and companion frequency (filled squares) as a function of the mass M_1 of the primary in a system. The multiplicity fraction for Sun-like stars (with $M_1 = 1.0 \pm 0.2\,M_\odot$) is $F_m \approx 0.4$; that is, of solar-mass stars that are not themselves bound to a more massive star, about 40% have at least one lower-mass stellar companion. The multiplicity fraction for O stars (with $M_1 \sim 28\,M_\odot$) is $F_m \approx 0.95$; nearly all O stars have at least one stellar companion. The companion frequency similarly increases with stellar mass, from $F_c \approx 0.5$ for Sun-like stars to $F_c \approx 2$ for O stars.

Once a binary system has been detected, some of its properties can be measured. For instance, the periodic behavior of eclipsing, spectroscopic, and astrometric binaries makes it possible to determine the orbital period $\mathcal{P}$. The orbital period is dictated by the Newtonian version of Kepler's Third Law (Equation 1.16), which can be scaled to either very small or very large orbits:

$$\mathcal{P} = 0.116\,\mathrm{d} \left(\frac{a}{1\,R_\odot}\right)^{3/2} \left(\frac{M_1 + M_2}{1\,M_\odot}\right)^{-1/2}$$

$$= 1.0\,\mathrm{Myr} \left(\frac{a}{10^4\,\mathrm{au}}\right)^{3/2} \left(\frac{M_1 + M_2}{1\,M_\odot}\right)^{-1/2}. \tag{12.3}$$

---

[2] A warning: the vocabulary used to discuss multiple-star systems is not always uniform. We follow Offner *et al.* 2022 in our distinction between the "multiplicity fraction" and "companion frequency."

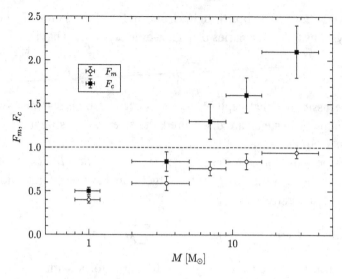

Figure 12.1 Dependence of the multiplicity fraction F_m (open circles) and the companion frequency F_c (filled squares) on the mass of the primary star in a stellar system. [Data from Moe and Di Stefano 2017]

A very long period binary, with an orbital size a comparable to the distance between stellar systems in its neighborhood, will be subject to disruption by passing stars. (Proxima Centauri's orbit about the α Centauri AB binary has $a \sim 9000$ au and $\mathcal{P} \sim 0.55$ Myr.)

At the other end of the range of orbital periods, we require that the stars not collide with each other at periapsis (the point of closest approach). At periapsis, the distance between the stars' centers is $r_{pe} = a(1-e)$, where e is the eccentricity of their mutual orbit. The requirement that this distance be greater than $R_1 + R_2$, the sum of the stellar radii, means that the minimum orbital period is

$$\mathcal{P}_{min}(e) = \frac{0.116\,d}{(1-e)^{3/2}} \left(\frac{M_1 + M_2}{1\,M_\odot}\right)^{-1/2} \left(\frac{R_1 + R_2}{1\,R_\odot}\right)^{3/2}. \qquad (12.4)$$

For example, consider a binary consisting of two main sequence stars, with masses $M_1 = 1\,M_\odot$ and $M_2 = 0.5\,M_\odot$. Since the sum of their radii is $R_1 + R_2 \approx 1.5\,R_\odot$ (from Equation 1.35), the minimum possible orbital period for this binary is $\mathcal{P}_{min} \approx 0.17\,d$, which assumes a circular orbit. However, when the two stars were brand new PMS stars on the deuterium-burning birthline, the sum of their radii was $R_1 + R_2 \approx 12\,R_\odot$. As PMS stars, then, their minimum possible orbital period was $\mathcal{P}_{min} \approx 4\,d$, assuming a circular orbit. After ~ 10 Gyr on the main sequence, the $1\,M_\odot$ primary will swell into a red giant; just before its helium flash, its radius will be $R_1 \sim 170\,R_\odot$ and its mass will be trimmed down to $M_1 \sim 0.8\,M_\odot$ by its stellar wind. At this point in time, the minimum possible orbital period will be $\mathcal{P}_{min} \sim 200\,d$.

Determining the true distribution of orbital periods for binaries is difficult. Each method of discovering binaries has different selection effects. For instance, astrometric binaries are most easily discovered when they have large physical separation a but are at a small distance d from the observer. By contrast, spectroscopic binaries are most easily discovered when they have large orbital speeds, implying a small separation a. Combining results from different methods, it is found that the distribution of $\mathcal{P}$ for a given primary mass is often well fitted by a lognormal distribution. For instance, if the primary is a G main sequence star ($M_1 \approx 1\,M_\odot$), the mean and standard deviation in $\log \mathcal{P}$ is 5.0 ± 2.3, when $\mathcal{P}$ is given in days. The distribution for M dwarfs is narrower and peaks at a shorter orbital period, with $\log \mathcal{P} = 3.2 \pm 1.3$. Although the peak in $\log \mathcal{P}$ corresponds to $\mathcal{P} = 10^5\,\mathrm{d} \approx 300\,\mathrm{yr}$ for G stars and $\mathcal{P} = 1600\,\mathrm{d} \approx 4\,\mathrm{yr}$ for M dwarfs, there is a tail extending to short periods. Some eclipsing binaries in which both members are on the main sequence have periods as short as $\mathcal{P} \sim 0.2\,\mathrm{d}$. Such a short period for a pair of main sequence stars implies that some mechanism has shrunk their orbit since they were large young PMS stars.

Determining the distribution of binary mass ratios, $q = M_2/M_1$, is difficult in the limit of small q, when the inconspicuous low-mass companion is not always detected. The distribution of q is customarily fitted with a power-law distribution:

$$f(q)dq \propto q^\gamma \, dq. \tag{12.5}$$

The value of γ is dependent on both the mass of the primary star and the orbital period of the system. For instance, consider binary systems whose primary is Sun-like, with $M_1 \approx 1.0\,M_\odot$. If we consider only systems with $q \geq 0.3$, the best slope is usually $\gamma = -0.5$, but steepens to $\gamma = -1.1$ for binaries with $\mathcal{P} > 10^6\,\mathrm{d}$. If the primary is an O star with $M_1 > 16\,M_\odot$, the slope gradually steepens from $\gamma = -0.5$ when $\mathcal{P} \approx 10\,\mathrm{d}$ to $\gamma = -2.0$ when $\mathcal{P} \approx 10^7\,\mathrm{d}$. All these slopes are shallower than we would expect if the masses of stars in binary slopes were drawn randomly from a Salpeter initial mass function. If the stellar mass function were $N(M) \propto M^{-2.35}$ down to a minimum cutoff mass M_{cut}, then a random draw of masses would produce $\gamma = -2.35$ as long as $M_1 \gg M_{cut}$. Some physical mechanism causes binaries to prefer more nearly equal stellar masses, especially for short-period systems. In fact, the distribution of q can often only be fitted adequately by assuming an "excess twin fraction" $\mathcal{F}_{twin}$ of binaries with $q > 0.95$. For solar-mass primaries, the excess twin fraction is $\mathcal{F}_{twin} \approx 0.3$ at $\mathcal{P} \approx 10\,\mathrm{d}$, decreasing to $\mathcal{F}_{twin} < 0.03$ at $\mathcal{P} \approx 10^7\,\mathrm{d}$.

Stars in short-period binaries are significantly affected by tides. For instance, binaries with $\mathcal{P} \leq 10\,\mathrm{d}$ usually have orbits of low eccentricity, as a result of **tidal circularization**. Figure 12.2 shows the orbital eccentricity e as a function of orbital period $\mathcal{P}$ for binary systems in which the primary is similar to the Sun. Although binaries with $\mathcal{P} > 10\,\mathrm{d}$ have a wide range of eccentricities, binaries with $\mathcal{P} < 10\,\mathrm{d}$ are nearly all of low eccentricity. To see how tidal effects can

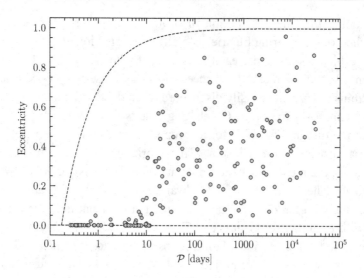

Figure 12.2 Eccentricity versus period for a sample of spectroscopic binaries in which the primary is a G main sequence star. Curved line: eccentricity for which a system of total mass $M = 1.5\,M_\odot$ has a periapsis distance $r_{pe} = a(1 - e) = 1.5\,R_\odot$. [Data from Pourbaix *et al.* 2004]

cause short-period binaries to have circular orbits, consider a binary system in which the orbit of the secondary relative to the primary has a semimajor axis a and an eccentricity e. As the stars orbit, their tidal distortion is greatest at periapsis and least at apoapsis. The resulting tidal flexure causes tidal heating of the two stars. The energy dissipation rate, averaged over a complete orbit, goes as $L_{\text{tide}} \propto e^2/\mathcal{P}^5 \propto e^2/a^{15/2}$; thus, short-period eccentric orbits are most effective at tidally heating the stars in a binary. The energy that heats the stars comes from their orbital energy,

$$E = -\frac{GM_1M_2}{2a}. \tag{12.6}$$

Since tidal effects don't change the mass of the two stars, the binary can lose energy only by shrinking its orbit. This looks like it will result in a runaway contraction as the energy dissipation rate L_{tide} becomes bigger as the orbit becomes smaller. However, we have to take into account the orbital angular momentum of the binary,

$$J = \frac{2\pi}{\mathcal{P}}\frac{M_1M_2}{M_1 + M_2}a^2\sqrt{1 - e^2}$$

$$= \left[G\frac{M_1^2M_2^2}{M_1 + M_2}a(1 - e^2)\right]^{1/2}. \tag{12.7}$$

If the orbital angular momentum J and the stellar masses M_1 and M_2 are conserved, then $a(1 - e^2)$ remains constant as the orbit shrinks, and a reaches a minimum value when $e = 0$. The empirical finding that short-period binaries with $P < P_{circ} \approx 10\,d$ are circularized implies that close binaries with $a < a_{circ}$ are circularized, with

$$a_{circ} = 19.5\,R_\odot \left(\frac{P_{circ}}{10\,d}\right)^{2/3} \left(\frac{M_1 + M_2}{1\,M_\odot}\right)^{1/3}. \tag{12.8}$$

The observed frequency of circular binaries in young open clusters indicates that the circularization occurs on a timescale $t_{circ} \sim 100\,Myr$.

In addition to producing circular orbits, the tidal effects within close binaries also tend to lock the stars into synchronous rotation, with the rotation period P_{rot} of each star in the binary being equal to the orbital period P. This tidal locking explains why gyrochronology breaks down as a method for determining the age of stars in close binary systems.

12.2 Close Binaries

Consider a close binary system; because of the short timescale for circularization, it is safe to assume that the stars are on circular orbits about the barycenter. The stars are separated by a distance a, the total mass of the system is $M \equiv M_1 + M_2$, and the mass ratio is $q \equiv M_2/M_1 \leq 1$. The reduced mass of the system is

$$M_r \equiv \frac{M_1 M_2}{M_1 + M_2} = \frac{q}{(1+q)^2} M. \tag{12.9}$$

The angular speed with which the stars orbit has the value (compare to Equation 12.3)

$$\Omega_0 = \frac{2\pi}{P} = \left(\frac{GM}{a^3}\right)^{1/2}. \tag{12.10}$$

The barycenter of the system is on the line connecting the centers of the two stars, at a distance $a_2 = a/(1 + q)$ from the secondary and $a_1 = aq/(1 + q)$ from the primary. To study this binary system, it is useful to set up a rotating coordinate system with its origin at the barycenter and z axis perpendicular to the orbital plane of the binary. The rotation of the coordinate system about the z axis has an angular speed Ω_0 equal to that of the binary, and with the same sense of rotation. Thus, the two stars are at rest in the rotating frame.

In the rotating frame, the effective potential of the binary is

$$\Phi_{eff}(\vec{r}) = \Phi_1(\vec{r}) + \Phi_2(\vec{r}) - \frac{1}{2}\Omega_0^2 R^2, \tag{12.11}$$

where R is the distance from the z axis, Φ_1 is the gravitational potential of the primary, and Φ_2 is the gravitational potential of the secondary. Figure 12.3 shows the equipotential contours (that is, the curves where the surfaces of constant Φ_{eff}

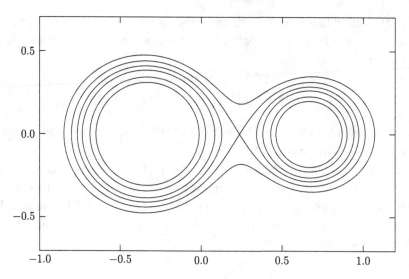

Figure 12.3 Equipotential contours, in the orbital plane of the rotating frame of reference, for a pair of point masses with mass ratio $q = M_2/M_1 = 0.5$. Units are chosen such that $a = 1$, $G(M_1 + M_2) = 1$, and thus $\Omega_0 = 1$.

intersect the orbital plane) for a binary with mass ratio $q = 0.5$; the primary is on the left. This figure assumes that the two stars are point masses. This isn't true for real stars, of course, but as we saw in Section 10.1, the "point mass core" approximation gives a reasonable fit for the equipotential contours that lie outside the dense central core of a star.

The L_1 Lagrangian point, a saddle point in the effective potential, lies between the two stars in the binary. For an equal-mass binary, with $q = 1$, the L_1 point lies exactly at the midpoint between the two stars. When $0.1 < q < 1$, the distance of the L_1 point from the center of the primary is well approximated by the formula $r_1 \approx a(0.5 - 0.1 \ln q)$. The two teardrop-shaped equipotential surfaces that just touch each other at the L_1 point are the **Roche lobes** of the effective potential. As we saw in Section 10.2 when discussing rapid rotators, the density and pressure of a star will be constant on surfaces of constant Φ_{eff}. If a star is much smaller than its Roche lobe, it will not suffer much tidal distortion in its shape, as illustrated by the nearly circular contours well inside the Roche lobes in Figure 12.3. However, when a star nearly fills its Roche lobe, it will have tidal bulges toward and away from its companion star. When it overflows its Roche lobe, then material can flow through the L_1 point onto the companion star.

The possibility of a star filling its Roche lobe leads to different classes of close binary systems, as illustrated in Figure 12.4. In a **detached binary**, both stars are well inside their Roche lobes. In a **semidetached binary**, one star is filling its Roche lobe; in this configuration, mass can be transferred from the lobe-filling star to its companion. In a **contact binary**, both stars fill their Roche

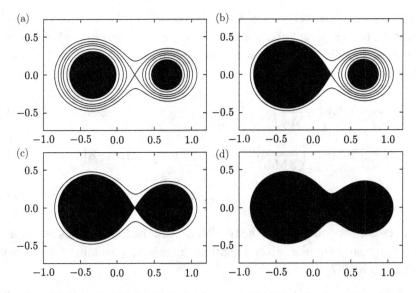

Figure 12.4 (a) Detached binary, with both stars within their Roche lobes. (b) Semidetached binary, with one star filling its Roche lobe. (c) Contact binary, with both stars filling their Roche lobes. (d) Common envelope, with gas from both stars outside the Roche lobes.

lobes. Finally, in a **common envelope event**, both stars significantly overflow their Roche lobes, and an outer envelope forms consisting of mingled gas from both stars. (Note that this is called a common envelope "event," or common envelope "phase," to emphasize the fact that such a configuration is short-lived.)

There are three stages in a star's career when it swells in physical radius and thus has the possibility of filling its Roche lobe. First, when a star is on the main sequence, its radius increases gradually with time. A star with $M_\star = 4\,M_\odot$, for example, swells from $R_\star \approx 2.3\,R_\odot$ on the zero age main sequence to $R_\star \approx 6.4\,R_\odot$ when it leaves the main sequence. A star that fills its Roche lobe while still on the main sequence is said to undergo "case A" mass transfer. The second stage when a star increases in radius is when it ascends the red giant branch for the first time. For a star with $M_\star = 4\,M_\odot$, the star's radius increases from $R_\star \approx 6.4\,R_\odot$ when it turns off the main sequence, to $R_\star \approx 23\,R_\odot$ at the base of the red giant branch, to $R_\star \approx 63\,R_\odot$ when helium ignition takes place. A star that fills its Roche lobe while traversing the Hertzsprung gap or ascending the red giant branch is undergoing "case B" mass transfer. The third stage when a star swells in radius is when it ascends the asymptotic giant branch. A star with initial mass $M_\star = 4\,M_\odot$ has a radius $R_\star \approx 41\,R_\odot$ when it ends its core helium-burning stage, but its radius increases to $R_\star \approx 280\,R_\odot$ by the time carbon ignition takes place. A star that fills its Roche lobe for the first time while ascending the asymptotic giant branch is undergoing "case C" mass transfer.

If we scale to a relatively massive, and thus rapidly evolving, binary system, we find that the orbital period of a semidetached system ranges from about a day for case A mass transfer, to about a year for case C mass transfer. Scaled to relevant stellar radii,

$$\mathcal{P} \sim 1\,\text{day} \left(\frac{M}{8\,M_\odot} \right)^{-1/2} \left(\frac{a}{8\,R_\odot} \right)^{3/2} \qquad \text{[case A]}, \qquad (12.12)$$

$$\sim 1\,\text{month} \left(\frac{M}{8\,M_\odot} \right)^{-1/2} \left(\frac{a}{80\,R_\odot} \right)^{3/2} \qquad \text{[case B]}, \qquad (12.13)$$

$$\sim 1\,\text{year} \left(\frac{M}{8\,M_\odot} \right)^{-1/2} \left(\frac{a}{400\,R_\odot} \right)^{3/2} \qquad \text{[case C]}. \qquad (12.14)$$

This timescale is comparable to the freefall time of the lobe-filling star. If the two stars are on nearly circular orbits, the orbital angular momentum of the binary is, from Equation 12.7,

$$J = \frac{q}{(1+q)^2} [GM^3 a]^{1/2}. \qquad (12.15)$$

As mass is transferred from one star to the other, the separation between the stars must then obey the relation

$$a = \frac{J^2}{GM^3} \frac{(1+q)^4}{q^2}. \qquad (12.16)$$

If the orbital angular momentum J and total mass M of the system are conserved during mass transfer, then

$$\frac{da}{dq} = \frac{J^2}{GM^3} \frac{2(1+q)^3(q-1)}{q^3}, \qquad (12.17)$$

which is negative when $q < 1$. If mass is transferred from the primary to the secondary, thus increasing q, the separation a between the stars decreases. The value of a reaches a minimum when $q = 1$, and the stars are of equal mass. If mass is transferred from the less massive star to the more massive, thus decreasing q, continuing the transfer is made more difficult by the fact that the two stars are receding from each other.

The transfer of mass between two stars in a close binary can profoundly affect their evolution. Consider, as an example, the binary system Algol.[3] The Algol system is at a distance $d \approx 29\,\text{pc}$, and has an apparent magnitude $m_V = 2.1$ when out of eclipse. Its brightness makes it a prime target for interferometric study, with the results shown in Figure 12.5. The more luminous star in the binary, Algol A, is eclipsed by the less luminous star, Algol B, once per orbital period, $\mathcal{P} = 2.8673\,\text{d}$. The two stars are on circular orbits with separation $a = 13.3\,R_\odot$. Algol A is a main sequence star of spectral type B8, with $M_A = 3.2\,M_\odot$ and

---

[3] Actually, Algol is a triple system, but we may safely ignore the very distant tertiary star.

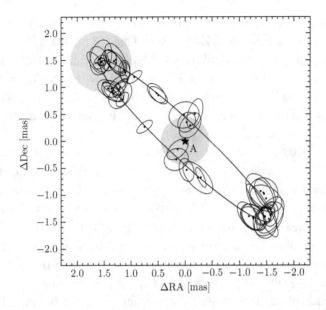

Figure 12.5 Positions of Algol B relative to Algol A measured by the CHARA interferometer. The best-fitting orbit is shown as the large ellipse. Smaller ellipses indicate the locations of the center of Algol B at the times of observation. Filled gray circles show the angular diameters of Algol A (0.88 mas) and Algol B (1.12 mas). [Data from Baron *et al.* 2012]

$R_A = 2.7\,R_\odot$. Algol B is a K2 subgiant, at the base of the red giant branch; it has $M_B = 0.7\,M_\odot$ and $R_B = 3.5\,R_\odot$. Thus, the less massive star, Algol B, has already evolved off the main sequence while the more massive star, Algol A, is still on the main sequence. However, we saw in Section 1.3 that the main sequence lifetime is shorter for more massive stars. This apparent contradiction gives rise to the **Algol paradox**: in close binaries, the less massive star is sometimes the more evolved star.

The Algol paradox is resolved by the existence of mass transfer in close binaries. On the zero age main sequence, Algol B was the *primary* of the system, with initial mass $M_{B,0} \approx 2.8\,M_\odot$ and radius $R_{B,0} \approx 2.3\,R_\odot$; Algol A was the secondary, with $M_{A,0} \approx 1.1\,M_\odot$ and $R_{A,0} \approx 1.1\,R_\odot$. The separation of the stars was $a_0 \approx 7.0\,R_\odot$ and the orbital period was $\mathcal{P}_0 \approx 1.1$ d, assuming conservation of orbital angular momentum during the later mass transfer phase. During its main sequence lifetime, Algol B was thus inside its Roche lobe, with a distance $r_1 \approx 0.56 a_0 \approx 3.9\,R_\odot \approx 1.7 R_{B,0}$ from the center of Algol B to the L_1 point of the system. Then, after a main sequence lifetime of $t_{MS} \sim 0.5$ Gyr, Algol B began traversing the Hertzsprung gap on its way to becoming a red giant. However, a funny thing happened on the way to the red giant branch. An isolated $2.8\,M_\odot$ star has about the same luminosity when it reaches the base of the red giant branch as it did as a zero age main sequence star; at both stages of its evolution, $L \sim 90\,L_\odot$.

However, its temperature as a fledgling red giant ($T \sim 5000\,\text{K}$) is only half the effective temperature it had as a ZAMS star. Thus, to maintain its luminosity, its radius must increase by a factor of four. If Algol B were an isolated star, its radius would have expanded from $R_B \approx 2.3\,R_\odot$ as a ZAMS star to $R_B \approx 9\,R_\odot$ at the base of the red giant branch. However, the proximity of Algol A meant that when Algol B was still a subgiant with radius $R_B \sim 3.9\,R_\odot$, its outer envelope extended beyond the Roche lobe and poured onto Algol A.

When $\Delta M \sim 0.85\,M_\odot$ of material was transferred to Algol A, the two stars were of equal mass, with $M_B = M_A \approx 1.95\,M_\odot$. At this moment, the distance between the two stars had its minimum value, $a_{\min} \approx 4.6\,R_\odot$, with the L_1 point being at a distance $r_1 = r_2 \approx 2.3\,R_\odot$ from each star. Although the two stars were of equal mass at this instant, their interior structures were quite different. Algol B had a giant's heart, with a core depleted of hydrogen. Algol A was a main sequence star at heart, with an extra layer of hydrogen-rich gas dumped on top. Once mass equality was reached, the two stars in the binary started to move away from each other. This slowed the transfer of mass from Algol B, but did not immediately cut it off. The main epoch of mass transfer, which took place on the timescale $t \sim 1\,\text{Myr}$ required to traverse the Hertzsprung gap, resulted in the transfer of $\Delta M \approx 2.1\,M_\odot$ of gas. Algol A, initially a G0 main sequence star with a lifetime of $t_{\text{MS}} \sim 8\,\text{Gyr}$, now finds itself a far more luminous B8 main sequence star, with its lifetime trimmed to $t_{\text{MS}} \sim 0.3\,\text{Gyr}$. At the moment, there is only a trickle of mass transfer in the Algol system: the flow from Algol B transfers $\sim 10^{-8}\,M_\odot\,\text{Myr}^{-1}$ to Algol A.

Key steps in the evolution of an Algol-like system are shown in Figure 12.6. Looking into the future of the Algol system itself, $\sim 300\,\text{Myr}$ from now Algol A will evolve off the main sequence and initiate another episode of significant mass transfer, this time in reverse, from Algol A back to Algol B. By this time, Algol B will have evolved all the way to become a white dwarf. Thus, the Algol system will eventually involve mass transfer from a star to a white dwarf.

We deduce, from the current properties of the Algol system, that it started as a short-period binary, with initial separation $a_0 \approx 7\,R_\odot$. What would happen to a system with a larger initial separation? In that case, the Roche lobe overflow would not occur until the primary evolves far up the red giant branch ("case B" mass transfer). The response of a red giant to mass loss is different from that of a main sequence star undergoing "case A" mass transfer. Lower-mass main sequence stars have smaller radii ($R_\star \propto M_\star$ for $M_\star < 2\,M_\odot$, for instance). As mass is lost from a main sequence star, it maintains its equilibrium by becoming smaller; this tends to throttle back mass transfer. However, at a given luminosity, lower-mass red giants have lower temperatures and thus bigger radii. As mass is lost from a red giant, it maintains its equilibrium by becoming bigger; this creates a strong positive feedback loop, as mass loss leads to a red giant overflowing its

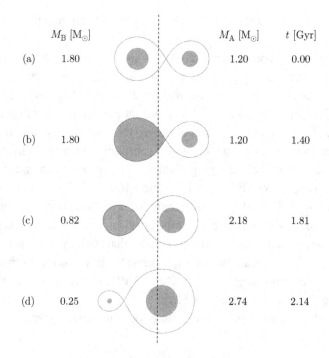

	M_B [M$_\odot$]			M_A [M$_\odot$]	t [Gyr]
(a)	1.80			1.20	0.00
(b)	1.80			1.20	1.40
(c)	0.82			2.18	1.81
(d)	0.25			2.74	2.14

Figure 12.6 Evolutionary phases of an Algol-like system with initial mass $M = 3\,\mathrm{M}_\odot$; vertical dashed line marks the barycenter. (a) Star B is the more massive star on the zero age main sequence. (b) Star B starts to overflow its Roche lobe. (c) Star B is a stripped red giant. (d) Star B is a white dwarf; star A is a red giant. [After Willems and Kolb 2004]

Roche lobe even more. The resulting rapid transfer of mass can lead to a common envelope event, where the companion star is engulfed by the donor red giant.

During the common envelope event, the red giant primary consists of a very dense degenerate core surrounded by a low-density envelope. If the secondary is a main sequence star, it is also significantly denser than the envelope of the red giant. Thus, in a common envelope event, there are two dense cores orbiting on a short timescale inside an extended low-density envelope. To simulate the effect on the envelope, use an eggbeater on a shallow plate of water.[4] The envelope is ejected by the "eggbeater" motion, while the orbit of the cores decays due to friction with the remaining envelope. The process ends either when the secondary is dissolved or when the envelope is fully ejected. If the secondary survives, it remains orbiting the exposed core of the former red giant. If the exposed core is not massive enough to ignite helium burning, it becomes a helium white dwarf. If the core is massive enough to ignite helium but not carbon, it becomes a carbon/oxygen white dwarf. If it is massive enough to ignite carbon, it can become a neutron star with a close companion, like the "black widow" pulsars mentioned

[4] You might want to try this experiment outside, while wearing a swimsuit.

in Section 9.2. An extremely massive core may even become a black hole with a close companion.

It is observed that some compact objects are in circular binary orbits with extremely short periods; as mentioned in Section 9.2, some binary pulsars have orbital periods as short as $\mathcal{P} \sim 1$ hr, requiring an orbital size $a < 1\,\mathrm{R}_\odot$. To be on such small orbits, the compact object and its companion must have disposed of orbital angular momentum by some means. One way to lose angular momentum, and thus spiral in to a smaller orbit, is through a common envelope event. However, there is another interesting channel for shrinking orbits, related to the magnetized winds that we discussed in Section 10.6.

For binaries with $\mathcal{P} < 10\,\mathrm{d}$, the timescale for circularization and synchronization is quite short. If one star in a short-period binary has a magnetized wind, its slower rotation will lead to tidal friction that swiftly brings it back into synchronization. In this way, the angular momentum loss comes primarily from the orbital angular momentum, $J_{\mathrm{orb}} \propto \Omega a^2 \propto a^{1/2}$, rather than from the smaller rotational angular momentum, $J_{\mathrm{rot}} \propto \Omega R_\star^2$.

In general, the rate of angular momentum loss due to a magnetized wind is

$$\dot{J}_{\mathrm{wind}} = \dot{M}_{\mathrm{wind}} \Omega r_{\mathrm{A}}^2 = \Omega \frac{B_{\mathrm{phot}}^2 R_\star^4}{v_{\mathrm{wind}}}, \qquad (12.18)$$

making use of Equation 10.46 for the definition of the Alfvén radius r_{A}. For a more specific example, let's take the Sun and spin it up to a rotational angular speed Ω greater than its current value $\Omega_\odot = 2.97 \times 10^{-6}\,\mathrm{s}^{-1}$. This will not greatly affect the Sun's radius and wind speed as long as Ω is small compared to the breakup speed Ω_{BU}. However, as long as the magnetic field is still unsaturated, we expect $B_{\mathrm{phot}} \propto \Omega$, giving

$$\dot{J}_{\mathrm{wind}} = \frac{B_\odot^2 \Omega_\odot^2 R_\odot^4}{v_\odot} \left(\frac{\Omega}{\Omega_\odot} \right)^3$$

$$\approx 1.2 \times 10^{-13} \Omega_\odot^2 M_\odot R_\odot^2 \left(\frac{\Omega}{\Omega_\odot} \right)^3, \qquad (12.19)$$

scaling to $B_\odot = 3\,\mathrm{G}$ at $\Omega = \Omega_\odot$, and assuming $v_\odot = 600\,\mathrm{km\,s}^{-1}$.

Suppose, however, that the magnetized Sun is rotating rapidly because it is in a close synchronized binary. Then, its rotational angular speed must equal the orbital angular speed, which Equation 12.10 tells us is

$$\Omega = \left(\frac{GM}{a^3} \right)^{1/2} = 2.4\,\Omega_\odot \left(\frac{M}{2\,M_\odot} \right)^{1/2} \left(\frac{a}{25\,R_\odot} \right)^{3/2}, \qquad (12.20)$$

where $M > 1\,\mathrm{M}_\odot$ is the total mass of the Sun plus the close companion we have given it. Thus, if we put the Sun in a binary close enough to be circularized and synchronized ($a < 25\,\mathrm{R}_\odot$ for a $2\,\mathrm{M}_\odot$ system), it would have to rotate faster than it does now, and thus have a stronger magnetic field and a larger value of $\dot{J}$. Substituting Equation 12.20 into Equation 12.19, we find that the resulting angular momentum loss from the magnetized wind would be

$$\dot{J}_{\text{wind}} = 1.7 \times 10^{-12}\Omega_\odot^2 M_\odot R_\odot^2 \left(\frac{M}{2\,M_\odot}\right)^{3/2}\left(\frac{a}{25\,R_\odot}\right)^{-9/2}, \tag{12.21}$$

with a very strong dependence on the radius of the orbit.

The angular momentum in a synchronized, circularized binary is primarily in its orbital angular momentum,

$$J_{\text{orb}} = \left[G\frac{M_1^2 M_2^2}{M_1 + M_2}\right]^{1/2} a^{1/2}$$

$$\approx 750\,\Omega_\odot M_\odot R_\odot^2 \left(\frac{M}{2\,M_\odot}\right)^{3/2}\left(\frac{a}{25\,R_\odot}\right)^{1/2}, \tag{12.22}$$

assuming a roughly equal-mass binary. This represents a far larger store of angular momentum than that provided by the Sun's current rotation, $J_\odot \approx 0.07\Omega_\odot M_\odot R_\odot^2$.

Combining Equations 12.21 and 12.22, the timescale for losing orbital angular momentum and spiraling inward is

$$t_J = \frac{J_{\text{orb}}}{\dot{J}_{\text{wind}}} \approx 4.4 \times 10^{14}\Omega_\odot^{-1}\left(\frac{a}{25\,R_\odot}\right)^5 \approx 4700\,\text{Gyr}\left(\frac{a}{25\,R_\odot}\right)^5. \tag{12.23}$$

For an orbital separation of $a = 25\,R_\odot$, corresponding to $\mathcal{P} \sim 10\,$d, this is a timescale much longer than the age of the universe; thus, angular momentum loss from magnetized winds doesn't significantly affect the longest-period synchronized binaries. However, because of the steep a^5 dependence, the value of t_J drops to 10 Gyr when $a \sim 8\,R_\odot$, corresponding to an orbital period of $\mathcal{P} \sim 2\,$d.

Loss of angular momentum from a close synchronized binary can be made more effective by having a pair of stars with magnetized winds, or by having unusually strong magnetic fields in one or both stars. However, it is generally true that magnetized winds are effective at bringing binary systems with a period of two to three days down to an orbital period of $\mathcal{P} < 1\,$d, permitting a main sequence star to overflow its Roche lobe. In very short period systems ($\mathcal{P} < 1.5\,$hr), gravitational waves can effectively carry away angular momentum; this provides a mechanism by which binary systems consisting of compact objects spiral inward.

12.3 Cataclysmic Variables

Loss of angular momentum in a common envelope event, or as the result of magnetized winds, can result in a close binary consisting of a star and a compact object such as a white dwarf, neutron star, or black hole. Mass transfer from the star to its compact companion can result in a **cataclysmic variable**, if the compact object is a white dwarf, or an **X-ray binary** if the compact object is a neutron star or black hole. As the name "cataclysmic variable" implies, the transfer of gas onto a white dwarf is prone to recurrent violent events, during which the luminosity of the binary system increases dramatically.[5] In extreme cases, the luminosity can increase by a factor of 10^6 or more. Cataclysmic variables are also known as **novae**. The name "nova" can be traced back to Tycho Brahe's book *De nova et nullius aevi memoria prius visa stella* (Of the star, new and never before seen in the life or memory of anyone). Although the "new star" that Tycho Brahe saw in 1572 turns out to have been a supernova, the term "nova stella," or "nova" for short, became applied to any star that underwent a sudden large increase in luminosity.

There are three main classes of novae. **Classical novae** increase in brightness by more than 10 magnitudes during an outburst, and have no recurrence of the outburst during historical times. After a classical nova reaches maximum luminosity, it decreases in brightness with a characteristic time $t \sim 50\,\text{d}$. **Recurrent novae** brighten by ~ 8 magnitudes during an outburst, with recurrences every $\sim 50\,\text{yr}$. **Dwarf novae** brighten by ~ 3 magnitudes during an outburst, with recurrences every $\sim 1\,\text{yr}$. It seems that any particular nova can slip from one of these classes into another. For instance, GK Persei had an extreme outburst in 1901, brightening by 13 magnitudes, which puts it into the classical nova category. However, from the late twentieth century it started undergoing outbursts of ~ 3 magnitudes in amplitude, once every $\sim 3\,\text{yr}$. It isn't clear what triggered the switch from being a classical nova to being a dwarf nova.

Why are cataclysmic variables prone to cataclysms? The most extreme classical novae are a million times more luminous at maximum than in quiescence, and even a dwarf nova is ten times more luminous than usual during an outburst. First, let's look at the observed properties of a cataclysmic variable when it is quiescent, in between outbursts. The white dwarfs in classical and recurrent novae are observed to have an average mass $M_{\text{WD}} \approx 1\,M_\odot$, corresponding to a radius (Equation 9.7)

$$R_{\text{WD}} \approx 0.013\,\text{R}_\odot \left(\frac{M_{\text{WD}}}{1\,M_\odot}\right)^{-1/3} \approx 9000\,\text{km} \left(\frac{M_{\text{WD}}}{1\,M_\odot}\right)^{-1/3}. \qquad (12.24)$$

This is much smaller than the white dwarf's Roche lobe, so as gas from the companion star falls toward the white dwarf, angular momentum conservation prevents it from falling straight to the white dwarf's surface. Instead, it forms an **accretion disk** around the white dwarf. The gas in the accretion disk slowly moves

---

[5] In the 1930s, Cecilia Payne-Gaposchkin and Sergei Gaposchkin pondered whether "catastrophic variable" or "cataclysmic variable" was the better term. They finally decided that "catastrophe" (originally meaning the denouement of a play) implied finality, while "cataclysm" (originally meaning a great flood) implied possible repetition.

inward as it loses angular momentum (probably by magnetorotational instability, but that's a topic for a different book).

A typical accretion rate in a cataclysmic variable is $\dot{M} \sim 10^{-3}\,M_\odot\,\mathrm{Myr}^{-1} \sim 10^{17}\,\mathrm{g\,s}^{-1}$. As the gas moves toward the white dwarf, its gravitational potential energy is converted to thermal energy in the hot disk, and then to photon energy. The accretion luminosity is (compare to Equation 7.46)

$$L_{\mathrm{acc}} \sim \frac{GM_{\mathrm{WD}}\xi\dot{M}}{R_{\mathrm{WD}}}, \tag{12.25}$$

where ξ is the efficiency with which gravitational potential energy is converted to photon energy. For a $1\,M_\odot$ white dwarf, the accretion luminosity is

$$L_{\mathrm{acc}} \sim 3\,\mathrm{L}_\odot \left(\frac{\xi\dot{M}}{10^{-3}\,M_\odot\,\mathrm{Myr}^{-1}} \right). \tag{12.26}$$

Much of the luminosity comes from the inner region of the accretion disk, close to the white dwarf's surface; this is because a gram of gas loses as much potential energy going from $r = 2R_{\mathrm{WD}}$ to $r = 1R_{\mathrm{WD}}$ as it does in going from infinity to $r = 2R_{\mathrm{WD}}$. We can make an estimate of the effective temperature T_{eff} of the accretion disk in this inner region by looking at the luminosity of the annulus whose outer radius is $2R_{\mathrm{WD}}$ and whose inner radius is $1R_{\mathrm{WD}}$:

$$2\pi \left[(2R_{\mathrm{WD}})^2 - R_{\mathrm{WD}}^2 \right] \sigma_{\mathrm{SB}} T_{\mathrm{eff}}^4 \sim \frac{1}{2} \frac{GM_{\mathrm{WD}}\xi\dot{M}}{R_{\mathrm{WD}}}. \tag{12.27}$$

(If you've spotted an extra factor of two in the above equation, it's because the annulus has two sides.) The effective temperature of the hot inner region of the disk is then

$$T_{\mathrm{eff}} \sim 50\,000\,\mathrm{K} \left(\frac{M_{\mathrm{WD}}}{1\,M_\odot} \right)^{1/4} \left(\frac{\xi\dot{M}}{10^{-3}\,M_\odot\,\mathrm{Myr}^{-1}} \right)^{1/4} \left(\frac{R_{\mathrm{WD}}}{9000\,\mathrm{km}} \right)^{-3/4}. \tag{12.28}$$

A blackbody spectrum at this temperature peaks at a photon energy of $h\nu \approx 2.82kT_{\mathrm{eff}} \sim 12\,\mathrm{eV}$. This is in the far ultraviolet range of the spectrum, corresponding to a wavelength of $\lambda \sim 1000\,\text{Å}$. Thus, when a cataclysmic variable is in its quiescent state, its spectrum will have two components: far UV light from the white dwarf's accretion disk, with $L \sim 3\,\mathrm{L}_\odot$, and near IR and optical light from the donor star, which may be either more or less luminous than the accretion disk. Figure 12.7 shows the far UV spectrum of the cataclysmic variable HR Delphini, in the quiescent state after its 1967 outburst. In this spectrum, note (in addition to the broad maximum around $\lambda \sim 1300\,\text{Å}$) the P Cygni line from C IV at $\lambda = 1550\,\text{Å}$. This is the signature of a strong outflow of gas after the 1967 outburst.

In the quiescent state of a cataclysmic variable, the far UV light is powered by the gravitational potential energy of the infalling gas. However, the gas working

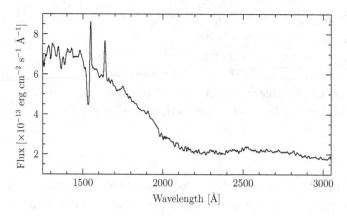

Figure 12.7 The far UV spectrum of HR Del ($d \approx 930\,\mathrm{pc}$). This spectrum is an average of multiple *IUE* observations during the interval 1979–1992. [Selvelli and Gilmozzi 2013]

its way through the accretion disk onto the white dwarf's surface is the hydrogen-rich outer envelope of the donor star. If hydrogen in the gas is fused into helium, that can provide a potent energy source for nova outbursts. The efficiency of hydrogen burning is

$$\eta_{\mathrm{fus}} = \frac{E_{\mathrm{fus}}}{mc^2} = 0.007\,12, \tag{12.29}$$

where E_{fus} is the energy released when a mass m of hydrogen is converted to helium. By contrast, the efficiency of dropping the same mass of hydrogen onto a white dwarf is

$$\eta_{\mathrm{WD}} = \frac{GM_{\mathrm{WD}}}{R_{\mathrm{WD}}c^2} \approx 1.6 \times 10^{-4} \left(\frac{M_{\mathrm{WD}}}{1\,M_\odot}\right) \left(\frac{R_{\mathrm{WD}}}{9000\,\mathrm{km}}\right)^{-1}. \tag{12.30}$$

Fusing the hydrogen into helium will thus release ~ 40 times more energy than was emitted as the hydrogen spiraled inward through the accretion disk.

But how do you coax hydrogen into fusion when it is near the surface of a white dwarf? We are used to thinking of fusion as something that occurs deep within a star, in the core or a thin shell not too high above the core. However, the gravitational acceleration at the surface of a white dwarf is high, with $g_{\mathrm{WD}} \sim 10^4 g_\odot$. Thus, the infalling gas is compressed into a thin dense layer. As more gas piles on, the base of the gas layer reaches a degenerate state, where the density is high but the temperature is not high enough to lift the degeneracy. Since degenerate gas is a superb heat conductor, the accreted gas comes to the same temperature as the degenerate core of the white dwarf. If this temperature is $T_c \geq 10^7$ K, then hydrogen burning begins with a **hydrogen flash**, similar to the helium flash with which the triple alpha process begins in relatively low-mass red giants. However, unlike the helium flash, which is buried below the thick envelope of a red giant, the hydrogen flash in a cataclysmic variable is readily visible.

Suppose that a cataclysmic variable spends a time t_{qui} in a quiescent state between outbursts, accreting gas at a rate $\dot{M}$. The amount of energy released by fusion of the accreted hydrogen is then

$$E_{nova} \approx 0.007\,12\,X\dot{M}t_{qui}c^2, \tag{12.31}$$

where X is the hydrogen mass fraction of the accreted material. If we assume $X \approx 0.7$, then

$$E_{nova} \approx 4.4 \times 10^{44}\,\text{erg}\left(\frac{\dot{M}}{10^{-3}\,M_\odot\,\text{Myr}^{-1}}\right)\left(\frac{t_{qui}}{50\,\text{yr}}\right). \tag{12.32}$$

This is about 70 times the amount of energy that the Sun radiates during the time interval t_{qui}. However, the nova's energy E_{nova} is mostly radiated away during a time interval $t_{burst} \sim 50\,\text{d}$. The average nova luminosity during that interval will be

$$L_{nova} \approx \frac{E_{nova}}{t_{burst}} \tag{12.33}$$

$$\approx 27\,000\,L_\odot\left(\frac{\dot{M}}{10^{-3}\,M_\odot\,\text{Myr}^{-1}}\right)\left(\frac{t_{qui}}{50\,\text{yr}}\right)\left(\frac{t_{burst}}{50\,\text{d}}\right)^{-1}.$$

This is larger by a factor of $\sim 10^4$ than the UV luminosity of the accretion disk in the system's quiescent state, and would represent a brightening by 10 magnitudes if E_{nova} all went into photons.

It is known that not all of E_{nova} takes the form of photons. Some of it goes into the kinetic energy of gas that is blown away from the white dwarf during the nova outburst. An important question, and one that is not yet fully answered, is whether the mass M_{ej} that is ejected during a nova outburst is greater than or less than the amount $t_{qui}\dot{M}$ that is accreted between outbursts. If more is ejected than is accreted, then the mass of the white dwarf decreases with time. However, if more is accreted than is ejected, then the mass of the white dwarf increases with time. This leads to the possibility that the white dwarf could be nudged over the Chandrasekhar limit, and become a thermonuclear supernova.

Let's consider what happens when the donor star pours gas onto a neutron star instead of a white dwarf; this constitutes an X-ray binary rather than a cataclysmic variable. With a mass $M_{NS} \approx 1.5\,M_\odot$ and radius $R_{NS} \approx 12\,\text{km}$, the accretion luminosity of an X-ray binary is

$$L_{acc} \sim 2700\,L_\odot\left(\frac{\xi\dot{M}}{10^{-3}\,M_\odot\,\text{Myr}^{-1}}\right), \tag{12.34}$$

about a thousand times greater than for a white dwarf with the same accretion rate. Most of the luminosity, again, will come from the region of the accretion disk close to the accreting compact object. In the case of accretion onto a neutron star, the effective temperature of this hot inner region will be (comparing to Equation 12.28)

$$T_{\text{eff}} \sim 8 \times 10^6 \, \text{K} \left(\frac{M_{\text{NS}}}{1.5 \, \text{M}_\odot} \right)^{1/4} \left(\frac{\xi \dot{M}}{10^{-3} \, \text{M}_\odot \, \text{Myr}^{-1}} \right)^{1/4} \left(\frac{R_{\text{NS}}}{12 \, \text{km}} \right)^{-3/4}. \quad (12.35)$$

A blackbody spectrum at this temperature will peak at a photon energy of $h\nu_{\text{peak}} \approx 2.82 k T_{\text{eff}} \sim 2 \, \text{keV}$. This is in the soft X-ray range of the spectrum, and explains how X-ray binaries got their name.

Although X-ray binaries show some variability in their flux, they are not prone to the same high-amplitude fusion-powered outbursts as cataclysmic variables. The efficiency of hydrogen burning is $\eta_{\text{fus}} \approx 0.007\,12$. The efficiency of dropping gas onto a neutron star is (Equation 12.30)

$$\eta_{\text{NS}} = \frac{GM_{\text{NS}}}{R_{\text{NS}} c^2} \sim 0.2 \left(\frac{M_{\text{NS}}}{1.5 \, \text{M}_\odot} \right) \left(\frac{R_{\text{NS}}}{12 \, \text{km}} \right). \quad (12.36)$$

In other words, since the radius of a neutron star is not much bigger than its Schwarzschild radius, the amount of energy you can get from dropping a mass m onto its surface is not much less than mc^2. If hydrogen dropped onto the surface of a neutron star undergoes fusion, that represents just a 4% addition to the energy that was released as the hydrogen spiraled in through the accretion disk.

12.4 Banging Stars Together

In a close binary, stars come close enough to transfer large amounts of gas. On occasion, stars and stellar remnants can come close enough to merge with each other. If two main sequence stars merge, the end result can be a **blue straggler** star. If two white dwarfs merge, the end result can be a thermonuclear supernova. If two neutron stars merge, the end result can be a black hole, with the emission of a burst of gravitational waves and a spray of r-process elements. Although neutron star mergers are interesting, their detailed general relativistic analysis takes us beyond the range of this book. Let's concentrate first on blue stragglers and then on thermonuclear supernovae.

A blue straggler can be most readily identified if it's within a globular cluster or open cluster, comprised of stars that all have nearly the same age. In this case, a blue straggler is observationally defined as a main sequence star that is hotter and brighter than the main sequence turnoff for that cluster. The term blue straggler was first used in print by Margaret Burbidge and Allan Sandage in 1958, when they wrote that "bright blue 'stragglers' are known in the Coma Berenices cluster and in Praesepe." As our example of a cluster containing blue stragglers, consider the globular cluster NGC 6397, whose lithium abundances we examined

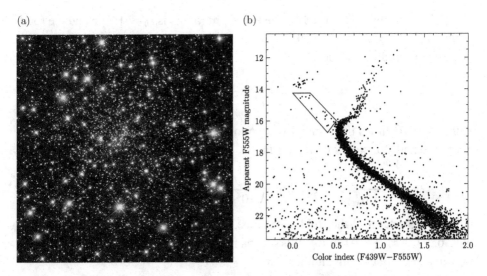

Figure 12.8 (a) Central regions of the globular cluster NGC 6397. Field of view is 2.12 × 2.18 arcmin, corresponding to ∼1.4 × 1.4 pc at the distance of NGC 6397. [HST/NASA/ESA] (b) H–R diagram of NGC 6397. The quadrilateral encloses the blue stragglers. [Data from Piotto *et al.* 2002]

in Figure 8.6. The brightest stars in NGC 6397, as seen in Figure 12.8(a), are red giants; at the cluster's estimated age of $t \approx 13.4$ Gyr, stars with initial mass $M_\star \sim 0.9\,M_\odot$ are just now ascending the red giant branch. The brightest blue stars in NGC 6397 are horizontal branch stars. However, in the H–R diagram of NGC 6397, shown in Figure 12.8(b), there is a smattering of stars between the horizontal branch at $m \sim 13.5$ and the main sequence turnoff at $m \sim 16.5$; these are the blue stragglers.

One way to make a blue straggler is by mass transfer in a close binary. The Algol system, for instance, is at least 500 Myr old; that's how long it took Algol B to evolve off the main sequence. However, the main sequence lifetime of Algol A is now a scant 300 Myr. Thus, if Algol were in a cluster with other 500 Myr-old stars, then Algol A would stand out as a blue straggler. However, some blue stragglers show no signs of being in a binary system. Isolated blue stragglers can result if one star in a binary completely cannibalizes the other. However, they can also result if two stars collide with each other. To see how often stellar collisions occur in a globular cluster, consider a simplified cluster in which all stars have nearly the same radius $R_\star$. In that case, stars will collide and merge when their centers come within a distance $r \sim 2R_\star$ of each other, and the collisional cross section will then be

$$\sigma_\star = \pi(2R_\star)^2 \approx 13\,\mathrm{R}_\odot^2 \left(\frac{R_\star}{1\,\mathrm{R}_\odot}\right)^2. \tag{12.37}$$

In the globular cluster NGC 6397, the half-light radius is $r_h \sim 1.7\,\mathrm{pc}$ and the total number of stars is $N_\star \sim 4 \times 10^5$. Thus, the mean number density of stars within the half-light radius will be

$$n_\star \approx \frac{3(N_\star/2)}{4\pi r_h^3} \sim 10^4\,\mathrm{pc}^{-3}, \tag{12.38}$$

which is $\sim 10^5$ times the number density of stars in the solar neighborhood. The mean free path for stellar collisions is

$$\lambda_{\mathrm{mfp}} \approx \frac{1}{n_\star \sigma_\star} \approx 16\,000\,\mathrm{Mpc} \left(\frac{n_\star}{10^4\,\mathrm{pc}^{-3}}\right)^{-1} \left(\frac{R_\star}{1\,\mathrm{R_\odot}}\right)^{-2}, \tag{12.39}$$

a distance greater than the Hubble distance. The central velocity dispersion within NGC 6397 is $\sigma_v \sim 10\,\mathrm{km\,s^{-1}}$; this means that the typical time that passes before a given star collides with another star is

$$t_{\mathrm{coll}} \approx \frac{\lambda_{\mathrm{mfp}}}{\sigma_v} \tag{12.40}$$

$$\approx 1.5 \times 10^6\,\mathrm{Gyr} \left(\frac{n_\star}{10^4\,\mathrm{pc}^{-3}}\right)^{-1} \left(\frac{R_\star}{1\,\mathrm{R_\odot}}\right)^{-2} \left(\frac{\sigma_v}{10\,\mathrm{km\,s^{-1}}}\right)^{-1},$$

which is $\sim 100\,000$ times the age of the globular cluster. Since any particular star in the cluster's central region has 1 chance in 100 000 of having undergone a collision, and since there are $\sim 200\,000$ stars in the central region, we would expect approximately one blue straggler to form by this mechanism during the lifetime of NGC 6397.

Since there are multiple blue stragglers in NGC 6397 today, straightforward collisions between single stars are unlikely to be the sole formation mechanism for its blue stragglers. A more fruitful mechanism is through binary/single or binary/binary encounters. If a single star comes within a distance $\sim a$ of a binary with semimajor axis a, an intricate three-body dance will ensue. One possible outcome of the encounter is the ejection of a high-speed star, with the remaining two stars forming an extremely close binary. Sometimes the close binary is so close that the two stars merge to form a blue straggler. Numerical simulations of clusters with plausible binary populations reveal that if the central density of the cluster is greater than $n_\star \sim 1000\,\mathrm{pc}^{-3}$, binary/binary and binary/single encounters are the dominant mechanism for making blue stragglers. At lower cluster densities, mass transfer in close binaries is the dominant mechanism.

If two white dwarfs collide, rather than two ordinary stars, the result can be spectacular, since colliding white dwarfs is one mechanism for making supernovae. A thermonuclear supernova occurs when a carbon/oxygen white dwarf approaches its Chandrasekhar limit, $M_{\mathrm{Ch}} = 1.46\,\mathrm{M_\odot}$. As the degenerate electrons become more nearly relativistic, the pressure of the white dwarf no longer suffices to support it against gravity, and it collapses and becomes denser. As it does so, the carbon and oxygen nuclei increase in temperature, and carbon burning begins with a thermonuclear runaway.

In the **single degenerate** scenario, a white dwarf is nudged toward the Chandrasekhar limit by having gas dumped onto it by a donor star in a close binary with the white dwarf. One problem with this scenario, which we mentioned above, is that such a system constitutes a cataclysmic variable. It's not certain whether the matter gained during the quiescent accretion phase of a cataclysmic variable is greater, on average, than the matter blown away during the subsequent nova outburst. Moreover, evidence for surviving stellar companions of thermonuclear supernovae is lacking.

In the **double degenerate** scenario for triggering a thermonuclear supernova, two white dwarfs collide, making a single white dwarf near or above the Chandrasekhar limit. One problem with this scenario is getting two white dwarfs to collide. The collisional cross section for two white dwarfs is smaller by a factor of $\sim 10^{-4}$ than the cross section for two Sun-like stars. Thus, since collisions between single stars are implausible, even in dense globular clusters like NGC 6397, collisions between single white dwarfs are implausible times ten thousand, assuming similar numbers of stars and white dwarfs in the cluster. In the blue straggler case, it is found that binary/binary and binary/single encounters dominate over encounters between single stars. This is also true for encounters between white dwarfs. However, most thermonuclear supernovae are not in globular clusters, galactic nuclei, or other regions with high stellar density. Thus, to produce thermonuclear supernovae in low-density regions, there must be another way of coaxing two white dwarfs into merging.

One way of merging white dwarfs is through the Lidov–Kozai mechanism in a triple system where the inner close binary consists of a pair of white dwarfs. (The outer perturbing tertiary body can be either a star or stellar remnant.) The Lidov–Kozai mechanism, in brief, explains how the eccentricity and inclination of the inner binary's orbit can be altered by the gravitational perturbation of the outer tertiary body. Let e be the eccentricity of the inner binary's orbit, and i be the inclination of the binary's orbital plane relative to the orbital plane of the tertiary about the binary's barycenter. As the outer tertiary tugs on the inner binary, a conserved property is

$$f_z \equiv \sqrt{1 - e^2} \cos i. \tag{12.41}$$

If the inner binary is initially on a nearly circular orbit ($e \approx 0$) and the outer perturbing tertiary is initially highly inclined ($\cos i \ll 1$), then the Lidov–Kozai mechanism can drive the eccentricity of the inner binary to very high values. If the periapse distance $a(1 - e)$ is then a few times R_{WD}, then tidal circularization will shrink the orbital size a. If the periapse distance $a(1 - e)$, through a combination of Lidov–Kozai oscillations and tidal circularization, becomes smaller than $2R_{WD}$, the two white dwarfs will merge. Numerical analysis reveals that white dwarf mergers can take place over times shorter than the Hubble time if the initial white dwarf separation is small and the perturbing tertiary is initially on a high-inclination orbit at a relatively small separation. If the tertiary is initially at a distance $d > 50a$, it requires fine tuning of the initial values of i and a to result in a merger in less than the Hubble time.

Fusing a mix of carbon and oxygen all the way to the iron peak releases energy with an efficiency

$$\eta_{\text{fus}} = \frac{E_{\text{fus}}}{mc^2} \approx 0.0009. \tag{12.42}$$

The energy released by fusing a $1.46\,M_\odot$ carbon/oxygen white dwarf into a mix of iron and nickel is then

$$E_{\text{fus}} \approx (0.0009)1.46\,M_\odot c^2 \approx 2.3 \times 10^{51}\,\text{erg}, \tag{12.43}$$

or $E_{\text{fus}} \approx 2.3$ foe. By contrast, as we saw in Equation 12.32, a recurrent nova outburst produces only half a microfoe. A supernova really is *super* compared to a mere nova. The energy released in the supernova explosion should be compared to the gravitational potential energy of a white dwarf just below the Chandrasekhar limit:

$$W_{\text{WD}} \approx -\frac{GM_{\text{WD}}^2}{R_{\text{WD}}} \approx -0.7 \times 10^{51}\,\text{erg} \approx -0.3 E_{\text{fus}}. \tag{12.44}$$

Thus, it is energetically possible for fusion to completely unbind the white dwarf, blowing the resulting iron and nickel into the interstellar medium without leaving a dense central object behind. This is in contrast to core collapse supernovae, which leave behind a central neutron star or black hole.

As it turns out, Tycho's "nova stella" of 1572 was a thermonuclear supernova. A *Chandra* image taken more than four centuries later (Figure 12.9) reveals an expanding supernova remnant rich in ionized iron. The iron is mixed back into the interstellar medium, ready for the creation of a new generation of higher-metallicity stars. And so the circle of stellar life continues . . .

Exercises

12.1 AGB stars experience strong mass loss with a slow wind speed, $v_{\text{wind}} \leq 10\,\text{km s}^{-1}$ or less. Suppose that the primary in a system, with mass M_1, is an AGB star, and the secondary, with mass M_2, is on the main sequence. The distance between the stars is a. What fraction of the ejected mass from an AGB star would the companion be able to accrete? Can you think of any observable consequences of such an accretion process for the secondary star?

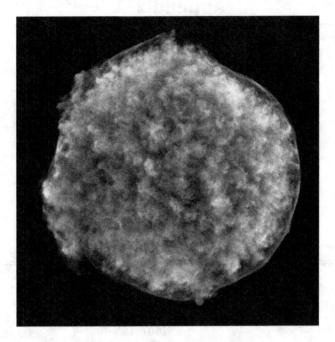

Figure 12.9 Remnant of Tycho's supernova ($d \sim 3\,\mathrm{kpc}$) as seen in X-rays (1.6–6 keV photon energy). Field of view is 10×10 arcmin, corresponding to $\sim 9 \times 9\,\mathrm{pc}$ at the distance of the remnant. [NASA/CXC/Rutgers/J. Warren *et al.*]

12.2 Consider a "twin" binary ($M_1 = M_2 = M_\star$) on a circularized, synchronized orbit. For what orbital period $\mathcal{P}(M_\star)$ will the timescale for them to spiral in and merge be the same as their main sequence lifetime?

12.3 Sketch out an evolutionary scenario that could produce a binary neutron star on a close orbit (similar to the binary pulsars discussed in Chapter 9). What will the outcome be if the neutron stars merge?

12.4 What fraction of $1\,\mathrm{M}_\odot$ stars would you expect to have a significant mass transfer (either gain or loss) from a binary companion at some point during their lifetime?

Appendix A

Constants and Units

Table A.1 Fundamental physical constants

Name	Symbol	Value	Units
Speed of light in vacuum	c	$2.99792458 \times 10^{10}$	$\mathrm{cm\,s^{-1}}$
Gravitation constant	G	6.67430×10^{-8}	$\mathrm{cm^3\,g^{-1}\,s^{-2}}$
Planck constant	h	$6.62607015 \times 10^{-27}$	$\mathrm{erg\,s}$
		$4.135667696 \times 10^{-15}$	$\mathrm{eV\,s}$
Reduced Planck constant	$\hbar$	$1.054571817 \times 10^{-27}$	$\mathrm{erg\,s}$
		$6.582119569 \times 10^{-16}$	$\mathrm{eV\,s}$
Boltzmann constant	k	1.380649×10^{-16}	$\mathrm{erg\,K^{-1}}$
		$8.61733326 \times 10^{-5}$	$\mathrm{eV\,K^{-1}}$
Elementary charge	e	$1.602176635 \times 10^{-19}$	C
Stefan–Boltzmann constant	σ_{SB}	$5.670375519 \times 10^{-5}$	$\mathrm{erg\,cm^{-2}\,s^{-1}\,K^{-4}}$
Thomson cross section	σ_e	$6.6524587321 \times 10^{-25}$	$\mathrm{cm^2}$
Atomic mass unit	m_{AMU}	$1.66053906660 \times 10^{-24}$	g
Proton mass	m_p	$1.67262193269 \times 10^{-24}$	g
Electron mass	m_e	$9.1093837015 \times 10^{-28}$	g
Proton magnetic moment	μ_p	$1.40160679736 \times 10^{-23}$	$\mathrm{erg\,G^{-1}}$
Electron magnetic moment	μ_e	$-9.2847647043 \times 10^{-21}$	$\mathrm{erg\,G^{-1}}$
Bohr radius	a_0	$5.29177210903 \times 10^{-9}$	cm

Source: 2018 CODATA recommended values.

Table A.2 Astronomical constants

Name	Symbol	Value	Units
Astronomical unit	au	$1.495978707 \times 10^{13}$	cm
Solar radius	$R_\odot$	6.957×10^{10}	cm
Solar luminosity	$L_\odot$	3.828×10^{33}	$\mathrm{erg\,s^{-1}}$
Solar mass parameter	$GM_\odot$	1.3271244×10^{26}	$\mathrm{cm^3\,s^{-2}}$
Solar mass	$M_\odot$	1.9884×10^{33}	g
Solar effective temperature	$T_\odot$	5772	K

Source: IAU 2012 resolution B2 on the redefinition of the astronomical unit of length; IAU 2015 resolution B3 on recommended nominal conversion constants for solar and planetary properties.

Table A.3 Astronomical and physical units

Name	Symbol	Value	Units
parsec	pc	206 264.8063	au
		$3.085\,677\,58 \times 10^{18}$	cm
day	d	86 400	s
Julian year	yr	365.25	d
		$3.155\,76 \times 10^7$	s
angstrom	Å	10^{-8}	cm
electron volt	eV	$1.602\,176\,634 \times 10^{-12}$	erg
standard atmosphere	atm	$1.013\,25 \times 10^6$	dyn cm^{-2}

Source: IAU 2012 resolution B2; IAU 2015 resolution B3; 2018 CODATA recommended values.

Appendix B

Properties of Example Stars

Table B.1 Properties of stars used as examples in the text

Name	Parallax [mas]	M_V	$B - V$	Spectral class	Mass [M$_\odot$]	Radius [R$_\odot$]	Luminosity [L$_\odot$]
Betelgeuse	5.5	−5.88	1.85	M2Ib	18	800	100 000
P Cygni	0.7	−5.85	0.42	B1Ia	30	75	600 000
Canopus	10.6	−5.61	0.15	A9II/F0Ib	9	65	11 000
Polaris Aa	7.3	−3.67	0.60	F8Ib	6	46	2400
Alcyone	8.1	−2.59	−0.09	B7III	6	9.3	2000
VY CMa	0.8	−2.45	2.24	M5Ia	20	1400	270 000
Regulus A	41.1	−0.53	−0.16	B8IV	3.8	3.9	300
Asterope	7.6	0.15	−0.04	B8V	2.9	2.7	100
Capella Aa	76.2	0.29	0.93	G8III	2.57	12	79
Vega	130	0.60	0.00	A0V	2.14	2.6	40
Pollux	96.5	1.06	1.00	K0III	1.9	9.1	33
Sirius A	374.5	1.41	0.00	A1V	2.14	1.74	25.5
R Doradus	18.3	1.71	1.58	M8III	0.8	340	5600
Mira A	11	1.72	1.10	M5-9III	1.1	300	7000
Altair	195	2.21	0.22	A7V	1.8	1.89	11
α Cen A	751	4.39	0.71	G2V	1.08	1.22	1.5
Sun	—	4.83	0.66	G2V	1.00	1.00	1.00
α Cen B	751	5.71	0.88	K1V	0.91	0.86	0.50
Lalande 21185	393	10.49	1.44	M2V	0.39	0.39	0.023
Kapteyn's Star	254	10.88	1.58	M1VI	0.28	0.29	0.012
Proxima Cen	768	15.56	1.82	M5.5V	0.12	0.14	0.0017

Basic data from SIMBAD [simbad.u-strasbg.fr; see references therein]. Masses of stars without known companions are approximate. Radii of oblate stars are volumetric radii. Properties of variable stars are time averages.

Further Reading, Bibliography, and Figure Credits

A l'alta fantasia qui mancò possa;
ma già volgeva il mio disio e il velle,
sì come rota ch'igualmente è mossa,
l'amor che move il sole e l'altre stelle.

Dante, Paradiso, Canto XXXIII, 142–145

Here ceased the powers of my high fantasy.
  Already were all my will and my desires
    turned – as a wheel in equal balance – by
The Love that moves the sun and the other stars.

Anthony Esolen, translator

Further Reading

Hansen, C. J., Kawaler, S. D., and Trimble, V. 2004, *Stellar Interiors: Physical Principles, Structure, and Evolution* (2nd edition), Springer

Kippenhahn, R., Weigert, A., and Weiss, A. 2012, *Stellar Structure and Evolution* (2nd edition), Springer

Prialnik, D. 2010, *An Introduction to the Theory of Stellar Structure and Evolution* (2nd edition), Cambridge University Press

Weiss, A., Willebrandt, W., Thomas, H.-C., *et al.* 2006, *Cox & Giuli's Principles of Stellar Structure* (3rd edition), Cambridge Scientific Publishers

Bibliography

In our reference list, we use compact abbreviations for the most frequently cited journals in astronomy. These abbreviations, as used by the SAO/NASA Astrophysics Data System, are:

- A&A = Astronomy and Astrophysics
- AJ = The Astronomical Journal
- ApJ = The Astrophysical Journal
- ApJS = The Astrophysical Journal Supplement
- ARA&A = Annual Review of Astronomy and Astrophysics
- MNRAS = Monthly Notices of the Royal Astronomical Society
- PASP = Publications of the Astronomical Society of the Pacific

Aerts, C., Mathis, S., and Rogers, T. M. 2019, "Angular momentum transport in stellar interiors," ARA&A, 57, 35

Akeson, R., Beichman, C., Kervella, P., *et al.* 2021, "Precision millimeter astrometry of the α Centauri AB system," AJ, 162, 14

Aston, F. W. 1920, "The mass-spectra of chemical elements," Philos. Mag., 39, 611

Baade, W., and Zwicky, F. 1934, "Cosmic rays from super-novae," Proc. Natl. Acad. Sci. USA, 20, 259

Bahcall, J. N., and Pinsonneault, M. H. 2004, "What do we (not) know theoretically about solar neutrino fluxes?" Phys. Rev. Lett., 92, 121301

Baron, F., Monnier, J. D., Pedretti, E., *et al.* 2012, "Imaging the Algol triple system in the *H* band with the CHARA interferometer," ApJ, 752, 20

Basri, G., Walkowicz, L. M., Batalha, N., *et al.* 2010, "Photometric variability in *Kepler* target stars: The Sun among stars – a first look," ApJL, 713, L155

Bedding, T. R., Zijlstra, A. A., von der Lühe, O., *et al.* 1997, "The angular diameter of R Doradus: A nearby Mira-like star," MNRAS, 286, 957

Bethe, H. A. 1939, "Energy production in stars," Phys. Rev., 55, 434

Brahe, T. 1573, *De nova et nullius aevi memoria prius visa stella*, Copenhagen: Lorentz Benedicht, printer

Burbidge, E. M., and Sandage, A. 1958, "Properties of two intergalactic globular clusters," ApJ, 127, 527

Burrows, A., Hubbard, W. B., Saumon, D., *et al.* 1993, "An expanded set of brown dwarf and very low mass star models," ApJ, 406, 158

Burrows, A., Marley, M., Hubbard, W. B., *et al.* 1997, "A nongray theory of extrasolar giant planets and brown dwarfs," ApJ, 491, 856

Cannon, A. J., and Pickering, E. C. 1901, "Spectra of bright southern stars photographed with the 13-inch Boyden telescope as a part of the Henry Draper Memorial," Annals of Harvard College Observatory, 28, 129

Cantat-Gaudin, T., Jordi, C., Vallenari, A., *et al.* 2018, "A Gaia DR2 view of the open cluster population in the Milky Way," A&A, 618, A93

Carnot, S. 1824, *Réflexions sur la puissance motrice du feu et sur les machines propres a développer cette puissance*, Paris: Bachelier

Chabrier, G. 2003, "Galactic stellar and substellar mass initial mass function," PASP, 115, 763

Chandrasekhar, S. 1931, "The highly collapsed configurations of a stellar mass," MNRAS, 91, 456

Chandrasekhar, S. 1931, "The maximum mass of ideal white dwarfs," ApJ, 74, 81

Claytor, Z. R., van Saders, J. L., Santos, Â. R. G., *et al.* 2020, "Chemical evolution in the Milky Way: Rotation-based ages for APOGEE–Kepler cool dwarf stars," ApJ, 888, 43

Cowan, J. J., Lawler, J. E., Sneden, C., *et al.* 2006, "Nucleosynthesis: Stellar and solar abundances and atomic data," Proceedings of the 2006 NASA Laboratory Astrophysics Workshop, p. 82

Cowling, T. G. 1941, "The non-radial oscillations of polytropic stars," MNRAS, 101, 367

Cranmer, S. R., and Saar, S. H. 2011, "Testing a predictive theoretical model for the mass loss rates of cool stars," ApJ, 741, 54

Cummings, J. D., Kalirai, J. S., Tremblay, P.-E., *et al.* 2018, "The white dwarf initial–final mass relation for progenitor stars from 0.85 to 7.5 $M_\odot$," ApJ, 866, 21

Deheuvels, S., Ballot, J., Eggenberger, P., *et al.* 2020, "Seismic evidence for near solid-body rotation in two *Kepler* subgiants and implications for angular momentum transport," A&A, 641, A117

Eddington, A. S. 1920, "The internal constitution of the stars," Nature, 106, 14

Eddington, A. S. 1924, "On the relation between the masses and luminosities of the stars," MNRAS, 84, 308

Eddington, A. S. 1926, *The Internal Constitution of the Stars*, Cambridge University Press

Eker, Z., Bakış, V., Bilir, S., *et al.* 2018, "Interrelated main-sequence mass–luminosity, mass–radius, mass–effective temperature relations," MNRAS, 479, 5491

Emden, R. 1907, *Gaskugeln: Anwendungen der mechanischen Wärmetheorie auf kosmologische und meteorologische Probleme*, B. Teubner

Fellgett, P. 1995, "Simple stars," Observatory, 115, 93 [source of epigraph for Chapter 2]

Gamow, G. 1928, "Zur Quantentheorie des Atomkernes," Z. Phys., 51, 204

García, R. A., Ceillier, T., Salabert, D., *et al.* 2014, "Rotation and magnetism of Kepler pulsating solar-like stars. Toward asteroseismically calibrated age–rotation relations," A&A, 572, A34

Godoy-Rivera, D., Pinsonneault, M. H., and Rebull, L. M. 2021, "Stellar rotation in the Gaia era: Revised open clusters' sequences," ApJS, 257, 46

Guillochon, J., Parrent, J., Kelley, L. Z., *et al.* 2017, "An open catalog for supernova data," ApJ, 835, 64

Hayashi, C. 1961, "Stellar evolution in early phases of gravitational contraction," PASJ, 13, 450

Hayashi, C., and Hoshi, R. 1961, "The outer envelope of giant stars with surface convection zone," PASJ, 13, 442

Henyey, L. G., LeLevier, R., and Levée, R. D. 1955, "The early phases of stellar evolution," PASP, 67, 154

Heyl, J., Caiazzo, I., and Richer, H. 2022, "Reconstructing the Pleiades with *Gaia* EDR3," ApJ, 926, 132

Holmes, A. 1913, *The Age of the Earth*, Harper & Bros., p. 157

Hoskin, M. 2014, "William Herschel and the planetary nebulae," JHA, 45, 219 [quotes Herschel's reason for the name "planetary nebula"]

Howe, R., Christensen-Dalsgaard, J., Hill, F., *et al.* 2000, "Dynamic variations at the base of the solar convection zone," Science, 287, 2456

Hoyle, F., Dunbar, D. N. F., Wenzel, W. A., *et al.* 1953, "A state in C^{12} predicted from astrophysical evidence," Phys. Rev. 92, 1095

Jacoby, G. H., Hunter, D. A., and Christian, C. A. 1984, "A library of stellar spectra," ApJS, 56, 257

Jeans, J. H. 1902, "I. The stability of a spherical nebula," Phil. Trans. A, 199, 1

Kastner, J. H., Montez Jr., R., Balick, B., *et al.* 2012, "The Chandra X-ray survey of planetary nebulae (ChanPlaNS): Probing binarity, magnetic fields, and wind collisions," AJ, 144, 58

Kelvin, Lord 1899, "The age of the Earth as an abode fitted for life," Science, 9, 665

Kilic, M., Bergeron, P., Kosakowski, A., *et al.* 2020, "The 100 pc white dwarf sample in the SDSS footprint," ApJ, 898, 84

Kraft, R. P. 1967, "Studies of stellar rotation. V. The dependence of rotation on age among solar-type stars," ApJ, 150, 551

Kramers, H. A. 1923, "On the theory of x-ray absorption and of the continuous x-ray spectrum," Philos. Mag., 46, 836

Lane, J. H. 1870, "On the theoretical temperature of the Sun, under the hypothesis of a gaseous mass maintaining its volume by its internal heat, and depending on the laws of gases as known to terrestrial experiment," Am. J. Sci. Arts, series 2, 50, 57

Lind, K., Primas, F., Charbonnel, C., *et al.* 2009, "Signatures of intrinsic Li depletion and Li–Na anti-correlation in the metal-poor globular cluster NGC 6397," A&A, 503, 545

Lodders, K. 2021, "Relative atomic solar system abundances, mass fractions, and atomic masses of the elements and their isotopes, composition of the solar photosphere, and compositions of the major chondritic meteorite groups," Space Science Reviews, 217, 44

Lu, J. R., Do, T., Ghez, A. M., *et al.* 2013, "Stellar populations in the central 0.5 pc of the Galaxy. II. The initial mass function," ApJ, 764, 155

Maeder, A. 1971, "The Schönberg–Chandrasekhar limit and rotation," A&A, 14, 351

Metcalfe, T. S., Chaplin, W. J., Appourchaux, T., *et al.* 2012, "Asteroseismology of the solar analogs 16 Cyg A and B from *Kepler* observations," ApJL, 748, L10

Michaud, G., Richer, J., and Vick, M. 2011, "Sirius A: Turbulence or mass loss?" A&A, 534, 18

Moe, M., and Di Stefano, R. 2017, "Mind your p's and q's: The interrelation between period ($\mathcal{P}$) and mass-ratio (q) distributions of binary stars," ApJS, 230, 15

Monnier, J. D., Zhao, M., Pedretti, E., *et al.* 2007, "Imaging the surface of Altair," Science, 317, 342

Montargès, M., Cannon, E., Lagadec, E., *et al.* 2021, "A dusty veil shading Betelgeuse during its Great Dimming," Nature, 594, 365

Morgan, W. W. 1937, "On the spectral classification of the stars of type A to K," ApJ, 85, 380

Najarro, F., Diller, D. J., and Stahl, O. 1997, "A spectroscopic investigation of P Cygni. I. H and He I lines," A&A, 326, 1117 [source for properties of stellar wind of P Cygni]

Narayanan, G., Heyer, M. H., Brunt, C., *et al.* 2008, "The Five College Radio Astronomy Observatory CO mapping survey of the Taurus Molecular Cloud," ApJS, 177, 341

Offner, S. S. R., Moe, M., Kratter, K. M., *et al.* 2022, "The origin and evolution of multiple star systems," in *Protostars and Planets VII* [arXiv:2203.10066]

Öpik, E. 1938, "Stellar structure, source of energy, and evolution," Publications de l'Observatoire Astronomique de l'Université de Tartu, 30C, 1

Özel, F., and Freire, P. 2016, "Masses, radii, and the equation of state of neutron stars," ARA&A, 54, 401

Pecaut, M. J., and Mamajek, E. E. 2013, "Intrinsic colors, temperatures, and bolometric corrections of pre-main-sequence stars," ApJS, 208, 9 [color–$T_{\rm eff}$ relation of Equation 1.23 is a fit to data from this paper]

Pickering, E. C. 1890, "The Draper Catalogue of stellar spectra photographed with the 8-inch Bache telescope as a part of the Henry Draper Memorial," Annals of Harvard College Observatory, 27, 1 [Pickering writes "the greater portion of this work . . . has been in charge of Mrs. M. Fleming."]

Pinsonneault, M., *et al.* 2023, in preparation

Piotto, G., King, I. R., Djorgovski, S. G., *et al.* 2002, "HST color–magnitude diagrams of 74 galactic globular clusters in the HST F439W and F555W bands," A&A, 391, 945

Pourbaix, D., Tokovinin, A. A., Batten, A. H., *et al.* 2004, "SB9: The ninth catalogue of spectroscopic binary orbits," A&A, 424, 727

Rosseland, S. 1924, "Note on the absorption of radiation within a star," MNRAS, 84, 525

Roxburgh, I. W. 2004, "2-dimensional models of rapidly rotating stars. I. Uniformly rotating zero age main sequence stars," A&A, 428, 171

Russell, H. N. 1925, "The problem of stellar evolution," Nature, 116, 209

Saha, M. N. 1921, "On a physical theory of stellar spectra," Proc. Roy. Soc. Lond., 99, 135

Salpeter, E. E. 1952, "Nuclear reactions in the stars. I. Proton–proton chain," Phys. Rev., 88, 547

Salpeter, E. E. 1955, "The luminosity function and stellar evolution," ApJ, 121, 161

Schmidt, D., Gorceix, N., Goode, P. R., *et al.* 2017, "*Clear* widens the field for observations of the Sun with multi-conjugate adaptive optics," A&A, 597, L8

Schönberg, M., and Chandrasekhar, S. 1942, "On the evolution of the main-sequence stars," ApJ, 96, 161

Schwarzschild, K. 1906, "Ueber das Gleichgewicht der Sonnenatmosphäre," Nachrichten von der Königlichen Gesellschaft der Wissenschaften zu Göttingen, Mathematisch-physikalische Klasse aus dem Jahre 1906, 41

Schwarzschild, M. 1941, "Overtone pulsations for the standard model," ApJ, 94, 245

See, T. J. J. 1893, "On the orbit of α Centauri," MNRAS, 54, 102

Selvelli, P., and Friedjung, M. 2003, "The active quiescence of HR Del (Nova Del 1967)," A&A, 401, 297

Selvelli, P., and Gilmozzi, R. 2013, "An archive study of 18 old novae. I. The UV spectra," A&A, 560, 49

Udalski, A., Szymanski, M., Kubiak, M., *et al.* 1999, "The Optical Gravitational Lensing Experiment. III. Period–luminosity–color and period–luminosity relations of classical Cepheids," Acta Astronomica, 49, 201

Uitenbroek, H., Dupree, A. K., and Gilliland, R. L. 1998, "Spatially resolved *Hubble Space Telescope* spectra of the chromosphere of α Orionis," ApJ, 116, 2501

Vasiliev, E., and Baumgardt, H. 2021, "*Gaia* EDR3 view on galactic globular clusters," MNRAS, 505, 5978

von Zeipel, H. 1924, "The radiative equilibrium of a rotating system of gaseous masses," MNRAS, 84, 48

Wang, M., Huang, W. J., Kondev, F. G., *et al.* 2021, "The AME2020 atomic mass evaluation (II). Tables, graphs, and references," Chinese Phys. C, 45, 030003

Willems, B., and Kolb, U. 2004, "Detached white dwarf main-sequence star binaries," A&A, 419, 1057

Xu, Y., Takahashia, K., Gorielya, S., *et al.* 2013, "NACRE II: An update of the NACRE compilation of charged-particle-induced thermonuclear reaction rates for nuclei with mass number $A < 16$," Nucl. Phys. A, 918, 61

Figure Credits

Astronomical images were obtained from a variety of sources with a preference for public-domain images or those licensed under Creative Commons. When we used figures from peer-reviewed scientific papers, we followed the publishers' guidelines for securing the necessary permissions.

Graphs composed by the technical editor were created using the Python matplotlib package in Jupyter notebooks.

Artwork and schematics created by the authors were composed using Microsoft PowerPoint.

Dedication: Cat and dog from a Book of Hours (use of Amiens), late fifteenth century AD. Abbéville Bibliothèque municipale MS 16, fol. 23v (bvmm.irht.cnrs.fr).

Figure 1.1: Intensitygram from the Helioseismic and Magnetic Imager on NASA's *Solar Dynamics Observatory* on May 20, 2022, 19:52:38 UTC. Image courtesy of NASA/SDO and the HMI science team.

Figure 1.2: Plotted by the technical editor, using data from the Interactive Solar Irradiance Data Center of the Laboratory for Atmospheric and Space Physics [University of Colorado Boulder]. Data are the 24-hour average of measurements made by the *Solar Radiation and Climate Experiment* (SORCE).

Figure 1.3: Plotted by the technical editor, using the ASTM E-490-00 solar spectrum normalized to $S_\odot = 1.361 \times 10^6 \, \mathrm{erg \, s^{-1} \, cm^{-2}}$.

Figure 1.4: Plotted by the technical editor, using data from Lodders 2021.

Figure 1.5: (a) ESO public image eso2003b. Scaling determined from Montargès *et al.* 2021, cropped and rescaled to a 180 mas field. © ESO. (b) Reproduced from Figure 12 of Uitenbroek *et al.* 1998, cropped to a 180 mas field. © AAS.

Figure 1.6: Plotted by the technical editor, using data from See 1893.

Figure 1.7: Drawn by the authors and technical editor.

Figure 1.8: Plotted by the technical editor, using data from Jacoby *et al.* 1984.

Figure 1.9: Plotted by the technical editor, using data from the GALAH Data Release 3, provided by Emily Griffith.

Figure 1.10: Plotted by the technical editor, using data from the *Gaia Catalogue of Nearby Stars*. Bolometric corrections, and conversions from *Gaia* EDR3 $BP - RP$ color index to T_{eff}, are from the PARSEC database (stev.oapd.inaf.it/YBC/). Solar metallicity is assumed.

Figure 1.11: Plotted by the technical editor, using data from Cranmer and Saar 2011 and references therein. Mass loss estimate for Sirius A is from Michaud *et al.* 2011.

Figures 1.12, 1.13: Plotted by the technical editor, using data from Eker *et al.* 2018. Data were provided in electronic format by Zeki Eker.

Figure 1.14: Plotted by the technical editor, using data from (a) Heyl *et al.* 2022 and (b) Godoy-Rivera *et al.* 2021.

Figures 2.1, 2.2: Drawn by the authors.

Figures 3.1–3.3: Computed and plotted by the authors and technical editor.

Figure 4.1: Computed and plotted by the authors and technical editor.

Figure 4.2: Plotted by the technical editor, using data provided by Franck Delahaye.

Figure 4.3: Plotted by the technical editor, using data from the Opacity Project OPserver (opacities.osc.edu/rmos.shtml).

Figure 4.4: Reproduced from upper left panel of Figure 1 in Schmidt *et al.* 2017, with permission. © ESO.

Figure 4.5: Computed and plotted by the authors and technical editor.

Figure 4.6: Drawn by the authors.

Figure 5.1: Plotted by the technical editor, using data from the AME2020 atomic mass evaluation, Wang *et al.* 2021.

Figure 5.2: Plotted by the technical editor, using data from Xu *et al.* 2013. Data were provided in electronic format by Kohji Takahashi.

Figure 5.3: Computed and plotted by the authors and technical editor.

Figure 5.4: Plotted by the technical editor, using data from the BP04 solar model.

Figure 5.5: Drawn by the authors.

Figure 5.6: Plotted by the technical editor, using data from the BP04+ solar model.

Figure 5.7: Drawn by the authors.

Figure 5.8: Computed and plotted by the authors and technical editor.

Figure 5.9: Plotted by the technical editor, using data from the BP04 solar model.

Figure 6.1: Plotted by the technical editor, from calculations made with the assistance of the Poly-Web tool by Rich Townsend (University of Wisconsin).

Figure 6.2: Plotted by the technical editor. The $n = 3$ polytrope data are taken from Figure 6.1; the solar model is that of BP04.

Figures 6.3, 6.4: Plotted by the technical editor, using data from the BP04 solar model.

Figures 6.5–6.8: Plotted by the technical editor, using numerical results from Modules for Experiments in Stellar Astrophysics (MESA).

Figure 7.1: Reproduced from Narayanan *et al.* 2008.

Figure 7.2: Plotted by the technical editor, using numerical results from MESA.

Figure 7.3: Computed and plotted by the authors and technical editor.

Figure 7.4: Plotted by the technical editor, using data from Burrows *et al.* 1993 and 1997 (astro.princeton.edu/~burrows).

Figure 8.1: Plotted by the technical editor, using data from Cantat-Gaudin *et al.* 2018.

Figures 8.2–8.5: Plotted by the technical editor, using numerical results from MESA.

Figure 8.6: Plotted by the technical editor, using data from Lind *et al.* 2009.

Figures 8.7, 8.8: Plotted by the technical editor, using numerical results from MESA.

Figure 8.9: Plotted by the technical editor, using data from Vasiliev and Baumgardt 2021.

Figures 8.10, 8.11: Drawn by the authors.

Figure 8.12: Plotted by the technical editor, using data from Cummings *et al.* 2018.

Figure 8.13: Image from Kastner *et al.* 2012 based on observations made with *Hubble Space Telescope*, and obtained from the *Hubble* Legacy Archive (chandra.harvard.edu/photo/2012/pne/).

Figure 9.1: Plotted by the technical editor, using data from Kilic *et al.* 2020.

Figure 9.2: Plotted by the technical editor, using data compiled by Paulo Freire (www3.mpifr-bonn.mpg.de/staff/pfreire/NS_masses.html). See Özel and Freire 2016 and references therein.

Figure 9.3: Plotted by the technical editor with assistance from Patrick J. Vallely, using data from the Open Supernova Catalog (sne.space). See Guillochon *et al.* 2017 and references therein.

Figure 9.4: (a) Image heic0609a, from *Hubble Space Telescope* ACS. NASA, ESA, and the Hubble Heritage (STScI/AURA)-ESA/Hubble Collaboration; R. Fesen (Dartmouth) and J. Long (ESA/Hubble). (b) Image from *Chandra* ACIS (NASA/CXC/SAO).

Figure 9.5: Plotted by the technical editor, using data from Cowan *et al.* 2006.

Figure 10.1: Reproduced from supporting material for Monnier *et al.* 2007. © AAAS.

Figure 10.2: Reproduced from Figure 1 of Roxburgh 2004, with permission. © ESO.

Figure 10.3: Reproduced from Figure 1 of Howe *et al.* 2000. © AAAS.

Figure 10.4: Plotted by the technical editor, using data from the supplementary material of Aerts *et al.* 2019 and Deheuvels *et al.* 2020.

Figure 10.5: Plotted by the technical editor, using numerical results from MESA.

Figure 10.6: Drawn by the authors.

Figure 10.7: Plotted by the technical editor, following Figure 2 of Basri *et al.* 2010. Data were provided by Gibor Basri.

Figure 10.8: Plotted by the technical editor, using data from (a) García *et al.* 2014 and (b,c) Godoy-Rivera *et al.* 2021.

Figure 10.9: Plotted by the technical editor, using data from Claytor *et al.* 2020.

Figure 11.1: Drawn by the technical editor with assistance from Radek Poleski; ZAMS taken from Figure 6.6.

Figure 11.2: Plotted by the technical editor, using data from Udalski *et al.* 1999.

Figure 11.3: Plotted by the technical editor, using data from Schwarzschild 1941 (in which Schwarzschild acknowledges "the help of the punched-card machines of the Thomas J. Watson Astronomical Computing Bureau").

Figure 11.4: Plotted by the technical editor, using data from the OGLE Atlas of Variable Star Light Curves (ogle.astrouw.pl/atlas/).

Figure 11.5: Drawn by the authors.

Figures 11.6, 11.7: Computed and plotted by the authors and technical editor.

Figure 11.8: Plotted by the technical editor, using frequencies tabulated by Metcalfe *et al.* 2012, and spectral data downloaded from the FAMED pipeline tutorial (github.com/EnricoCorsaro/FAMED).

Figure 11.9: Plotted by the technical editor with assistance from Mathieu Vriard.

Figure 11.10: Plotted by the technical editor, using data from Pinsonneault *et al.* 2023.

Figure 12.1: Plotted by the technical editor, using data from Lu *et al.* 2013.

Figure 12.2: Plotted by the technical editor, using data from the Ninth Catalogue of Spectroscopic Binary Orbits (sb9.astro.ulb.ac.be). See Pourbaix *et al.* 2004.

Figures 12.3, 12.4: Computed and plotted by the authors and technical editor.

Figure 12.5: Plotted by the technical editor, using data from Baron *et al.* 2012.

Figure 12.6: Plotted by the technical editor, following Figure 1 of Willems and Kolb 2004.

Figure 12.7: Plotted by the technical editor following Figure 1 of Selvelli and Friedjung 2003, using archival process spectra of Selvelli and Gilmozzi 2013.

Figure 12.8: (a) Image opo0321a from *Hubble Space Telescope* WFPC2. NASA/ESA and the Hubble Heritage Team (AURA/STScI). (b) CMD data from Piotto *et al.* 2002, replotted by the authors and technical editor.

Figure 12.9: Image from *Chandra* ACIS. NASA/CXC/Rutgers/J. Warren and J. Hughes *et al.* (Obs. ID 3832).

Further Reading header: Dante and Beatrice conversing, from Canto IV of the *Paradiso*. Bodleian Library, ms. Holkham misc. 48, p. 119, *c.*AD 1350–75, artist unknown.

Index

Printed in the United States
by Baker & Taylor Publisher Services